# Methods in Molecular Biology

*Series Editor*
John M. Walker
School of Life and Medical Sciences
University of Hertfordshire
Hatfield, Hertfordshire, UK

For further volumes:
http://www.springer.com/series/7651

For over 35 years, biological scientists have come to rely on the research protocols and methodologies in the critically acclaimed *Methods in Molecular Biology* series. The series was the first to introduce the step-by-step protocols approach that has become the standard in all biomedical protocol publishing. Each protocol is provided in readily-reproducible step-by-step fashion, opening with an introductory overview, a list of the materials and reagents needed to complete the experiment, and followed by a detailed procedure that is supported with a helpful notes section offering tips and tricks of the trade as well as troubleshooting advice. These hallmark features were introduced by series editor Dr. John Walker and constitute the key ingredient in each and every volume of the *Methods in Molecular Biology* series. Tested and trusted, comprehensive and reliable, all protocols from the series are indexed in PubMed.

# Cryptosporidium

## Methods and Protocols

### Second Edition

Edited by

**Jan R. Mead**

*Department of Pediatrics, Emory University, Atlanta, GA, USA*

**Michael J. Arrowood**

*Division of Foodborne, Waterborne, and Environmental Diseases, National Center for Emerging and Zoonotic Infectious Diseases, Centers for Disease Control and Prevention (Retired), Atlanta, GA, USA*

Humana Press

*Editors*
Jan R. Mead
Department of Pediatrics
Emory University
Atlanta, GA, USA

Michael J. Arrowood
Division of Foodborne, Waterborne, and Environmental Diseases, National Center for Emerging and Zoonotic Infectious Diseases
Centers for Disease Control and Prevention (Retired)
Atlanta, GA, USA

ISSN 1064-3745 ISSN 1940-6029 (electronic)
Methods in Molecular Biology
ISBN 978-1-0716-4823-0 ISBN 978-1-0716-4824-7 (eBook)
https://doi.org/10.1007/978-1-0716-4824-7



This Humana imprint is published by the registered company Springer Science+Business Media, LLC, part of Springer Nature.
The registered company address is: 1 New York Plaza, New York, NY 10004, U.S.A.

# Preface

*Cryptosporidium* is a ubiquitous opportunistic protozoan parasite with significant implications for public health, veterinary medicine, and environmental science. It can infect the intestinal tract causing diarrheal illness that may be life-threatening in immunocompromised and immunodeficient individuals. It is also recognized as an important disease in young children in developing countries where it may be responsible for growth deficits and cognitive problems. The need for accurate clinical and environmental detection, effective treatments, and a deeper understanding of its biology has driven decades of research. This book brings together established and cutting-edge methodologies and protocols that span the detection, characterization, and genomic analysis of *Cryptosporidium*, providing researchers with a comprehensive resource for advancing studies in this field.

The aim of this book is to encompass broad aspects of *Cryptosporidium* research. As with the first edition, the second edition includes established methods reported in journal publications but provided in more detail and with helpful notes not typically included in a standard scientific paper. These methods have been updated to outline the latest modifications that are currently used in laboratories.

The first section of this volume focuses on diagnostic assays, encompassing traditional microscopy-based techniques, nucleic acid-based molecular detection, and immunoassay-based methodologies. These foundational approaches remain essential for clinical and epidemiological investigations. Advanced methods are presented for isolating and detecting *Cryptosporidium* in environmental samples, particularly from water sources, which are critical for understanding transmission dynamics and public health risk assessments. Methods for the purification of cryptosporidial oocysts from clinical specimens and from experimentally infected animal models are also included.

Advancements in genomic analysis have transformed our ability to study *Cryptosporidium* at the species and subtype level. Several chapters in this book detail whole genome sequencing methodologies, genotyping approaches, and novel techniques such as hybridization bait capture analyses. These technologies provide powerful insights into the genetic structure and evolution of *Cryptosporidium*, facilitating epidemiological tracking and improved disease surveillance.

A significant portion of this book is dedicated to functional genomics and experimental models. Methodologies and tools include those for genetic manipulation (including CRISPR/cas9), gene silencing, and host-pathogen interactions. A new protocol describes labeling cryptosporidial proteins for tracking and functional studies. Established animal models, including mouse and calf models, are discussed in detail along with in vitro culture systems that simulate the intestinal environment. These models may be instrumental in evaluating therapeutic interventions and understanding *Cryptosporidium's* pathogenic mechanisms.

The book concludes with a focus on parasite disinfection and on drug discovery and testing, highlighting high-content screening methods, phenotypic assays, and efficacy testing of potential disinfectants and therapeutics. These chapters underscore the urgent need for effective anti-*Cryptosporidium* drugs, given the limited treatment options currently available.

By compiling these methodologies in a single volume, we aim to provide a valuable resource for researchers, clinicians, and public health professionals engaged in the study of *Cryptosporidium*. We hope that the protocols outlined in this book will facilitate new discoveries and contribute to the ongoing efforts to control and mitigate the impact of cryptosporidiosis worldwide.

We extend our gratitude to all contributing authors, whose expertise and dedication have made this book possible. Their collective efforts represent foundational and innovative *Cryptosporidium* research, and we trust that their insights will inspire future advancements in this field.

*Atlanta, GA, USA* — ***Jan R. Mead***
*Atlanta, GA, USA* — ***Michael J. Arrowood***

# Contents

## Contributors


FIIFI AGYABENG-DADZIE • *Department of Genetics, University of Georgia, Athens, GA, USA; Center for Tropical and Emerging Global Diseases, University of Georgia, Athens, GA, USA*

MICHAEL J. ARROWOOD • *Division of Foodborne, Waterborne, and Environmental Diseases, National Center for Emerging and Zoonotic Infectious Diseases, Centers for Disease Control and Prevention (Retired), Atlanta, GA, USA; Retired, Peachtree Corners, GA, USA*

MALINA A. BAKOWSKI • *Calibr-Skaggs Institute for Innovative Medicines, La Jolla, CA, USA*

MEGAN S. BEAUDRY • *Department of Environmental Health Science, College of Public Health, University of Georgia, Athens, GA, USA; Center for Tropical and Emerging Global Diseases, University of Georgia, Athens, GA, USA; Daicel Arbor Biosciences, Ann Arbor, MI, USA*

SEEMA BHALCHANDRA • *Tufts Medical Center, Boston, MA, USA*

MOHAMMAD IMTIAJ UDDIN BHUIYAN • *Department of Environmental Health Science, College of Public Health, University of Georgia, Athens, GA, USA*

CARRIE F. BROOKS • *Center for Tropical and Emerging and Global Diseases, University of Georgia, Athens, GA, USA*

JESSICA H. BYERLY • *Department of Pathobiology, School of Veterinary Medicine, University of Pennsylvania, Philadelphia, PA, USA*

DAVIEL CARDENAS • *Tufts Medical Center, Boston, MA, USA*

ALEJANDRO CASTELLANOS-GONZALEZ • *Infectious Disease Division, Department of Internal Medicine, University of Texas Medical Branch, Galveston, TX, USA*

RACHEL M. CHALMERS • *Cryptosporidium Reference Unit, Public Health Wales Microbiology and Health Protection, Singleton Hospital, Swansea, UK; Cryptosporidium Reference Unit, Public Health Wales Microbiology and Health Protection, Singleton Hospital, Swansea, Wales, UK*

JACKSON F. CHEN • *National Reference Centre for Parasitology, Research Institute of the McGill University Health Centre, Montreal, QC, Canada; Infectious Diseases and Immunity in Global Health program, Research Institute of the McGill University Health Centre, Montreal, QC, Canada; Division of Experimental Medicine, Faculty of Medicine, McGill University, Montreal, QC, Canada*

XIAN-MING CHEN • *Department of Microbial Pathogens and Immunity, Rush University Medical Center, Chicago, IL, USA*

KRISTIN ELWIN • *Cryptosporidium Reference Unit, Public Health Wales Microbiology and Health Protection, Singleton Hospital, Swansea, Wales, UK*

MATTHEW FREESE • *Department of Molecular Microbiology, Washington University School of Medicine, St Louis, MO, USA*

TRAVIS C. GLENN • *Department of Environmental Health Science, College of Public Health, University of Georgia, Athens, GA, USA; Department of Genetics, University of Georgia, Athens, GA, USA; Institute of Bioinformatics, University of Georgia, Athens, GA, USA*

MARION L. GRAHAM • *Department of Microbial Pathogens and Immunity, Rush University Medical Center, Chicago, IL, USA*

Valentin Greigert • *Department of Molecular Microbiology, Washington University School of Medicine, St Louis, MO, USA; Laboratoire ESCAPE, UFR Médecine, Université de Reims Champagne-Ardenne, Reims, France*
Michael E. Grigg • *Laboratory of Parasitic Diseases, Division of Intramural Research, NIAID, NIH, Bethesda, MD, USA*
Wanyi Huang • *State Key Laboratory for Animal Disease Control and Prevention, Center for Emerging and Zoonotic Diseases, College of Veterinary Medicine, South China Agricultural University, Guangzhou, China*
Amy M. Kahler • *Division of Foodborne, Waterborne, and Environmental Diseases, National Center for Emerging and Zoonotic Infectious Diseases, Atlanta, GA, USA*
Jessica C. Kissinger • *Department of Genetics, University of Georgia, Athens, GA, USA; Center for Tropical and Emerging Global Diseases, University of Georgia, Athens, GA, USA; Institute of Bioinformatics, University of Georgia, Athens, GA, USA*
Hymlaire Lamisere • *Tufts Medical Center, Boston, MA, USA; Tufts University Graduate School of Biomedical Sciences, Boston, MA, USA*
Melissa S. Love • *Calibr-Skaggs Institute for Innovative Medicines, La Jolla, CA, USA*
Mia C. Mattioli • *Division of Foodborne, Waterborne, and Environmental Diseases, National Center for Emerging and Zoonotic Infectious Diseases, Atlanta, GA, USA*
Case W. McNamara • *Calibr-Skaggs Institute for Innovative Medicines, La Jolla, CA, USA*
Jan R. Mead • *Veterans Affairs Medical Center, Decatur, GA, USA; Department of Pediatrics, Emory University, Atlanta, GA, USA*
Jennifer L. Murphy • *Division of Foodborne, Waterborne, and Environmental Diseases, National Center for Emerging and Zoonotic Infectious Diseases, Centers for Disease Control and Prevention, Atlanta, GA, USA*
Momar Ndao • *National Reference Centre for Parasitology, Research Institute of the McGill University Health Centre, Montreal, QC, Canada; Infectious Diseases and Immunity in Global Health program, Research Institute of the McGill University Health Centre, Montreal, QC, Canada; Division of Experimental Medicine, Faculty of Medicine, McGill University, Montreal, QC, Canada; Division of Infectious Diseases, Department of Medicine, Faculty of Medicine, McGill University, Montreal, QC, Canada*
Mattie C. Pawlowic • *Wellcome Trust Centre for Anti-Infectives Research, University of Dundee, Dundee, UK*
Kelly N. Petersen • *Odum School of Ecology, University of Georgia, Athens, GA, USA*
Soumya Ravindran • *Department of Molecular Microbiology, Washington University School of Medicine, St Louis, MO, USA*
Michael W. Riggs • *School of Animal and Comparative Biomedical Sciences, The University of Arizona, Tucson, AZ, USA*
Harriet Risby • *Cryptosporidium Reference Unit, Public Health Wales Microbiology and Health Protection, Singleton Hospital, Swansea, Wales, UK*
Guy Robinson • *Cryptosporidium Reference Unit, Public Health Wales Microbiology and Health Protection, Singleton Hospital, Swansea, UK; Cryptosporidium Reference Unit, Public Health Wales Microbiology and Health Protection, Singleton Hospital, Swansea, Wales, UK*
Dawn M. Roellig • *Division of Foodborne, Waterborne, and Environmental Diseases, National Center for Emerging and Zoonotic Infectious Diseases, Atlanta, GA, USA*
Adam Sateriale • *Francis Crick Institute, London, UK*
Deborah A. Schaefer • *School of Animal and Comparative Biomedical Sciences, The University of Arizona, Tucson, AZ, USA*

MATTHEW H. SEABOLT • *Division of Foodborne, Waterborne, and Environmental Diseases, National Center for Emerging and Zoonotic Infectious Diseases, Atlanta, GA, USA*

SEBASTIAN SHAW • *Department of Pathobiology, School of Veterinary Medicine, University of Pennsylvania, Philadelphia, PA, USA*

L. DAVID SIBLEY • *Department of Molecular Microbiology, Washington University School of Medicine, St. Louis, MO, USA; Department of Molecular Microbiology, Washington University School of Medicine, St Louis, MO, USA*

KARINE SONZOGNI-DESAUTELS • *National Reference Centre for Parasitology, Research Institute of the McGill University Health Centre, Montreal, QC, Canada; Infectious Diseases and Immunity in Global Health program, Research Institute of the McGill University Health Centre, Montreal, QC, Canada*

THADDEUS STAPPENBECK • *Department of Pathology and Immunology, Washington University School of Medicine, St Louis, MO, USA*

BORIS STRIEPEN • *Department of Pathobiology, School of Veterinary Medicine, University of Pennsylvania, Philadelphia, PA, USA*

AMANDA H. SULLIVAN • *Department of Environmental Health Science, College of Public Health, University of Georgia, Athens, GA, USA; Institute of Bioinformatics, University of Georgia, Athens, GA, USA*

PIOTR T. TUCZAPSKI • *Department of Environmental Health Science, College of Public Health, University of Georgia, Athens, GA, USA; Department of Plant Biology, University of Georgia, Athens, GA, USA*

SUMITI VINAYAK • *Department of Pathobiology, College of Veterinary Medicine, University of Illinois Urbana-Champaign, Urbana, IL, USA*

CHENCHEN WANG • *State Key Laboratory for the Diagnosis and Treatment of Severe Infectious Diseases, Institute of Zoonosis, and College of Veterinary Medicine, Jilin University, Jilin, China*

DONGQIANG WANG • *State Key Laboratory for the Diagnosis and Treatment of Severe Infectious Diseases, Institute of Zoonosis, and College of Veterinary Medicine, Jilin University, Jilin, China*

TIANPENG WANG • *State Key Laboratory for Animal Disease Control and Prevention, Center for Emerging and Zoonotic Diseases, College of Veterinary Medicine, South China Agricultural University, Guangzhou, China*

YI WANG • *Department of Pathology and Immunology, Washington University School of Medicine, St Louis, MO, USA*

HONORINE D. WARD • *Tufts Medical Center, Boston, MA, USA; Tufts University Graduate School of Biomedical Sciences, Boston, MA, USA*

GEORGIA WILKE • *Department of Molecular Microbiology, Washington University School of Medicine, St Louis, MO, USA*

WILLIAM WITOLA • *Department of Pathobiology, University of Illinois College of Veterinary Medicine, Urbana, IL, USA*

LIHUA XIAO • *State Key Laboratory for Animal Disease Control and Prevention, Center for Emerging and Zoonotic Diseases, College of Veterinary Medicine, South China Agricultural University, Guangzhou, China; State Key Laboratory for Animal Disease Control and Prevention, Center for Emerging and Zoonotic Diseases, College of Veterinary Medicine, South China Agricultural University, Guangzhou, China*

RUI XU • *State Key Laboratory for Animal Disease Control and Prevention Center for Emerging and Zoonotic Diseases, College of Veterinary Medicine, South China Agricultural University, Guangzhou, China*

HAILI ZHANG • *Department of Veterinary Pathobiology, College of Veterinary Medicine & Biomedical Sciences, Texas A&M University, College Station, TX, USA; GenoBioTX, Sugar Land, TX, USA*
GUAN ZHU • *State Key Laboratory for the Diagnosis and Treatment of Severe Infectious Diseases, Institute of Zoonosis, and College of Veterinary Medicine, Jilin University, Jilin, China; Department of Veterinary Pathobiology, College of Veterinary Medicine & Biomedical Sciences, Texas A&M University, College Station, TX, USA*

## Abbreviations

| | |
|---|---|
| 3D | Three dimensional |
| 3T3 | Mouse fibroblast cells (3T3) |
| ABSL | Animal Biosafety Level |
| AID | Auxin-inducible degron |
| ALI | Air-liquid interface |
| AMO | Anti-miRNA antisense inhibitor oligoribonucleotide (AMO) technology |
| *Ase* I | *Ase*I |
| BCA | Bicinchoninic acid |
| BCIP/NBT | 5-bromo-4-chloro-3-indolyl phosphate/nitroblue tetrazolium |
| BKI | Bumped kinase inhibitor |
| BSA | Bovine serum albumin |
| Caco-2 | Human colorectal adenocarcinoma cells |
| CDF | Chlorine demand-free |
| CMGF- | Complete media without growth factors |
| CMGF+ | Complete media with growth factors |
| CNV | Copy number variation |
| Cp15 | 15-kDa antigen |
| Cp17 | 17-kDa antigen |
| Cp27 | 27-kDa antigen |
| CRISPR | Clustered regularly interspaced short palindromic repeats |
| $C_T$ | Cycle threshold |
| CV | Coefficient of variation |
| DABCO | 1,4-Diazabicyclo-[2.2.2]octane |
| DAPI | 4', 6-Diamidino-2-phenylindole (DAPI) |
| DEUF | Dead-end ultrafiltration |
| $dH_2O$ | Deionized water |
| DI | Deionized |
| DPD | N,N-diethyl-p-phenylenediamine |
| DTT | Dithiothreitol |
| EC numbers | Enzyme commission numbers |
| ECS | Extra-capillary space |
| EDAC | 1-ethyl-3-(3-dimethylaminopropyl) carbodiimide |
| EDTA | Ethylenediaminetetraacetic acid |
| EIA | Enzyme immunoassay |
| EST | Express sequence tag |
| FBS | Fetal bovine serum |
| FDU | 5-Fluoro-2'-deoxyuridine |
| FITC | Fluorescein isothiocyanate |
| GST | Glutathione-S-transferase |
| hAgo2 | Human Argonaute |
| HCS | High-content screening |
| HCT-8 | Human ileocecal adenocarcinoma cells |
| HEPES | 4-(2-Hydroxyethyl)-1-piperazineethanesulfonic acid |
| HFF | Human foreskin fibroblast |
| HIEs | Human intestinal enteroids |

| | |
|---|---|
| H-InMyoFib | Primary human intestinal myofibroblasts |
| HPA lectin | *Helix pomatia* lectin |
| HRM | High-resolution melt |
| HTS | High-throughput screening |
| IACUC | Institutional Animal Care and Use Committee |
| ICLF | Immunochromatographic lateral flow |
| ICS | Intra-capillary space |
| IECs | Intestinal epithelial cells |
| IFA | Immunofluorescence assay |
| IFNg-KO | Interferon gamma knock-out mice |
| IFNgR-KO | Interferon gamma receptor knock-out mice |
| IL-12 KO | IL-12 knock-out mice |
| IMS | Immunomagnetic separation |
| IPTG | Isopropylthio-b-D-galactoside |
| IPTG | Isopropyl β-d-1-thiogalactopyranoside |
| kDa | Kilodaltons |
| LD decay | Linkage disequilibrium decay |
| $Lgr5^+$ | Leucine-rich repeat-containing G-protein coupled receptor 5 cells |
| LSU | Large subunit |
| mAID | Mini-AID |
| *Mbo*II | *Mbo*II |
| MDCK | Madin Darby canine kidney |
| MEGA | Molecular Evolutionary Genetics Analysis |
| MEGA7 | MEGA7 |
| miRNA | MicroRNA |
| MLFT | Multi-locus fragment-sizing typing |
| MLST | Multi-locus sequencing typing |
| MLVA | Multi-locus variable number of tandem repeats analysis |
| MSDS | Mucous |
| NaOCl | NaOCl (sodium hypochlorite) |
| NCBI | National Center for Biotechnology Information |
| NGS | Next-generation sequencing |
| NKCC1 | Sodium potassium chlorine co-transporter (Na-K-Cl co-transporter) |
| OCT | Optimal cutting temperature |
| ORF | Open reading frame |
| PCA | Principle component analysis |
| PCR | Polymerase chain reaction |
| PCR-RFLP | Polymerase chain reaction restriction fragment length polymorphism |
| PDMS | Polydimethylsiloxane |
| PMSF | Phenymethylsulfonyl fluoride |
| PPE | Personal protective equipment |
| PTR | Protein transfection reagents |
| Pur | Puromycin |
| PVA | Polyvinyl alcohol |
| PVDF | Polyvinylidene fluoride |
| qPCR | Quantitative polymerase chain reaction |
| qRT-PCR | Real-time quantitative PCR |
| RankL | Receptor activator of NFκB ligand |
| RFLP | Restriction fragment length polymorphism |

| | |
|---|---|
| RFU | Relative fluorescence unit |
| RISC | RNA-induced silencing complex |
| RNAi | RNA interference |
| ROCK inhibitor | Rho kinase inhibitor |
| RT-PCR | Real-time PCR |
| RT-qPCR | Reverse transcription, quantitative PCR |
| SCID | Severe combined immune deficient mice |
| SDS | Sodium dodecyl sulfate (sodium lauryl sulfate) |
| siRNA | Small interfering RNA |
| SmGM-2 | Smooth Muscle Growth Medium-2 |
| SNAP23 | Synaptosome associated protein 23 |
| SNP | Single nucleotide polymorphism |
| SOP(s) | Standard operating procedure |
| SspI | *Ssp*I |
| ssRNA | Single strand RNA |
| SSU | Small subunit |
| SSU rDNA | Small subunit ribosomal DNA |
| SSU rRNA | Small subunit ribosomal RNA |
| Sulfo-NHS | *N*-Hydroxysulfosuccinimide |
| TAE | Tris acetate EDTA |
| TCA cycle | Tricarboxylic acid cycle |
| TDC | Taurodeoxycholic acid |
| TEER | Trans-epithelial electrical resistance |
| UV | Ultraviolet |
| VNTR | Variable number tandem repeats |
| WGA | Whole genome amplification |
| WGS | Whole genome sequencing |

# Chapter 1

# *Cryptosporidium* Diagnostic Assays: Microscopy


## Guy Robinson and Rachel M. Chalmers



## Abstract

Stained microscopy of fecal smears was the cornerstone of *Cryptosporidium* diagnosis for many years and provides a low-cost method for detecting oocysts. The development and commercialization of improved enzyme immunosorbent assays (EIAs) for coproantigen detection provided an automatable method for mass testing and rapid diagnostics when incorporated onto a cartridge format. Similarly, immunochromatographic lateral flow (ICLF) assays enable rapid diagnostics. Nevertheless, it is important that positive reactions by EIAs or ICLF assays are confirmed. Here, we describe microscopical methods using tinctorial stains for the diagnosis of acute cryptosporidiosis and using immunofluorescent reagents for diagnosis or for confirmation of EIA- or ICLF-positive reactions.



**Key words** *Cryptosporidium*, Oocyst detection, Microscopy, Ziehl–Neelsen, Auramine phenol, Immunofluorescent


## 1 Introduction

There are no international standard methods for the diagnosis of cryptosporidiosis, but, depending on the setting, both low- and high-tech options are available, and those based on stained microscopy of stools are described here. Historically, histological staining and microscopy of thin sections of gut biopsy material were used to diagnose *Cryptosporidium* infection, but nowadays, the oocyst stage of the life cycle, shed in feces, is the usual diagnostic target. Stained microscopy forms the basis of low-tech methods as oocysts are hard to identify without staining. It may be necessary to examine stool samples taken over several days until oocysts are seen, or an alternative diagnosis is made, if patients continue to have symptoms. Here, we describe the preparation of fecal smears and staining by acid-fast modified Ziehl–Neelsen (Kinyoun) and auramine phenol stains and their subsequent microscopic examination [1, 2, 3]. Prior concentration, preferably by formalin-ethyl acetate sedimentation, may be used but is not usually necessary for the diagnosis of acute infection unless the samples are submitted in


Jan R. Mead and Michael J. Arrowood (eds.), *Cryptosporidium: Methods and Protocols*, Methods in Molecular Biology, vol. 2978, https://doi.org/10.1007/978-1-0716-4824-7_1,

preservative. While the reagents are cheap, good microscopic examination is labor-intensive. The use of more expensive immunofluorescent stains improves both sensitivity and specificity and reduces examination time, which is also described here.

The use of antibodies raised against oocyst antigens has also been incorporated into commercially available enzyme immunosorbent assays (EIAs) (in either a 96-well plate or a cartridge format) and immunochromatographic lateral flow (ICLF) assays (in either cartridge or dip stick assays). These do not rely on microscopy skills, vary in their sensitivity and specificity, and should be assessed for suitability for the population being tested. Generally, EIA kits have greater sensitivity and specificity for *Cryptosporidium* than ICLF and modified Ziehl–Neelsen staining (see [4, 5] for comparative assessment of methods). They are not described here as the manufacturer's instructions should be followed, including whether preservatives are compatible. Generally, stools should not be concentrated prior to use of EIA or ICLF as antigens may be lost.

## 2 Materials

### 2.1 Preparation of Slides for Stained Microscopy

1. Glass microscope slides (modified Ziehl–Neelsen and auramine phenol staining) or single- or multiple-well glass microscopy slides (immunofluorescence staining).
2. Sample applicator sticks.
3. Distilled water.
4. Methanol.
5. Diamond marker or pencil.

### 2.2 Formalin-Ethyl Acetate Sedimentation for Concentration of Samples Submitted in Preservative

Commercial devices and reagents are available for the formalin-ethyl acetate concentration of parasites [6, 7], or the following reagents and protocol may be used.

1. 15-mL conical centrifuge tube.
2. Sample applicator sticks.
3. 10% formalin.
4. Sieve with 425-μm aperture, equivalent to 36-mesh British Standard (BS 410-86) or 40-mesh American Standard (ASTM E11-81).
5. Beaker.
6. Ethyl acetate (*see* **Note 1**).

### 2.3 Modified Ziehl–Neelsen Reagents

Commercial supplies are available for all reagents.

1. Strong carbol fuchsin (1–3% basic fuchsin): fully dissolve 20 g basic fuchsin in 200 mL absolute methanol on a magnetic

stirrer. Add 125 mL liquid phenol (general purpose reagent [GPR; 80% w/w in distilled water]) carefully until well mixed, and make up to the final volume with 1675 mL deionized water. Mix thoroughly and filter before use through Whatman No.1 filter paper to remove debris. Store the stock reagent in the dark in the range 20 ± 5 °C.

2. 1% acid methanol: carefully add 20 mL concentrated hydrochloric acid to 1980 mL of absolute methanol and mix. Store at ambient temperature.
3. 0.4% malachite green: add 2 g malachite green to 480 mL deionized water and mix on a magnetic stirrer before filtering through Whatman No.1 filter paper. Store at ambient temperature.

### 2.4 Auramine Phenol Stain Reagents

Commercial supplies are available for all reagents.

1. Auramine phenol: dissolve 3 g phenol in 100 mL deionized water and slowly add 0.3 g auramine O before filtering through Whatman No. 1 filter paper. Store in the range 20 ± 5 °C in a light-proof glass bottle with an airtight stopper.
2. 3% Acid methanol: carefully add 60 mL concentrated hydrochloric acid to 1940 mL absolute methanol and mix.
3. 0.1% potassium permanganate: add 0.5 g potassium permanganate to 499.5 mL deionized water and mix using a magnetic stirrer before filtering through Whatman No. 1 filter paper.

### 2.5 Immuno-fluorescent Microscopy Reagents

1. Commercially available kits may include: *Cryptosporidium* oocyst-positive and oocyst-negative controls, fluorescently labeled (e.g., with fluorescein isothiocyanate [FITC]) anti-*Cryptosporidium* monoclonal antibodies, and glycerol-based mounting medium containing a photobleaching (quenching) inhibitor.

### 2.6 Microscopy

1. Microscope with bright field and epifluorescence capability.
2. Fluorescence filter blocks: FITC (excitation filter 490 nm; emission filter 510 nm) or ultraviolet (UV; excitation filter 355 nm; emission filter 450 nm) filters for auramine phenol. For immunofluorescent staining, the required filter block is dependent on the fluorophore used (the FITC filter noted above is most common).
3. Objective lenses with 10X eye piece lenses: 40X–50X immersion objectives for screening modified Ziehl–Neelsen and auramine-phenol-stained samples with 100X immersion objective for examining finer detail; 20X for screening immunofluorescent-stained samples with 40X and 100X immersion for confirmatory examination.

## 3 Methods

Wear appropriate personal protective equipment (PPE) during all sample processing steps. Each of these methods is appropriate for examining both human and animal fecal samples.

### 3.1 Preparation of Slides for Stained Microscopy

1. Include a verified positive control on a separate slide each time this procedure is performed.
2. Mark the reference number of the specimen on a clean, dry, glass microscope slide with a diamond marker or pencil (or equivalent), and use a separate microscope slide for each specimen.
3. Fecal samples must be well mixed with the wooden end of a clean applicator stick. Non-diarrheic, very solid, or dry feces should be emulsified in oocyst-free, distilled water. Mucoid samples can be digested prior to the preparation of a smear using mucous liquefying agent (e.g., 1:1 (v/v) dithiothreitol (1% (w/v) in water; commercial preparations available) or 0.1% potassium hydroxide) and incubating for 15 min at ambient temperature.
4. *For modified Ziehl–Neelsen and auramine phenol staining:* use a clean applicator stick to prepare a medium to thick smear of the well-mixed stool sample, of approximately 4 × 1.5 cm, with areas of varying thickness, on a labeled, clean, dry, glass microscope slide. Allow to air-dry. Place slides on a staining rack over the sink and fix by flooding with methanol for 3 min. Drain off and allow to air dry.
5. *For immunofluorescent staining:* smear the homogenized sample, not too thickly (*see* **Note 2**), within the well boundary of a single-well microscope slide, leaving a gap of approximately 1 mm between the edge of the sample and the well. If the sample is fluid, add 5 μL of sample to the well. Allow to air dry. Fix by carefully applying sufficient methanol (approximately 20 μL) to cover each sample without spilling outside of the well. Allow to air-dry.

### 3.2 Formalin-Ethyl Acetate Sedimentation for Concentration of Stools

1. Using a sample applicator stick, transfer approximately 0.5–1 g of a well-mixed fecal sample (750 μL of liquid) to a clean 15-mL tube containing 7 mL of 10% formalin.
2. Homogenize the sample with the applicator stick, breaking up any lumps as necessary.
3. Filter the suspension through the sieve into a beaker (*see* **Note 3**) and pour back into the 15-mL tube.
4. Add 3 mL ethyl acetate and cap the tube before vigorously shaking for 30 s.

5. Release the built-up pressure by gently unscrewing before resealing and centrifuging the tube at 1100 × *g* for 2 min.
6. Using a sample applicator stick, loosen the fatty plug at the top of the sample.
7. Gently pour off the fatty plug and fluid column that is present above the required sample pellet, into a resealable liquid waste container. Allow one or two drops to run back into the tube over the sample pellet.
8. Resuspend the pellet by gentle agitation (*see* **Note 4**) and transfer a subsample onto a microscope slide for drying and staining.

### 3.3 Modified Ziehl–Neelsen Staining

1. Immerse or flood slides on a staining rack over the sink, with cold strong carbol fuchsin and allow to stain for 15 min.
2. Rinse slides thoroughly with tap water (*see* **Note 5**).
3. Immerse or flood slides with 1% acid methanol for 10–15 s to decolorize (*see* **Note 6**).
4. Rinse slides thoroughly with tap water (*see* **Note 5**).
5. Immerse or flood slides with 0.4% malachite green for 30 s to counterstain.
6. Rinse slides thoroughly in tap water (*see* **Note 5**).
7. Allow the slides to air-dry.
8. Examine for the presence of oocysts by scanning slides systematically using at least the 40X objective lens of a bright-field microscope. Confirm the presence of oocysts using a high-power (e.g., 100X) objective lens (*see* **Note** 7).
9. Measure the size and shape of the red-stained bodies using a calibrated eyepiece graticule or reticle (*see* **Notes 8** and **9**).

### 3.4 Auramine Phenol Staining

1. Immerse or flood slides on a staining rack over the sink, with auramine phenol stain for 10 min.
2. Rinse slides thoroughly with tap water (*see* **Note 5**).
3. Immerse or flood slides with 3% acid alcohol for 5 min to decolorize.
4. Rinse slides thoroughly in tap water (*see* **Note 5**).
5. Counterstain by immersing or flooding in 0.1% potassium permanganate for 30 s.
6. Rinse slides thoroughly with tap water to remove excess stain (*see* **Note 5**).
7. Allow the slides to air-dry at room temperature (*see* **Note 10**).
8. Examine for the presence of oocysts, using an epifluorescence microscope equipped with FITC (excitation filter 490 nm;

emission filter 510 nm) or UV (excitation filter 355 nm; emission filter 450 nm) filters (*see* **Note 11**), by scanning the slide systematically using at least the 40X objective lens. Confirm the presence of oocysts using a high-power (e.g., 100X) objective lens (*see* **Note 12**).

9. Measure the size and shape of the fluorescent bodies by slowly increasing the voltage (light intensity) of the bright-field light source so that both fluorescent and bright-field images can be seen concurrently and using a calibrated eyepiece graticule or reticle (*see* **Note 8**).

### 3.5 Immunofluorescent Staining

1. Stain and mount slides according to manufacturers' instructions.
2. Scan the slide for oocysts using a 20X objective lens and confirm using a 40X or greater objective lens of an epifluorescence microscope equipped with an appropriate filter set (e.g., for FITC-labeled stain, excitation wavelength 490 nm and emission wavelength 510 nm) (*see* **Note 13**).
3. Measure the size and shape of the fluorescent bodies using the 100X by slowly increasing the voltage (light intensity) of the bright-field light source so that both fluorescent and bright-field images can be seen concurrently and using a calibrated eyepiece graticule or reticle (*see* **Note 8**). If necessary, slides can be stored at 2–8 °C for up to 7 d without the loss of fluorescence.

## 4 Notes

1. Diethyl ether was historically used, but now replaced by ethyl acetate which although safer is still highly flammable. Caution must be taken to ensure no proximity to naked flames or sparks during use and to avoid prolonged inhalation or skin contact. All precautions outlined in the material safety data sheet (MSDS) or safety data sheet (SDS) must be observed.
2. The transparency should be sufficient to read the face of a watch or text of a newspaper through the smear.
3. The sieve should fit comfortably within the rim of the beaker. Both the sieve and beaker must be washed thoroughly in running tap water between each sample. Observe local waste disposal requirements (both chemical and biological). Disposable sieves reduce the risk of cross-contamination between samples.
4. A large pellet indicates that too much sample was processed, insufficient shaking with the ethyl acetate, or a too fast or long centrifuge step was used.

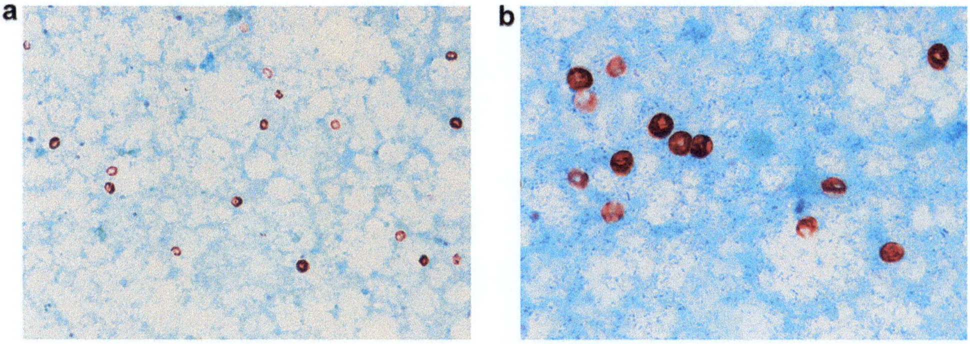


**Fig. 1** *Cryptosporidium* oocysts (~5 μm diameter) demonstrating typical features with modified Ziehl–Neelsen staining as seen using (**a**) 50X objective lens for screening and (**b**) 100X objective lens for confirmation. Often, the sporozoites lay toward the edge of oocysts which gives them the so-called doughnut or erythrocytic staining (denser around the outside and fainter or clear in the middle)

5. Depending on the nature of the feces, some of the sample could wash off the slide at this point; care must be taken to avoid this occurring by adjusting the flow of water to an appropriate level or gently moving the slide while submerged in a water container.
6. Over-decolorizing must be avoided by adhering to the specified timings.
7. The smear can be examined with or without a coverslip. *Cryptosporidium* spp. oocysts stain red on a pale green background (*see* Fig. 1). The degree and proportion of staining varies with individual oocysts. Internal structures take up the stain to varying degrees; some may appear amorphous, while others may contain the characteristic crescent shapes of up to four sporozoites. The presence of "ghost" (unstained) oocysts may be apparent as holes or gaps in the counterstain.
8. Most *Cryptosporidium* spp. oocysts in human samples (e.g., *Cryptosporidium parvum* and *Cryptosporidium hominis*) appear as round or slightly ovoid, 4–6 μm in diameter. In animal samples, larger *Cryptosporidium* oocysts (e.g., *Cryptosporidium andersoni*) may also be seen and appear more ovoid, ~7.5 μm in length by ~5.5 μm in width. Always refer to the positive control to ensure that the size, shape and color of the putative oocyst are consistent with those of the positive control. However, *Cryptosporidium* spp. cannot be differentiated by microscopy so report as "*Cryptosporidium* oocysts seen."
9. Yeasts and fecal debris can also stain red and correspond to the correct size range. However, the staining is usually more uniform across the body with often more variation in the size and shape of the stained bodies compared with *Cryptosporidium* (*see* Fig. 2). Some bacterial spores may also stain red, but these are too small to cause confusion.

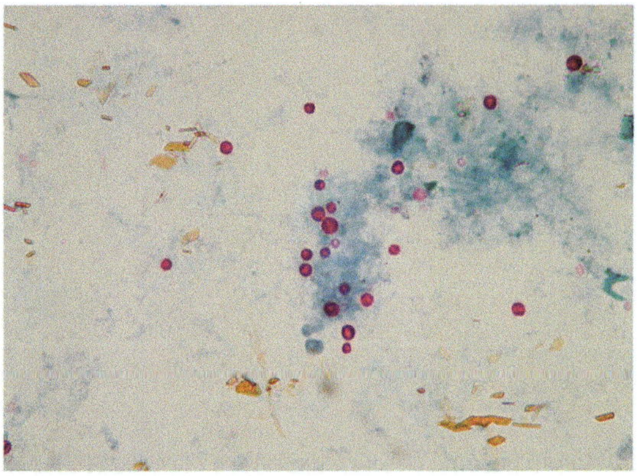

**Fig. 2** Non-*Cryptosporidium* artifacts exhibiting nonspecific modified Ziehl–Neelsen staining as seen using a 50X objective

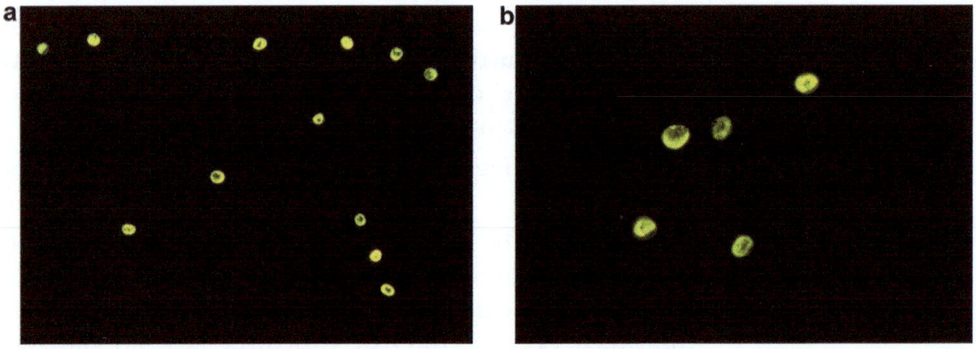

**Fig. 3** *Cryptosporidium* oocysts (~5 μm diameter) in feces demonstrating typical features with auramine phenol staining as seen using (**a**) 50X objective lens for screening and (**b**) 100X objective lens for confirmation. The oocyst wall is faintly visible around the internal contents that exhibit the stain. Often, the sporozoites lay toward the edge of oocysts which gives them the so-called doughnut or erythrocytic staining (denser around the outside and fainter or clear in the middle)

10. Do not blot slides dry, as some blotting papers contain fluorescent fibers.
11. Do not look at the fluorescent light (particularly with the UV filters) around the stage area of the microscope as this can be harmful to your eyes.
12. The smear can be examined with or without a coverslip, and *Cryptosporidium* spp. oocysts exhibit a characteristically yellow/green fluorescence against a dark background. It is the internal sporozoites that absorb the stain and fluoresce brightly, with individual crescent-shaped sporozoites wrapped around each other often visible [8] (*see* Fig. 3).

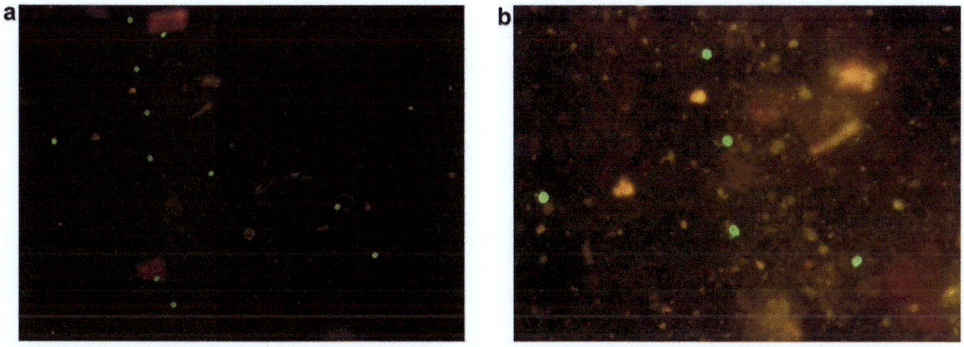


**Fig. 4** *Cryptosporidium* oocysts (~5 μm diameter) in feces demonstrating typical staining with FITC-labeled antibody as seen using a) 20X objective lens for screening and b) 40X objective lens for confirmation. The oocyst wall is brightly stained. Note an open oocyst (low center in image) demonstrating the "Pac-Man" appearance of oocysts that have been split open (excysted)

13. FITC-stained oocysts exhibit a bright apple-green fluorescence under the FITC filter set. Often the fluorescence has an increased intensity around the entire circumference of the oocyst, with no visible breaks in oocyst wall staining. If Evans blue or Eriochrome Black, which reduce nonspecific fluorescence, is included in the kit, the background fluorescence will be red (*see* Fig. 4a, b).

## Competing Interest Statement

The authors have no conflict of interests to declare that are relevant to the content of this chapter.

### References

1. Casemore DP, Armstrong M, Sands RL (1985) Laboratory diagnosis of cryptosporidiosis. J Clin Path 38:1337–1341
2. Smith H (2008) Diagnostics. In: Fayer R, Xiao L (eds) *Cryptosporidium* and cryptosporidiosis, 2nd edn. CRC Press, Boca Raton
3. Anon (2016) Chapter 2.9.4. cryptosporidiosis. In: Manual of diagnostic tests and vaccines for terrestrial animals. Available at: http://www.oie.int/fileadmin/Home/eng/Health_standards/tahm/2.09.04_CRYPTO.pdf
4. Chalmers RM, Campbell BM, Crouch N, Charlett A, Davies AP (2011) Comparison of the diagnostic sensitivity and specificity of seven *Cryptosporidium* assays used in the United Kingdom. J Med Microbiol 60:1598–1604
5. Chalmers RM, Katzer F (2013) Looking for *Cryptosporidium*: the application of advances in detection and diagnosis. Trends Parasitol 29: 237–251
6. Perry JL, Matthews JS, Miller GR (1990) Parasite detection efficiencies of five stool concentration systems. J Clin Microbiol 28:1094–1097
7. Saez AC, Manser MM, Andrews N, Chiodini PL (2011) Comparison between the Midi Parasep and Midi Parasep Solvent Free (SF) faecal parasite concentrators. J Clin Pathol 64:901–904
8. Hanscheid T, Melo Cristino J, Salgado MJ (2008) Screening of auramine-stained smears of all fecal samples is a rapid and inexpensive way to increase the detection of coccidial infections. Int J Infect Dis 12:47–50

# Chapter 2

# *Cryptosporidium* Diagnostic Assays: Molecular Detection and Species Determination by Real-Time PCR-Based Assays

Guy Robinson, Kristin Elwin, and Rachel M. Chalmers

## Abstract

Molecular diagnostic assays for *Cryptosporidium* are usually based on polymerase chain reaction (PCR) and may detect the entire genus or target-specified species. Of the ever-growing number of species, fewer than half have been reported from humans, and most human cases of cryptosporidiosis are caused by *Cryptosporidium parvum* or *Cryptosporidium hominis*. Here, we describe a real-time PCR for the detection of all *Cryptosporidium* spp. that can then be differentiated by sequencing the PCR amplicons and a duplex real-time PCR for the simultaneous detection and differentiation of *C. parvum* and *C. hominis*.

**Key words** *Cryptosporidium*, *C. parvum*, *C. hominis*, Detection, DNA extraction, Real-time PCR

## 1 Introduction

Molecular detection of *Cryptosporidium* spp. in feces, or other relevant clinical samples such as bile, small bowel aspirate, sputum, bronchoalveolar lavage, antral washout, or liver biopsy, is usually based on polymerase chain reaction (PCR) amplification of sporozoite deoxyribonucleic acid (DNA). Molecular methods offer improved sensitivity and detection rates compared with oocyst detection by tinctorial microscopy and antigen detection by enzyme immunoassays (EIA) or immunochromatographic lateral flow assays (ICLF) [1, 2]. A critical stage for molecular detection is the adequate disruption of oocysts for efficient DNA extraction from the sporozoites contained within. Oocysts may be partially purified first, using a flotation or sedimentation technique, or DNA may be extracted directly from feces [1]. Options for oocyst disruption include bead beating, freeze–thaw cycles, heating, or chemical/enzymatic treatments [1]. Options for subsequent DNA extraction and purification include commercial spin columns, Glassmilk, and Chelex resin [1, 3]. Many multipathogen gastrointestinal diagnostic PCR assays are now available in "black box"

Jan R. Mead and Michael J. Arrowood (eds.), *Cryptosporidium: Methods and Protocols*, Methods in Molecular Biology, vol. 2978, https://doi.org/10.1007/978-1-0716-4824-7_2,

format [4]. Here, we outline *Cryptosporidium* DNA extraction directly from stool and other sample types and describe two options for real-time PCR detection. The first is designed to detect any species in the genus *Cryptosporidium* using a real-time PCR targeting the small subunit ribosomal ribonucleic acid (SSU rRNA) gene (adapted from [5–7]). By sequencing the DNA amplicons, the infecting species can be determined. Of the ever-growing list of *Cryptosporidium* species, fewer than half have been reported infecting humans, and most cases of human cryptosporidiosis are caused by just two species, *Cryptosporidium parvum* (zoonotic) or *Cryptosporidium hominis* (anthroponotic). The second assay described here is therefore designed for the specific detection and differentiation of *C. parvum* and *C. hominis* using a real-time PCR developed from two previously published assays [8, 9].

## 2 Materials

### 2.1 DNA Extraction (See Note 1)

1. Commercially available kits suitable for the extraction of DNA from appropriate specimen types and preparations may vary (*see* **Note 2**). See kit inserts for materials required. Column-based commercial DNA extraction kits usually contain some sort of lysis buffer, proteinase K, wash buffers, and a membrane-based spin column.
2. If semi-purification of the sample is required, see the salt flotation method in Chap. 5 or the sedimentation method in Chap. 1.
3. Control material should be considered carefully (*see* **Note 2**).

### 2.2 Genus-Specific Real-Time PCR for all Cryptosporidium Species (Adapted from 5–7)

This assay is suitable for testing both human and animal samples, as all known species of *Cryptosporidium* are amplified by these primers. This assay was validated on human clinical and animal samples at the Cryptosporidium Reference Unit using the Rotor-Gene Q (Qiagen), with the specified reagents. If this assay is used as a primary diagnostic, an internal control to quality assure the process needs to be included in each sample (*see* **Note 3**).

1. Vortex mixer and microcentrifuge for 1.5-mL tubes, or a combination instrument (e.g., Grant Instruments Combi-Spin PCV-2400).
2. PCR workstation with ultraviolet (UV) decontamination.
3. Variable-volume pipettes.
4. Filter-protected pipette tips.
5. Real-time PCR thermocycler (e.g., Qiagen Rotor-Gene Q HRM; Qiagen) (*see* **Note 1**).
6. Type-it HRM PCR kit (Qiagen) (*see* **Note 1**).

7. 10-μM 18S forward primer (18SF): standard desalted purification, 5′-GGA AGG GTT GTA TTT ATT AGA TAA AG-3′ [5–7].
8. 10-μM 18S reverse primer (18SR): standard desalted purification, 5′-CTC ATA AGG TGC TGA AGG AGT A-3′ [5–7].
9. Nuclease-free water.

### 2.3 Duplex Real-Time PCR for Cryptosporidium hominis and Cryptosporidium parvum

This assay was validated on human clinical samples at the Cryptosporidium Reference Unit using both the Rotor-Gene 3000 and 6000 (Corbett Research) and the Rotor-Gene Q (Qiagen), with the specified reagents (*see* **Notes 4** and **5**). If this assay is used as a primary diagnostic, an internal control to quality assure the process needs to be included in each sample (*see* **Notes 2** and **3**).

1. Vortex mixer and microcentrifuge for 1.5-mL tubes, or a combination instrument (e.g., Grant Instruments Combi-spin PCV-2400).
2. PCR workstation with UV decontamination.
3. Variable-volume pipettes.
4. Filter-protected pipette tips.
5. Real-time PCR thermocycler (e.g., Qiagen Rotor-Gene Q HRM) (*see* **Note 1**).
6. TaqMan Environmental Master Mix 2.0 (Applied Biosystems, Foster City, CA) or a suitable alternative commercial real-time master mix if locally verified.
7. *C. hominis* A135 forward primer (CRUA135ChF): standard desalted purification, 10 μM, 5′-CAC CAA AGA TAA TGG ATG TTG TTG AT-3′ (2861-2886 XM-661774 TU502).
8. *C. hominis* A135 reverse primer (CRUA135ChR): standard desalted purification, 10 μM, 5′-AAT TGC TTC GAC ATC GTC CAA T-3′ (2964-2985 XM-661774 TU502).
9. *C. hominis* A135 MGB TaqMan probe (CRUA135ChTM): 10 μM, 6-FAM-5′-CAA ACG AGC TAT TAA AGG-3′ -MGB-NFQ (2903-2920 XM-661774 TU502).
10. Lib13 forward primer (CRULib13F): standard desalted purification, 10 μM, 5′-TCC TTG AAA TGA ATA TTT GTG ACT CG-3′ (458-483 B78618 IOWA).
11. *C. parvum* Lib13 reverse primer (CRULib13RCp): standard desalted purification, 10 μM, 5′-TTA ATG TGG TAG TTG CGG TTG AAC-3′ (600-623 B78618 IOWA).
12. *C. parvum* Lib13 MGB TaqMan probe (CRULib13CpTM): 10 μM, VIC-5′-TAT CTC TTC GTA GCG GCG TA-3′ -MGB-NFQ (569-588 B78618 IOWA).
13. Nuclease-free water.

## 3 Methods

### 3.1 DNA Extraction

Options for oocyst disruption prior to DNA extraction and purification include:

1. Bead beating: This method of oocyst disruption uses zirconia beads shaken very fast on an instrument such as the FastPrep-24 (MP Biomedicals) (*see* **Note 6**).
2. Freeze–thawing: This method usually cycles between freezing in liquid nitrogen and thawing at 95 °C. If liquid nitrogen is difficult to source, a dry ice/methanol slurry can be used as an alternative.
3. Heating: Incubating samples at 95–100 °C for 10 min to 1 h is sufficient for disrupting the oocysts prior to DNA extraction.
4. Chemical/enzymatic treatment: Most DNA extraction kits include a lysis stage that involves incubation with a buffer containing components for chemical or enzymatic digestion.

There are several options for *Cryptosporidium* DNA extraction and purification depending upon the nature of the sample being tested:

1. Directly from stool: Spin-column or magnet-based extraction, with constituents to counter the inhibitory substances found in feces, is a common means of extracting directly from stool samples due to their high performance in providing good-quality purified DNA (*see* **Note 7**).
2. From a semi-purified oocyst suspension (e.g., following flotation method in Chap. 5 or sedimentation method in Chap. 1) and other liquid sample types (e.g., bile, small bowel aspirate, sputum, bronchoalveolar lavage, antral washout): Column-based kits can also be used to extract DNA from these liquid samples (*see* **Note 8**).
3. From tissue samples: Commercially available tissue extraction kits/protocols are suitable for extracting *Cryptosporidium* DNA from tissue samples known or suspected to contain the parasite (e.g., liver biopsy).

### 3.2 Genus-Specific Real-Time PCR for All Cryptosporidium Species (Adapted from 5–7)

1. Allow the DNA samples and PCR reagents (*see* Table 1) to thaw separately at room temperature.
2. Momentarily vortex the DNA samples and PCR reagents before centrifuging briefly to remove contents from the tube lids.
3. In a DNA-free PCR workstation, prepare sufficient PCR master mix in a 1.5-mL microcentrifuge tube for the required number of samples, plus a no-template control (nuclease-free

**Table 1**
**Real-time SSU rDNA PCR master mix recipe, adapted from [5–7]**

| Master mix components | 1× reaction (μL) |
|---|---|
| Type-it HRM PCR kit (Qiagen) | 12.5 |
| 18SF | 2.5 |
| 18SR | 2.5 |
| nf$H_2O$ | 2.5 |
| [DNA] | [5] |
| Total reaction volume | 25 |

**Table 2**
**Conditions for the real-time SSU rDNA PCR**

| Stage | Temperature | Duration (s) | Comments |
|---|---|---|---|
| Hold (Hot start) | 95 °C | 300 | |
| Cycling (50 cycles) | 95 °C<br>62 °C<br>72 °C | 45<br>60<br>60 | <br><br>Acquiring on appropriate channel with calibration based on autogain optimization on first acquisition |

water), a *Cryptosporidium*-positive DNA control, and 20% extra master mix for wastage, based on the details for a single reaction (*see* Table 1).

4. Briefly vortex the master mix and spin down in a centrifuge for a few seconds to remove contents from the tube lid.
5. Pipette 20 μL of master mix into a labeled PCR tube for each sample.
6. Add 5 μL of nuclease-free water to the no-template control tube.
7. Move the tubes to a designated DNA template addition PCR workstation and UV decontaminate the workstation and tools used for reagent preparation.
8. Add 5 μL of each DNA sample to its corresponding tube.
9. Finally, to further reduce the risk of cross-contamination, add 5 μL of *Cryptosporidium*-positive DNA to the last tube.
10. Transfer the PCR tubes to a real-time thermocycler and run the following three-step cycling conditions (*see* Table 2):

11. Once the run has completed, analyze the results by applying a fluorescence threshold of 0.05, checking that the positive controls are positive (have a sigmoidal curve in the correct channel), observing the characteristic curve, and recording the cycle threshold (Ct) values from each sample.
12. The *Cryptosporidium* species present in positive samples is determined by Sanger sequencing the PCR amplicons (*see* **Note 9**). If this is not undertaken or sent for sequencing the same day, store samples at 2–8 °C for up to a week or − 20 °C for longer periods.

### *3.3 Duplex Real-Time PCR of* Cryptosporidium hominis *and* Cryptosporidium parvum

The two PCR assays designed to be specific for *C. hominis* and *C. parvum*, combined into the duplex real-time assay described here [10], are modified from the *C. hominis* assay described by Hadfield et al. [8] and *C. parvum* described by Tosini et al. [9] (*see* **Notes 4** and **10**).

1. In a DNA-free area, prepare sufficient PCR master mix for the number of samples required, plus a no-template control (nuclease-free water), and *C. hominis* and *C. parvum* DNA-positive controls, based on the details for a single reaction in Table 3.
2. Pipette 22 μL of master mix into each PCR tube and add 3 μL of test DNA.
3. Transfer the PCR tubes to a thermocycler and run the following two-step cycling conditions (*see* Table 4):
4. Once the run has completed, analyze the results by applying a fluorescence threshold of 0.05, checking that the positive

**Table 3**
***C. hominis* and *C. parvum* duplex real-time PCR master mix recipe**

| Master mix components | 1× reaction (μL) |
|---|---|
| Environmental master mix | 12.5 |
| CRUA135ChF | 2.25 |
| CRUA135ChR | 0.75 |
| CRUA135ChTM | 0.25 |
| CRULib13F | 2.25 |
| CRULib13RCp | 2.25 |
| CRULib13CpTM | 0.25 |
| nf$H_2O$ | 1.5 |
| [DNA] | [3] |
| Total reaction volume | 25 |

**Table 4**
**Conditions for *C. hominis* and *C. parvum* duplex real-time PCR**

| Stage | Temperature | Duration (s) | Comments |
|---|---|---|---|
| Hold (Hot start) | 95 °C | 600 | |
| Cycling (50 cycles) | 95 °C | 15 | |
| | 60 °C | 60 | Acquiring on appropriate channel with calibration based on autogain optimization on first acquisition |

controls appear in the correct PCR channel, observing the characteristic curves, and recording the Ct values from each sample (*see* **Note 11**).

## 4 Notes

1. Wear appropriate personal protective equipment (PPE). Extreme care must be taken to avoid contamination of PCR reagents and equipment with genomic and amplified DNA. Work in a low- to high-risk direction: reagent tubes and racks should be handled before sample tubes and racks. Gloves should be changed after touching sample tubes/racks or other potentially contaminated areas such as the discard box.
2. It is important to choose a kit that incorporates *Cryptosporidium* oocyst disruption. Alternatively, the method may be adapted to ensure adequate disruption of oocysts before sporozoite DNA can be extracted and purified, for example, by bead beating, freeze–thaw cycles, heating, or chemical/enzymatic treatments [1].
3. All tests should include a known and characterized *Cryptosporidium*-positive sample as a process-positive control, which will evidence adequate disruption of oocysts, extraction of DNA, and PCR amplification. In addition, for diagnostic tests, it is good practice to use an internal control within each sample, either by amplifying DNA also known to be present in clinical samples (e.g., human DNA targets) or spiking with exogenous DNA prior to extraction. This provides surety that a negative result is true, due to the absence of *Cryptosporidium* and not due to PCR inhibition or lack of extraction from within that sample. Several products are available commercially that provide this element of quality control. *Cryptosporidium*-positive and negative control DNA should also always be included in the PCR to demonstrate that it worked correctly and without

reagent contamination. If the same positive control DNA is used each time, assay results can be used for performance monitoring.

4. The assay is presented in duplex format here, but the targets can be run independently. If *C. hominis* only or *C. parvum* only is sought, the reagents and conditions remain the same but with only the primers specific for the species of interest. The missing reaction volume from removing the unwanted primers and probe must be made up with nuclease-free water. *C. parvum* only may be suitable for samples where only this zoonotic species is sought (e.g., young livestock).
5. Use of alternative reagents, conditions, or thermocyclers may alter the assay's performance and will need optimizing and validating to ensure satisfactory sensitivity and specificity.
6. A vortex shaker is not sufficiently vigorous.
7. There are several commercial kits available for the extraction of DNA from stool, feces, or soil, which can all be used for this sample type. Follow the manufacturer's instructions and ensure adequate disruption of oocysts, for example, by extended heating or chemical/enzymatic treatments [1].
8. For semi-purified suspensions of oocysts, the kits designed for the extraction of DNA from stool or soil are not required due to the much cleaner nature of the samples. Instead, blood or bodily fluid DNA extraction kits and protocols may be used, following adequate oocyst disruption.
9. Various commercial providers are available to provide Sanger sequencing by mailing the PCR amplicons and an aliquot of the primers at ambient temperature, with the sequencing data usually returned via email the following day. However, if facilities are available and/or a rapid turnaround is required, in-house sequencing can be performed according to the specific instrument manufacturer's instructions.
10. Specificity testing of this assay demonstrated slight amplification of *Cryptosporidium cuniculus* in the *C. hominis* assay and *Cryptosporidium equi* in the *C. parvum* assay, but the real-time profiles of all four species are distinguishable (*see* Figs. 1 and 2 and **Note 11**). Suspected *C. cuniculus* or *C. equi* isolates can be confirmed by PCR and sequencing the SSU rRNA gene as described above.
11. All curves should be sigmoidal and reach high fluorescence. Consider using a confirmation assay, such as the real-time SSU rDNA sequencing assay described above, if results are not typical. For example, if a sample tested using the A135 *C. hominis* assay produces a high Ct (or does not quite cross the threshold) and very low final fluorescence (green FAM

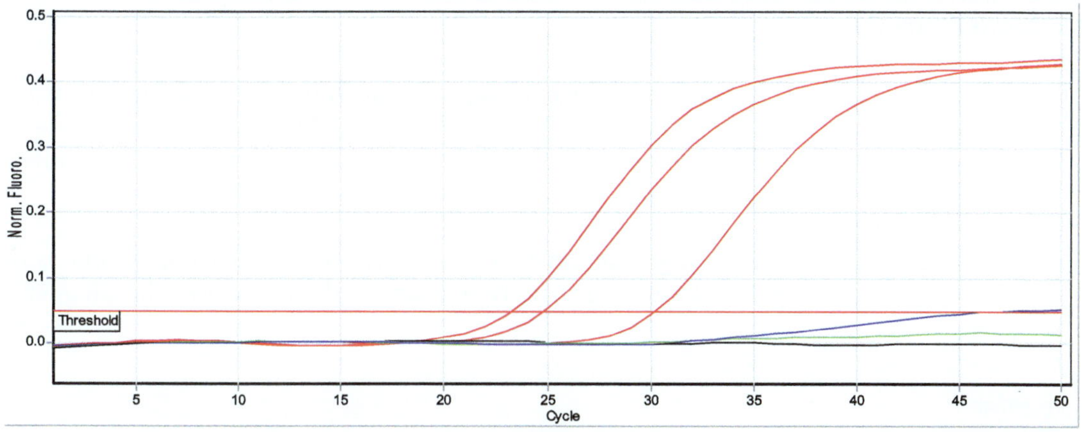


**Fig. 1** Amplification curves produced by the A135 *Cryptosporidium hominis* polymerase chain reaction (PCR) showing three characteristic *C. hominis* curves (red) and a single *C. cuniculus* (blue) curve. Also shown: threshold (red horizontal line), *C. parvum* (green), and no-template control (black)

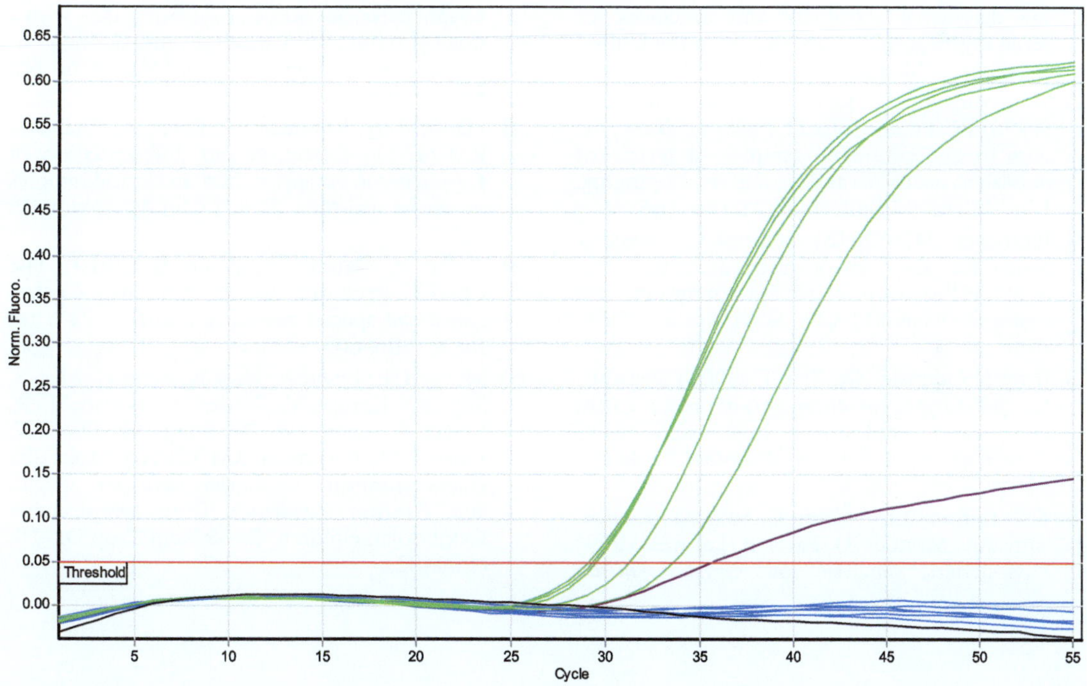


**Fig. 2** Amplification curves produced by the Lib13 *C. parvum* PCR showing five characteristic *C. parvum* curves (green) and a single *Cryptosporidium equi* (purple) curve. Also shown: threshold (red horizontal line), *C. hominis* (blue), and no-template control (black)

channel), this may indicate the presence of *C. cuniculus* (*see* Fig. 1). If a sample tested using the Lib13 *C. parvum* assay has a Ct within the normal range (less than 40) but very low final fluorescence (yellow VIC channel), this may indicate (extremely rarely) the presence of *C. equi* (*see* Fig. 2).

## Acknowledgments

The research leading to these methods has received funding from the European Union Seventh Framework Programme ([FP7/2007–2013] [FP7/2007–2011]) under Grant Agreement No. 311846.

**Competing Interest Statement** The authors have no conflicts of interest to declare that are relevant to the content of this chapter.

### References

1. Smith H (2008) Diagnostics. In: Fayer R, Xiao L (eds) *Cryptosporidium* and cryptosporidiosis, 2nd edn. CRC Press, Boca Raton
2. Chalmers RM, Campbell BM, Crouch N, Charlett A, Davies AP (2011) Comparison of the diagnostic sensitivity and specificity of seven *cryptosporidium* assays used in the United Kingdom. J Med Microbiol 60:1598–1604
3. Wiedenmann A, Kruger P, Botzenhart K (1998) PCR detection of *Cryptosporidium parvum* in environmental samples—a review of published protocols and current developments. J Ind Microbiol Biotechnol 21:150–166
4. Binnicker MJ (2015) Multiplex molecular panels for diagnosis of gastrointestinal infection: performance, result interpretation, and cost-effectiveness. J Clin Microbiol 53:3723–3728
5. Jiang J, Alderisio KA, Xiao L (2005) Distribution of *Cryptosporidium* genotypes in storm event water samples from three watersheds in New York. Appl Environ Microbiol 71:4446–4454
6. Xiao L, Escalante L, Yang C, Sulaiman I, Escalante AA, Montali RJ, Fayer R, Lal AA (1999) Phylogenetic analysis of *Cryptosporidium* parasites based on the small-subunit rRNA gene locus. Appl Environ Microbiol 65:1578–1583
7. Xiao L, Alderisio K, Limor J, Royer M, Lal AA (2000) Identification of species and sources of *Cryptosporidium* oocysts in storm waters with a small-subunit rRNA-based diagnostic and genotyping tool. Appl Environ Microbiol 66: 5492–5498
8. Hadfield SJ, Robinson G, Elwin K, Chalmers RM (2011) Detection and differentiation of *Cryptosporidium* spp. in human clinical samples by use of real-time PCR. J Clin Microbiol 49: 918–924
9. Tosini F, Drumo R, Elwin K (2010) The CpA135 gene as a marker to identify *Cryptosporidium* species infecting humans. Parasitol Int 59:606–609
10. Moore CE, Elwin K, Phot N, Seng C, Mao S, Suy K, Kumar V, Nader J, Bousfield R, Perera S, Bailey JW, Beeching NJ, Day NP, Parry CM, Chalmers RM (2016) Molecular characterisation of *Cryptosporidium* species and *Giardia duodenalis* from symptomatic Cambodian children. PLoS Negl Trop Dis 10: e0004822

Chapter 3

# Detection of *Cryptosporidium* Recovered from Large-Volume Water Samples Using Dead-End Ultrafiltration

**Amy M. Kahler and Mia C. Mattioli**


## Abstract

This procedure provides instructions for detection of *Cryptosporidium* from large-volume water samples. Water samples are collected by dead-end ultrafiltration in the field and processed in a laboratory. Microbes recovered from the filters are further concentrated and subjected to *Cryptosporidium* isolation or nucleic acid extraction methods for the detection of *Cryptosporidium* oocysts or *Cryptosporidium* deoxyribonucleic acid (DNA).


**Key words** *Cryptosporidium*, Dead-end ultrafiltration, IMS, Microscopy, Real-time PCR, Water

## 1 Introduction

This procedure provides instructions for detection of *Cryptosporidium* recovered from large-volume water samples. Water samples are collected using a hollow-fiber, dead-end ultrafiltration method (DEUF). The hollow-fiber ultrafilters used in this procedure are kidney dialysis filters which have pore sizes on the order of 10 nm (~30 kilodaltons [kDa]) and thus can be used to recover protozoan parasites, bacteria, viruses, and large toxins from water samples [1–3]. Dialysis filters are available in a wide range of filter fiber surface areas, but those with larger surface areas can process more water without pressure increases or clogging compared to filters with smaller surface areas. For this procedure, a filter with a fiber surface area of >2 $m^2$ is recommended, especially for surface waters. The DEUF method is designed to be performed in the field with minimal equipment and setup [4]. In the DEUF procedure, water flows into the ultrafilter, through the hollow-fiber membranes, and out of the ultrafilter pores while microbes are captured within the hollow fibers. The ultrafilters are capable of filtering 10–50 L of

Jan R. Mead and Michael J. Arrowood (eds.), *Cryptosporidium: Methods and Protocols*, Methods in Molecular Biology, vol. 2978, https://doi.org/10.1007/978-1-0716-4824-7_3,

turbid surface water or hundreds of liters of finished drinking water. The volume of water filtered will depend on water quality characteristics and the suspected concentrations of target microorganisms. After ultrafiltration, the ultrafilter is processed in a laboratory. The ultrafilter is backflushed with 500 mL of a solution containing 0.5% Tween 80, 0.01% sodium polyphosphate, and 0.001% Antifoam L-30 emulsion to recover the microbes from the ultrafilter. If the concentrations of the target microbes are sufficiently high, the resulting backflushed solution can be analyzed directly. If the concentrations of the target microbes are low or unknown, the backflushed solution can be further concentrated to achieve a volume that is amenable to downstream detection methods.

For *Cryptosporidium* detection, the sample concentrate may be subjected to immunomagnetic separation (IMS) and fluorescent antibody (FA) microscopy for observation of *Cryptosporidium* oocysts [5] and/or nucleic acid extraction and real-time polymerase chain reaction (PCR) for detection of *Cryptosporidium* deoxyribonucleic acid (DNA) [6, 7]. The choice of detection methods should be determined by the goals of the study and/or the unique characteristics of the water type being analyzed. The performance recovery efficiency of each methodological step (DEUF, secondary concentration, IMS, and nucleic acid extraction) may vary depending on filter type as well as water quality and composition [1, 2]. Therefore, it is recommended that the complete method be evaluated and validated before processing real-world samples to ensure that effective detection can be achieved [8].

## 2 Materials (*See* Note 1)

### 2.1 Dead-End Ultrafiltration

1. Insulated cooler.
2. Reusable freezer packs.
3. Kidney dialysis filter, 2.0–2.5 $m^2$ fiber area recommended (*see* **Note 2**).
4. Masterflex L/S 36 tubing (platinum-cured silicone or C-Flex ULTRA) (VWR, Radnor, PA).
5. Female DIN to 3/8″ barb fitting adapters, polypropylene (Molded Products, Inc., Harlan, IA).
6. SNP-8 acetal copolymer tubing clamps.
7. Blood port kidney storage cap (Molded Products).
8. Dialysate port cap, optional or required, depending on type of ultrafilter used (Molded Products).
9. Flow totalizing meter, 1/2″ schedule 80 polyvinyl chloride (PVC) National Pipe Thread (NPT) turbine, 3.8–38 L/min

(Omega Engineering, Inc. Swedesboro, NJ) fitted with two straight barbed to male NPT threaded adapters, 3/4″ National Pipe Taper Male (NPTM) × 3/8″ tubing inside diameter (ID).

10. 20-L collapsible sample collection containers (Cubitainer®, optional).
11. Long nose pliers.
12. Scissors.
13. 500-mL bottles (Nalgene).
14. Sodium thiosulfate.
15. Geotech Geopump peristaltic pump (Geotech Environmental Equipment, Inc., Denver, CO) with Easy-Load II pump head placed on the Series 2 drive (for nonpressurized water sources).
16. Alcohol wipe or equivalent.
17. 60-mL syringe.
18. 1/2″ ID hose × swivel female garden hose thread (FGHT) nylon swivel female insert.
19. SNP-12 acetal copolymer tubing clamps.
20. Masterflex I/P 89 platinum-cured silicone tubing (VWR).
21. Reducing connector 5/8″ to 3/8″.
22. SNP-28 acetal copolymer tubing clamps.
23. SNP-24 acetal copolymer tubing clamps.
24. SNP-19 acetal copolymer tubing clamps.

### 2.2 Backflushing

1. Masterflex L/S digital pump system with Easy-Load II pump head, 115/230 V (or use the same pump as for field collection) (Cole-Parmer)..
2. Lab jack with 10″ × 10″ top.
3. Ultra-flex support with base plate, 18″ arm length.
4. Platinum-cured silicone tubing (Masterflex L/S 36 tubing or C-Flex ULTRA) (VWR).
5. Blunt-edge forceps.
6. Face shield.
7. Long nose pliers.
8. SNP-12 or SNP-14 acetal copolymer tubing clamps.
9. 500-mL bottle.
10. 1-L bottle.
11. 10-mL plastic serological pipette.
12. Deionized or distilled (DI) water.
13. Tween 80.

14. Sodium hexametaphosphate.
15. Antifoam L-30 emulsion.

### 2.3 Secondary Concentration

1. Centrifuge capable of accepting 200- to 500-mL conical centrifuge tubes.
2. 200-mL bottles or 500-mL bottles.
3. 50-mL plastic serological pipette.
4. 0.01 M phosphate-buffered saline (PBS), pH 7.2–7.4 (1×) or equivalent buffer solution for resuspension.

### 2.4 Immunomagnetic Separation (IMS) and Fluorescent Antibody (FA) Microscopy

1. Glass Leighton tubes or Dynabeads L-10 tubes (Applied Biosystems Inc., Beverly, MA).
2. Deionized or distilled (DI) water.
3. Anti-*Cryptosporidium* magnetic beads (Dynabeads anti-*Cryptosporidium* or Dynabeads GC-Combo) (Applied Biosystems).
4. Rotary mixer for immunomagnetic beads (Dynabeads rotary mixer, Applied Biosystems).
5. Magnet for magnetic bead capture in Leighton tubes or 10- to 30-mm tubes (MPC-6 magnetic particle concentrator, Applied Biosystems).
6. 1-mL or 2-mL plastic serological pipettes.
7. 1.5-mL nuclease-free (NF) microcentrifuge tubes (Invitrogen Corp., Carlsbad, CA).
8. Magnet for magnetic bead capture in microcentrifuge tubes (MPC-S magnetic particle concentrator) (Applied Biosystems).
9. 0.01 M PBS, pH 7.2–7.4 (1×).
10. 0.1 N HCl.
11. 1 N NaOH.
12. Two-well microscope slides with adhesive coating (SuperStick slides, Waterborne Inc., New Orleans, LA).
13. Heat block or slide warmer.
14. EasyStain *Cryptosporidium* oocyst labeling reagent (bioMérieux Inc., Durham, NC).
15. Coverslips, 22 × 60 mm.
16. Clear nail polish.
17. Fluorescent microscope with fluorescein isothiocyanate (FITC) filter.

### 2.5 Nucleic Acid Extraction and Real-Time PCR

1. Molecular grade ethanol, 200 proof.
2. Nuclease-free water.

3. 0.2-mm zirconium oxide beads, $Y_2O_3$-stabilized, 95% (Union Process, Inc., Akron, OH).
4. 0.5-mm zirconium oxide beads, $Y_2O_3$-stabilized, 95% (Union Process, Inc.).
5. 0.1 N HCl.
6. Oven.
7. 0.5-mL nuclease-free mL tubes.
8. Double-ended micro-tapered stainless steel spatula.
9. Twist ties.
10. FastPrep-24 bead beater (MP Biomedicals, LLC, Santa Ana, CA).
11. FastPrep-compatible 2-mL empty bead beating tubes (MP Biomedicals).
12. FastPrep-compatible caps for 2-mL beating tubes (MP Biomedicals).
13. UNEX lysis buffer (Microbiologics, Inc., St. Cloud, MN).
14. Proteinase K, ≥600 mAU/mL.
15. Silica HiBind RNA minicolumn RNACOL (Omega Bio-tek, Inc., Norcross, GA).
16. OneStep PCR inhibitor removal kit (Zymo Research, Irvine, CA).
17. 2-mL collection tube.
18. 1.5-mL nuclease-free microcentrifuge tubes.
19. Tris-EDTA (TE) buffer, pH 8.0, molecular biology grade.
20. TaqMan Environmental Master Mix 2.0 (Life Technologies Corp., Carlsbad, CA).
21. Nuclease-free (NF) water.
22. Oligonucleotides.

    **Forward primer**: ATG ACG GGT AAC GGG GAA T.

    **Reverse primer**: CCA ATT ACA AAA CCA AAA AGT CC.

    **Probe**: 6FAM-CGC GCC TGC TGC CTT CCT TAG ATG-B HQ1.

23. T4 gene 32 protein (gp32).
24. Bovine serum albumin (BSA, molecular biology grade).
25. TaqMan Exogenous Internal Positive Control Kit (Life Technologies Corp.).
26. Aerosol barrier pipette tips.
27. Applied Biosystems 7500 real-time PCR thermocycler (Life Technologies Corp.).

28. MicroAmp optical 96-well reaction plate or MicroAmp optical 8-tube strip (Life Technologies Corp.).
29. Optical adhesive film (MicroAmp 96-well format or MicroAmp optical 8-cap strip) (Life Technologies Corp.).
30. Laminar flow cabinet or similar PCR-compatible workstation.
31. Plate or strip spinner.
32. RNase AWAY surface decontaminant (Thermo Fisher Scientific, Inc., Norcross, GA).

## 3 Methods

### 3.1 Reagent Preparation

1. Sodium thiosulfate solution (1% w/v): Add 5 g sodium thiosulfate to a 500-mL Nalgene bottle; add 500 mL DI water or ultrafilter effluent as described in Subheading 3.2, **step 15**.
2. Backflush solution (0.5% Tween 80, 0.01% sodium polyphosphate [NaPP], and 0.001% Antifoam L-30 emulsion): Add 10 mL DI water to a screw cap tube; add 1 g NaPP and 100 μL Antifoam (*see* **Note 3**) to the DI water (10% NaPP, 1% Antifoam), and shake vigorously to dissolve the NaPP (*see* **Note 4**). This stock solution can be kept at room temperature or at 4 °C. Add 500 mL DI water to a 500-mL bottle; add 2.5 mL Tween 80 and 500 μL of the NaPP/Antifoam stock solution to the DI water. Swirl to dissolve the Tween 80. If desired, the backflush solution can be prepared a day in advance and stored at 4 °C. Warm to room temperature before use.
3. BSA/gp32 solution (10 mg/mL BSA, 500 μg/mL gp32): Add 10 mL TE buffer to 100 mg of BSA and dissolve by swirling. Aliquot mL into 1.5-mL microcentrifuge tubes and store at 4 °C. Thaw gp32 and add 950 μL BSA solution and invert to mix; store at 4 °C.

### 3.2 Ultrafiltration (DEUF, per CDC/USEPA Method)

1. Prepare the flow totalizer: Screw in a tubing adapter to each end of the totalizer; take a length of L/S 36 tubing and cut in half; push tubing onto both sides of the meter; no tubing clamps are needed.
2. Label ultrafilter with an indelible marker.
3. Fill out pertinent information on chain of custody form or sample collection bench sheet: Pertinent information may include sample location and/or sample ID, measured water quality parameters, water type, site description, etc.
4. Remove one end port cap and screw in a DIN adapter (this is the influent port).

5. Push platinum-cured silicone (L/S 36) tubing onto the DIN adapter and secure with an SNP-8 clamp. The length of influent tubing needed will be determined by the distance to the water source.
6. Close off the side port closest to the influent port by pushing the cap until it clicks.
7. Screw a blood port cap into the end port opposite the influent port.
8. Remove the cap from the effluent port (furthest from the influent port) and save. Push the flow totalizer tubing onto the effluent port, making sure that the arrow on the flow totalizer points away from the ultrafilter. No tubing clamp is needed.
9. Record the initial flow totalizer meter reading.
10. Place the effluent tubing in an empty 20-L Cubitainer® or equivalent container, downstream of the water being collected, or near a floor drain. The effluent water will be free of microbial contamination, so it can be released into the environment as conditions allow (*see* **Note 5**).
11. Nonpressurized water sources (*see* Fig. 1): Place the influent tubing into the water source (or water that has been collected in a 20-L Cubitainer®). Ensure that the end of the tubing will stay below the surface of the water (*see* **Note 6**). Feed the influent tubing through the pump head and close the pump head using the lever. Plug in the "battery" power cord into the pump and the other end into the battery (*see* **Note 7**). The Geopump comes with alligator clips so that any external battery that is 12–18 V DC @ 70 watts or 90–260 V AC @ 47–65 Hz can be used. Place the battery in a location where it will not get wet. Position the toggle switch on the pump for the appropriate direction of flow. Ensure the speed dial is set to zero and turn on the pump. Gradually increase the speed to the maximum setting (*see* **Note 8**).

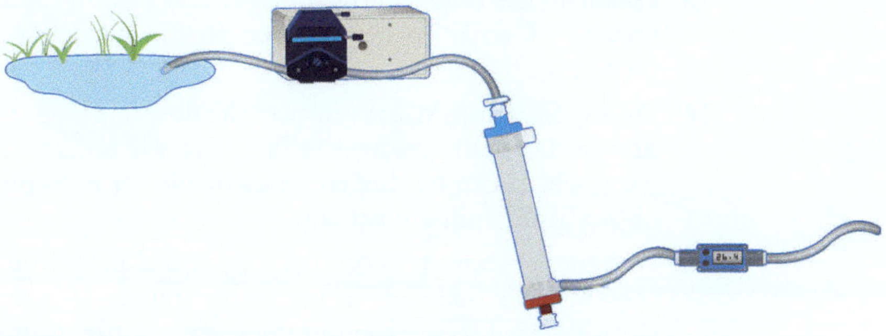

**Fig. 1** Dead-end ultrafiltration (DEUF) assembly for nonpressurized water sources

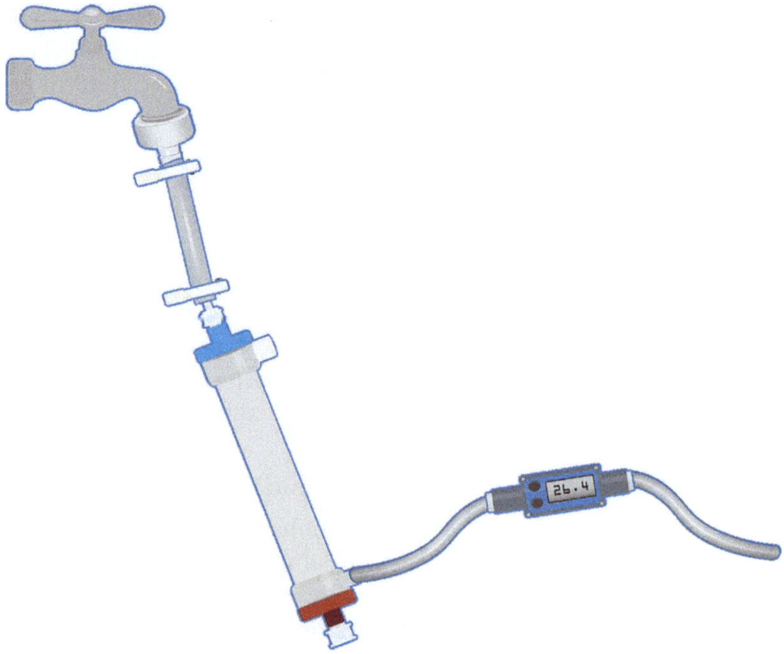


**Fig. 2** DEUF assembly for pressurized water sources

12. Pressurized water sources (*see* Fig. 2): Wipe the faucet with an alcohol wipe or other sanitizer. Turn on the faucet and purge the water for 2–3 min. Turn off the faucet before setting up the ultrafilter. If sampling from a standard garden faucet (hose bib), screw on an FGHT adapter. Push influent tubing onto the adapter and secure with an SNP-12 clamp. If sampling from a nonstandard faucet, push 4–6 inches of I/P 89 tubing over the faucet head and secure with an SNP-28 clamp. Use a reducing connector to connect the I/P 89 and L/S 36 influent tubing; secure both sides with SNP-19 and SNP-8 clamps, respectively. Turn on the faucet and gradually increase the flow until the desired flow rate is achieved (3–5 L/min) (*see* **Note 9**).
13. Calculate the desired flow totalizer end reading by adding the volume of water to be collected to the starting reading (*see* **Note 10**).
14. During filtration, visually inspect the flow rate from the effluent tubing. Dramatic changes in flow rate will indicate filter clogging, which can be due to water quality or entrapment of an object in the influent tubing.
15. Prepare the sodium thiosulfate solution if a chlorine residual is present (or suspected) in the water source. If water was not already added to the sodium thiosulfate bottle during preparation, fill the bottle to the 500-mL mark using the ultrafilter effluent. Shake to dissolve and set aside.

16. Once the desired volume of water has been filtered, turn off the pump or faucet.
17. For water sources with a chlorine residual, follow the chlorine quenching steps below. For nonchlorinated water sources, proceed to **step 20**.
18. Nonpressurized water sources: Place the influent tubing into the sodium thiosulfate solution. Turn on the pump to draw the sodium thiosulfate solution into the ultrafilter. Continue until the sodium thiosulfate solution has been drawn through most of the influent tubing, but do not allow air to be pumped into the ultrafilter. Turn off the pump and release the lever on the pump head.
19. Pressurized water sources: remove the influent tubing from the faucet and cut the tubing so that only 2–3 inches remain attached to the ultrafilter. Remove the plunger from the 60-mL syringe. Insert the tip of the syringe into the influent tubing and secure with an SNP-8 clamp. Pour 60 mL of the sodium thiosulfate solution into the syringe and insert the plunger to push the solution through the ultrafilter. Remove the SNP-8 clamp and take the syringe out of the tubing. Remove the plunger from the syringe. Pulling out the plunger before removing the syringe from the tubing will create negative pressure and draw liquid out of the ultrafilter. Repeat sodium thiosulfate injection (as described above) one more time.
20. Unscrew the DIN adapter with influent tubing. Screw a blood port cap into the influent port.
21. Place influent tubing and clamps into a waste bag for used supplies. Return to the lab for disposal or decontamination.
22. Remove the effluent tubing and re-cap the effluent port using the cap supplied with the ultrafilter. The effluent tubing and flow totalizer can be reused for subsequent samples.
23. Place the sealed ultrafilter in a cooler for shipment or transport to the laboratory. Use ice packs or bags of ice to keep the cooler chilled. The ideal temperature range is 2–8 °C.

### 3.3 Backflushing and Secondary Concentration

1. Prepare the backflush solution.
2. Assemble the system as shown in Fig. 3 (*see* **Note 11**): Set the pump on the lab jack; raise the lab jack so that the pump head will be at the same height as the side port of the ultrafilter where the backflush will enter. Clamp ultrafilter vertically into support base so that end ports are at the top and bottom, and side ports are pointed in the direction of the pump. Ensure the ultrafilter is high enough that a 1-L bottle can be placed under

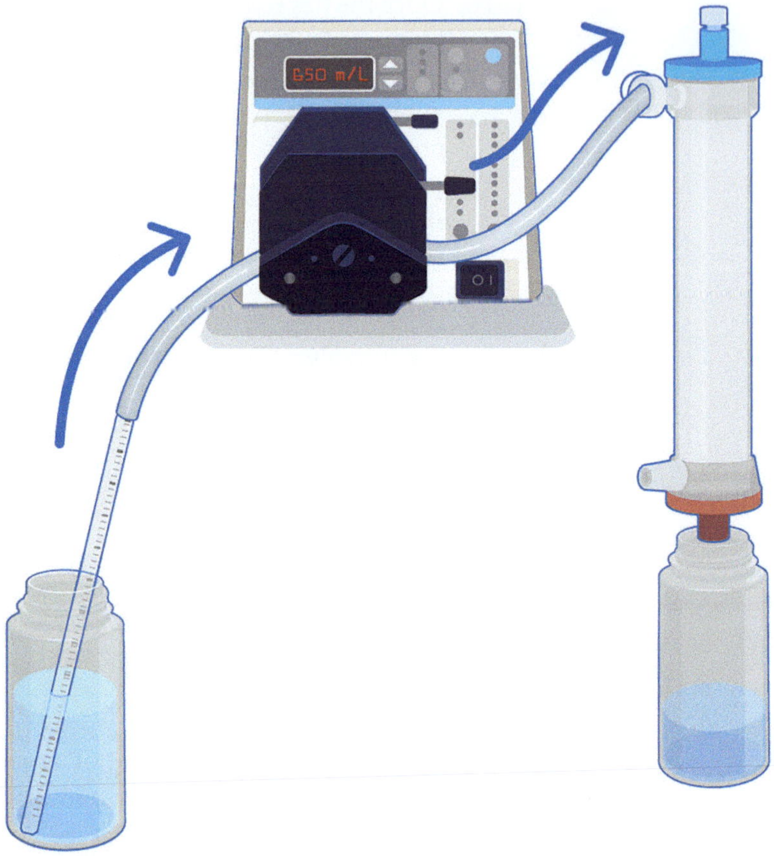


**Fig. 3** DEUF assembly for backflushing and recovery of captured sample concentrate

the end port. Ensure a blood port storage cap is screwed into the top port. Remove the cap from the highest side port. Ensure the remaining side port cap is secure.

3. Cut a 10–12″ length of platinum-cured silicone (L/S 36) tubing.
4. Push the tubing onto the open side port and secure with an SNP-12 or SNP-14 clamp.
5. Feed the tubing through the pump head.
6. Remove the cotton plug from a 10-mL serological pipette using a pair of forceps. Break off the conical tip of the plastic pipette (*see* **Note 12**).
7. Push the influent tubing onto the top of the pipette and place the pipette into the backflush solution.
8. Close the pump head and ensure the direction of flow is correct. Set the pump speed to 650 mL/min (*see* **Note 13**).

9. Remove the blood port storage cap from the lower end port and place a sterile 1-L bottle under the port.
10. Turn on the pump, and pump the entire backflush solution through the ultrafilter.
11. Turn off the pump when no backflush solution remains in the bottle and the flow out of the ultrafilter slows to a trickle. Do not continue pumping once the flow of liquid from the ultrafilter has stopped. Doing so will cause pressure to build up and increase the risk of the side port cap to burst off.
12. Open the pump head slowly to release the pressure in the influent tubing.
13. Remove the tubing clamp and save for subsequent ultrafilters.
14. Release the ultrafilter from the support base and cap the bottle containing the sample concentrate.
15. Unscrew the blood port caps and save for decontamination.
16. Measure and record the backflush concentrate volume, if necessary.
17. Centrifuge the backflush concentrate at 1500–2000× *g* for 15 min according to U.S. Environmental Protection Agency (USEPA) Method 1623.1 (*see* **Note 14**).
18. Carefully remove the supernatant using a 50-mL pipette, leaving 1–5 mL of supernatant above the pellet.
19. Measure and record the packed pellet volume and total concentrate volume.
20. Proceed to Subheadings 3.4 and 3.5 for *Cryptosporidium* detection by immunofluorescence assay microscopy or Subheadings 3.6 and3.7 for *Cryptosporidium* detection by real-time PCR. If both detection methods are utilized, split the centrifuge concentrate for processing by each method (*see* **Note 15**).

### 3.4 Immunomagnetic Separation (IMS, per USEPA Method 1623.1)

1. If the packed pellet volume is ≤0.5 mL, the entire sample concentrate, or up to 10 mL, can be processed by IMS. If the packed pellet volume is >0.5 mL, process no more than the equivalent of 0.5 mL. If desired, the sample concentrate can be split into subsamples in order to process the entire volume. If the packed pellet volume is ≤0.5 mL, but the total sample concentrate volume is >10 mL, the sample concentrate can be split into subsamples in order to process the entire volume, if desired.
2. Prepare a 1× dilution of SL-A buffer from the provided 10× buffer stock using DI water. For each sample processed, 1.5 mL of 1× SL-A buffer will be required.

3. Add the sample concentrate to an L-10 tube. Bring the volume up to 10 mL with DI water if processing <10 mL sample concentrate.
4. Add 1 mL each of the 10× SL-A and SL-B buffers to the L-10 tube.
5. Vortex the Dynabeads for 10 s and add 100 μL to the L-10 tube.
6. Place the L-10 tube on the rotary mixer and rotate at 18 rpm for 1 h at room temperature..
7. Place the L-10 tube in the MPC-6, ensuring the flat side of the tube is flush against the magnet.
8. Starting in an upright position, gently rock the tube by hand through 180° for 2 min, with approximately 1 tilt per s.
9. Remove the cap from the L-10 tube and pour the supernatant into a collection beaker. Allow more supernatant to settle and remove with a 1-mL or 2-mL serological pipette.
10. Remove the L-10 tube from the MPC-6 and resuspend the sample in 0.5 mL of 1× SL-A buffer by repeatedly releasing the liquid down the flat side of the tube. Do not vortex.
11. Transfer the solution to a 1.5-mL microcentrifuge tube. Repeat rinsing procedure two more times.
12. Place the microcentrifuge tube in the MPC-S with the magnet in place.
13. Starting in an upright position, gently rock the tube by hand through 180° for 1 min, with approximately 1 tilt per s.
14. Aspirate the supernatant from the tube using a 1000-μL micropipettor.
15. Add 1 mL of 0.01 M PBS to the tube, taking care not to disturb the beads at the back of the tube.
16. Remove the magnet from the MPC-S and gently resuspend the beads.
17. Replace the magnetic strip in the MPC-S.
18. Starting in an upright position, gently rock the tube by hand through 180° for 1 min, with approximately 1 tilt per s.
19. Aspirate the supernatant from the tube using a 1000-μL micropipettor.
20. Let the tube stand undisturbed for 1 min, then aspirate any remaining liquid from the tube.
21. Remove the magnet from the MPC-S and add 50 μL of 0.1 N HCl to the tube. Vortex on high for 50 s.
22. Allow the tube to stand undisturbed in an upright position for 10 min at room temperature.

23. Vortex on high for 30 s.
24. Tap the sample into the base of the tube and place in the MPC-S.
25. Replace the magnetic strip in the MPC-S and allow the tube to stand undisturbed for at least 10 s.
26. Add 5 μL NaOH to two sample wells on a two-well adhesive microscopy slide, or 10 μL to one well if the volume from the two dissociation steps will be combined to the same well.
27. While the tube is still in the MPC-S, transfer the sample volume from the tube to a microscopy slide well.
28. Repeat the dissociation process (*see* Subheading 3.4, **steps 21–27**); add the sample volume to a second sample well or combine with the volume from the first dissociation.
29. Proceed directly to fluorescent antibody microscopy or store the slides up to 24 h in a closed container at 4 °C.

### 3.5 Fluorescent Antibody (FA) Microscopy

1. Place the Fixing Buffer from the EasyStain kit into a beaker of ice water; hold at 4 °C until use.
2. Dry the sample suspension on the two-well adhesive microscope slides on a slide warmer or heat block (≤60 °C, usually ≤10 min). Alternatively, the slides can be dried inside a 37 °C incubator.
3. Add 50 μL EasyStain reagent to the well.
4. Incubate at room temperature for 30 min or in a box containing a moist tissue at 37 °C for 15 min (*see* **Note 16**).
5. Place the slide onto a paper towel and tilt the slide to absorb the stain from the well.
6. Add 300 μL ice-cold Fixing Buffer to the well, ensuring the entire well is covered. Let stand for 2 min.
7. Place the slide onto a Kimwipe or paper towel and tilt the slide to absorb the Fixing Buffer from the well.
8. Add one drop of Mounting Medium to the well and apply the coverslip (*see* **Note 17**).
9. Apply clear nail polish to the edges of the coverslip to seal the coverslip to the slide.
10. Examine the slide under 200X–400X magnification using a fluorescent microscope with a FITC filter. *Cryptosporidium* oocysts will appear as ~5 μm circles with a bright green fluorescent oocyst wall.

### 3.6 Nucleic Acid Extraction

1. Prepare 70% ethanol by combining 35 mL of 200-proof ethanol and 15 mL of nuclease-free (NF) water in a 50-mL conical tube. Store the solution at room temperature.

2. Prepare the extraction beads: Submerge the 0.2 mm and 0.5 mm beads separately in 0.1 N HCl in a 250-mL glass beaker and mix for 30 s. Let sit for 10 min, then decant the HCl. Rinse the beads by adding enough deionized or distilled water (DI water) to submerge the beads. Mix for 30 s and decant the DI water. Repeat the DI rinsing procedure four more times. Transfer the beads to separate clean 400-mL glass beakers. Bake the beads in a laboratory oven at 200 °C for 30 min (or until completely dry) and allow to cool to room temperature. Transfer the beads to a container with a cap, such as a 50-mL conical tube, and store at room temperature.
3. Cut the cap off a 0.5-mL microcentrifuge tube and secure with a twist tie onto the end of a double-ended spatula.
4. Scoop one capful each of the 0.2 mm and 0.5 mm acid-wash beads into a bead beating tube.
5. Add 750 μL water sample concentrate to the prepared bead beating tube (*see* **Note 18**). Add 750 μL UNEX lysis buffer and 75 μL Proteinase K, then vortex to mix.
6. Allow the sample to sit at room temperature for at least 15 min for Proteinase K activity.
7. Place the bead beating tube securely in the bead beater. Balance with a similarly weighted bead beating tube if processing one sample.
8. Beat the sample at 6.0 m/s for 60 s.
9. Centrifuge the bead beating tube at 10,000× *g* for 30 s.
10. Prepare the silica column by placing into a provided 2-mL collection tube.
11. Transfer 700 μL of supernatant from the bead beating tube into a prepared silica column.
12. Centrifuge the silica column at 10,000 × *g* for 1 min.
13. Pour the filtrate from the collection tube into an autoclave waste container; place the silica column back into the collection tube.
14. Repeat **steps 11–13** to process the remaining volume, being careful to avoid the beads and any pellet material while removing the supernatant.
15. Add 500 μL of 100% ethanol to the silica column.
16. Centrifuge the silica column at 10,000 × *g* for 1 min.
17. Pour the filtrate from the collection tube into an autoclave waste container; place the silica column back into the collection tube.
18. Add 500 μL of 70% ethanol to the silica column.
19. Centrifuge the silica column at 10,000 × *g* for 1 min.

20. Pour the filtrate from the collection tube into an autoclave waste container; place the silica column back into the collection tube.
21. Centrifuge the silica column at 10,000× *g* for 1 min. This is a dry spin to remove any residual ethanol.
22. Discard the collection tube and place the silica column into a labeled 1.5-mL microcentrifuge tube.
23. Add 80 μL TE buffer to the silica column.
24. Centrifuge the silica column at 9000× *g* for 1 min.
25. Insert a OneStep PCR inhibitor removal column into a provided collection tube.
26. Add 600 μL of the Prep Solution and centrifuge at 8000× *g* for 3 min.
27. Transfer the column to a labeled 1.5-mL microcentrifuge tube.
28. Transfer the eluted TE buffer from the silica column to the OneStep column.
29. Centrifuge the OneStep column at 8000× *g* for 1 min (*see* **Note 19**).
30. Discard the OneStep column and store the nucleic acid eluate at −20 or −80 °C until PCR analysis.

### 3.7 Real-Time PCR

1. Thaw-frozen reagents in a laminar flow or PCR-compatible cabinet.
2. Prepare the master mixture solution by adding the PCR reagents to a 1.5-mL microcentrifuge tube (*see* Table 1). Mix

**Table 1**
***Cryptosporidium* real-time PCR master mixture recipe**

| Reagent | Initial concentration | Volume per 50 μL reaction (μL) | Final concentration |
|---|---|---|---|
| Environmental master mix 2.0 | 2× | 25 | 1× |
| Forward primer | 10 μM | 1.25 | 250 nM |
| Reverse primer | 10 μM | 1.25 | 250 nM |
| Probe | 4 μM | 1.25 | 100 nM |
| BSA/gp32 | 10 mg/mL BSA; 500 μg/mL gp32 | 2.5 | 0.5 mg/mL BSA; 25 μg/mL gp32 |
| Internal control mix | 10× | 5 | 1× |
| Internal control DNA | 50× | 1 | 1× |
| NF water | NA | 7.75 | NA |

the solution inverting the tube several times or briefly vortexing.

3. Add 45 μL of the master mixture into each reaction tube in a 96-well plate or 8-tube strip.
4. Add 5 μL template DNA to the appropriate reaction tubes. Triplicate reaction tubes are recommended for environmental samples (*see* **Note 20**).
5. Add 5 μL nuclease-free water as a no-template control to the appropriate reaction tubes. Additional no-template controls using the internal control blank can also be analyzed.
6. Add 2–5 μL positive control, depending on the concentration and desired cycle threshold (Ct) value. If adding less than 5 μL, add additional NF water to bring the total reaction volume to 50 μL.
7. Briefly spin tube strips or 96-well plates to move liquid to the bottom of the tubes.
8. Run samples with the following cycling conditions: 95 °C for 10 min, followed by 45 cycles of 95 °C for 15 s and 55 °C for 1 min. Fluorescence acquisition is done during the 55 °C step. (*see* **Note 21**).
9. Analyze amplification curves visually to obtain detect/nondetect status of each sample. Calculate Ct values by manually or automatically setting the threshold level for the run. When comparing Ct values between real-time PCR runs, ensure the threshold level is the same for each run.

### *3.8 Decontamination and Disposal*

1. Discard the ultrafilter and side port caps in an autoclave waste container.
2. Remove tubing clamps, faucet adapters, reducing connectors, and DIN adapters from influent sample collection tubing. Discard the tubing in an autoclave waste container.
3. Inspect and clean the effluent tubing and flow totalizer. The setup can remain assembled for future use. Discard tubing that has become cracked or worn. Store in a cool, dry place.
4. Place reusable plastic DEUF components into a solution of 10% bleach for ≥1 h. Rinse thoroughly with DI water and air dry.
5. Discard centrifuge supernatant in a beaker containing 10% (v/v) sodium hypochlorite and dispose of according to the laboratory's standard operating procedures (SOPs).
6. Ultraviolet (UV) sterilize laminar flow or PCR-compatible cabinet and wipe down pipettes and equipment with RNase AWAY.

## 4 Notes

1. Catalog numbers are included for reference; substitutions can be made provided the substitute item has the same specifications and equivalent performance as the suggested item.
2. Filters with 2.0–2.5 $m^2$ fiber area are recommended for adequate throughput of water of varying composition and turbidity. Method performance for the chosen filter should be evaluated before use in the field, ideally with water of similar characteristics to those being tested.
3. In order to more easily pipette the Antifoam solution, cut off the tip of a 1000-μL pipette tip to create a larger opening.
4. NaPP requires considerable shaking to dissolve, and a wrist shaker is ideal for this purpose. Alternatively, the stock solution can be prepared in advance to allow time to dissolve or the volume can be scaled up to 100 mL, and the solution can be stirred in a beaker on a magnetic stir plate.
5. For sampling nonflowing bodies of water, the ultrafilter effluent should be directed away from the location where the influent tubing is drawing water.
6. It is important that the tubing stays submerged during filtration. A bit of air in the line will not cause a problem, but drawing in too much air will cause pumping to fail. When this occurs, the pump will not draw any water, even when the tubing is resubmerged. Turn off the pump and open the pump head to release the pressure. Pumping can resume once the tubing is resubmerged in the water. For natural water sources (e.g., lakes, rivers), it can be helpful to loosely attach the tubing to something in the water using a zip tie. If that is not possible, a rock or other object can be gently placed over the tubing to secure it, provided the tip of the tubing can be positioned such that it will not draw in sediment.
7. Charge pump batteries 24 h before use. Non-US outlets require a step-down converter with an outlet adapter.
8. If a Geotech pump (or equivalent) is not used, ensure the filtrate rate does not exceed 5 L/min.
9. Certain power sources may not provide enough power to achieve a flow rate of $\geq$3.8 L/min. When this is the case, the pump should be run at full speed to achieve the highest flow rate possible. The flow totalizer will provide a less accurate (but still useful) estimate of the volume passed through the ultrafilter as the flow rate decreases. It may also be useful to purge any air from the totalizer at the beginning of the sample collection by briefly orienting the totalizer so the outlet is pointed

"up" and visually noting the disappearance or absence of bubbles exiting the totalizer.

10. If you do not have a flow totalizer, you can manually calculate the flow rate using a graduated cylinder or 1-L bottle and a stopwatch.

11. There is no directionality to the ultrafilter. The end ports are color coded for hemodialysis, but there is no difference in functionality for water sampling.

12. The 10-mL pipette gives more control during the backflushing, but it is not necessary. If opting out of using the pipette, make the tubing long enough to place the end directly into the backflush bottle.

13. If the pump used for backflushing does not have a display for the speed setting, the desired pumping rate of 650 mL/min can be determined by pumping water through a length of tubing and measuring the flow rate. This setting can be marked on the pump for future reference. If the Geotech pump is used for backflushing, it requires going above the 650 mL/min setting to achieve pumping. Once pumping is initiated, the dial should immediately be turned back to the 650 mL/min setting.

14. A centrifugation speed up to 4000 × *g* can be used if there are other analytes in the sample that require higher *g* forces to pellet.

15. For potable or groundwater samples (i.e., small concentrate volume), the pellet can be split 50:50 for nucleic acid extraction and IMS. For surface water samples (i.e., larger concentrate volume), up to 750 μL can be used for nucleic acid extraction, and a portion of the remaining volume can be used for IMS.

16. An empty micropipette tip box makes a suitable storage container for slides. For staining, remove the pipette tip tray and place a folded Kimwipe or paper towel in the bottom of the box. Mist or squirt DI water to moisten the towel. Replace the pipette tip tray to hold the slide. After the slide is stained, remove the moist towel. If the pipette tip box lid is not opaque, wrap the lid with aluminum foil. Store the stained slide in the dark box until microscopy is performed.

17. It is important to minimize the formation of bubbles when placing the coverslip, as they will interfere with examination under the microscope. It is helpful to connect the coverslip with the mounting medium at an angle. Place the long-edge coverslip along one end of the slide and lower it down at an angle until it is resting on top. Alternatively, lower the slide down onto the coverslip at an angle.

18. For low-quality water samples, or when sample volume is limited, less than 750 μL can be extracted. Add an equal volume of UNEX lysis buffer and 10% Proteinase K (10% of sample volume, not total volume after UNEX buffer is added).
19. Some water concentrates may produce a nucleic acid extract that is not completely recovered from the OneStep column after 1 min. Subsequent spins (1 min at a time) can be performed to recover a larger volume.
20. Because PCR inhibitors are sometimes present, it may be useful to also test template samples diluted 1/10.
21. Cycling times will vary depending on the master mix and real-time thermocycler used. Follow the manufacturer's instructions for the appropriate cycling times.

## Acknowledgments

Use of trade names is for identification only and does not imply endorsement by the Public Health Service or by the U.-S. Department of Health and Human Services. The findings and conclusions in this report are those of the authors and do not necessarily represent the official position of the Centers for Disease Control and Prevention.

**Competing Interest Statement** The authors have no conflicts of interest to declare that are relevant to the content of this chapter.

### References

1. Mull B, Hill VR (2012) Recovery of diverse microbes in high turbidity surface water samples using dead-end ultrafiltration. J Microbiol Meth 91(3):429–433. https://doi.org/10.1016/j.mimet.2012.10.001
2. Smith CM, Hill VR (2009) Dead-end hollow-fiber ultrafiltration for recovery of diverse microbes from water. Appl Environ Microbiol 75(16):5284–5289. https://doi.org/10.1128/AEM.00456-09
3. Raphael BH, Lautenschlager M, Kahler A, Pai S, Parks BA, Kalb SR, Maslanka SE, Shah S, Magnuson M, Hill VR (2012) Ultrafiltration improves ELISA and Endopep MS analysis of botulinum neurotoxin type A in drinking water. J Microbiol Meth 90(3):267–272. https://doi.org/10.1016/j.mimet.2012.05.020
4. USEPA (2022) Protocol for the collection of water samples for detection of pathogens and bioterrorism agents. USEPA. https://nepis.epa.gov/Exe/ZyPURL.cgi?Dockey=P1017HMN.txt. Accessed 1 Jan 2025
5. USEPA (2012) Method 1623.1: *Cryptosporidium* and *Giardia* in water by filtration/IMS/FA. Office of Water, Washington, D.C.
6. Hill VR, Narayanan J, Gallen RR, Ferdinand KL, Cromeans T, Vinje J (2015) Development of a nucleic acid extraction procedure for simultaneous recovery of DNA and RNA from diverse microbes in water. Pathogens 4(2):335–354. https://doi.org/10.3390/pathogens4020335
7. Jothikumar N, da Silva AJ, Moura I, Qvarnstrom Y, Hill VR (2008) Detection and differentiation of *Cryptosporidium hominis* and

*Cryptosporidium parvum* by dual TaqMan assays. J Med Microbiol 57(Pt 9):1099–1105. https://doi.org/10.1099/jmm.0.2008/001461-0

8. USEPA (2017) Validation of U.-S. Environmental Protection Agency environmental sampling techniques that support the detection and recovery of microorganisms. USEPA https://nepis.epa.gov/Exe/ZyPURL.cgi?Dockey=P1017PXQ.txt. Accessed 1 Jan 2025

# Chapter 4

# Genomic Characterization of *Cryptosporidium* spp. via iNextEra Library Preparation and Hybridization Bait Capture

**Mohammad Imtiaj Uddin Bhuiyan, Fiifi Agyabeng-Dadzie, Amanda H. Sullivan, Megan S. Beaudry, Kelly N. Petersen, Piotr T. Tuczapski, Michael E. Grigg, Jessica C. Kissinger, and Travis C. Glenn**

## Abstract

Next-generation DNA sequencing (NGS) is used to study the genome sequences of *Cryptosporidium* spp., but NGS is challenging when pure *Cryptosporidium* oocysts are limited in number or not available. Varying levels of parasites present in fecal samples, combined with the abundance of host cells, bacterial and other microbial cells, and undigested food particles, often result in fecal DNA samples with ~0.1% *Cryptosporidium* DNA, making genome-scale sequencing of *Cryptosporidium* from such samples cost-prohibitive. DNA extractions from fecal samples are, however, widely available and commonly used for polymerase chain reaction (PCR)-based diagnostics which can detect fg levels of *Cryptosporidium* DNA in complex DNA mixtures. Here, we describe an Illumina NGS sample preparation protocol (iNextEra) that can generate libraries from a wide range of DNA input (<1 ng to >60 ng). We then use those libraries within a modified myBaits capture hybridization protocol using CryptoCap_100K baits to enrich *Cryptosporidium* genomic DNA from a complex DNA background to increase the percentage of generated sequence reads that map to target *Cryptosporidium* reference genome sequences. Thus, iNextEra libraries and capture hybridization facilitate genome-level sequencing of this critical pathogen from widely available samples with less cost, thereby opening new opportunities to understand the complex biology of this important pathogen.

**Key words** *Cryptosporidium* spp., Fecal DNA, Genome, Sequence capture, Illumina, Massively parallel sequencing

## 1 Introduction

Whole-genome sequencing (WGS) is a widely used approach to sequence microorganisms of public health importance. Cryptosporidiosis, the disease caused by the protozoan parasite *Cryptosporidium*, is the second leading cause of diarrhea in infants [1] and of diarrheal deaths in children under the age of five, in developing countries [2]. Despite having such a huge disease burden, WGS of

Jan R. Mead and Michael J. Arrowood (eds.), *Cryptosporidium: Methods and Protocols*, Methods in Molecular Biology, vol. 2978, https://doi.org/10.1007/978-1-0716-4824-7_4,

*Cryptosporidium* spp. is challenging due to some unique features of this organism. First, the most accessible life stage of *Cryptosporidium* spp. is oocysts, which are shed into feces. Separating the oocysts from other fecal material is possible [3], but the process is laborious and expensive and is not often amenable to automation or pooling while maintaining sample identity. Second, each individual oocyst contains only four haploid genomes totaling <40 fg of DNA [4]. Third, the range of *Cryptosporidium* oocysts in human fecal samples can vary significantly (e.g., 10,000 to 500,000 oocysts per gram of stool [5]). Fourth, oocysts from a single fecal sample are a population of many individual parasites which derive from an unknown number of founding individuals. Fifth, obtaining DNA from oocysts is challenging with substantial variation observed among different extraction protocols that impact the yield of *Cryptosporidium* DNA. In general, it is optimal to have millions of oocysts available for DNA extraction prior to WGS, but this is not feasible for most samples and impossible at later stages of infection, when relatively few oocysts are shed. Thus, it is difficult to obtain the quantities of pure *Cryptosporidium* DNA from clinical and environmental samples that are often required for WGS, especially without costly deep sequencing [6] where most of the data are from off-target organisms (hosts, bacteria, undigested food, etc.).

Instead of purifying oocysts from fecal material and then obtaining DNA from the purified oocysts, an alternative approach is to extract DNA directly from the entire fecal sample prior to detecting *Cryptosporidium* DNA within the mixed DNA sample. This latter approach is commonly used to diagnose *Cryptosporidium* infection via PCR-based methods. There are several PCR and quantitative polymerase chain reaction (qPCR) assays available for the detection and identification of *Cryptosporidium* species across various sample types [7]. Although PCR and qPCR have excellent sensitivity, qPCR gives relatively limited amounts of information from each sample, and sequencing of amplicons is difficult and expensive to scale to genome-wide assays [8]. Therefore, while obtaining whole-genome sequencing (WGS) data from fecal DNA samples is ideal, many current methods used for *Cryptosporidium* struggle to meet this challenge.

Capture hybridization is a widely used target-enrichment approach that is well-suited for WGS of *Cryptosporidium*. The general workflow is that an NGS library is prepared from a DNA sample; then, long biotinylated oligonucleotides of DNA or RNA (i.e., baits) are used to capture complementary target DNA of interest in a hybridization reaction and tether the baits and their hybridized/captured DNA to streptavidin beads while the unwanted DNA is washed away [9]. Capture hybridization can be used for a broad array of goals [10]. Recent work has demonstrated that multiple bait sets can capture large portions of the

*Cryptosporidium* genome [4, 12], but the publications lack some background and details that are important to consider when implementing these techniques.

We describe here a protocol for capture hybridization with two critical components to facilitate efficient collection of *Cryptosporidium* genomic DNA from fecal DNA samples. First, we describe the details of implementing a low-cost high-throughput library preparation protocol (iNextEra, derived from [11]) for WGS that works across a broad range of input DNA amounts (<1 ng to >60 ng), requires little bench time and no specialized equipment, and succeeds with limited PCR cycles. Second, we describe the details of a capture hybridization protocol (myBaits with CryptoCap_100K bait set) that can enrich human-infecting *Cryptosporidium* spp. from the complex background of DNA present in the generated iNextEra WGS library, thus reducing the amount of sequencing needed and the associated sequencing costs.

## 2 Materials

### *2.1 Sample DNA*

1. A variety of methods and commercial kits are available to isolate DNA from various sources of *Cryptosporidium*, including environmental, fecal, or purified oocyst samples [13, 14] (*see* **Note 1**).

### *2.2 DNA Quantification*

1. Qubit™ 1X dsDNA High Sensitivity Assay Kit (Thermo Fisher Scientific Inc., Waltham, MA).
2. Qubit™ Flex Fluorometer (Thermo Fisher Scientific) (*see* **Note 2**).
3. Qubit™ Flex 0.2-mL eight-well strip tubes with attached flat caps (Thermo Fisher Scientific). 0.2-mL eight-well strip PCR tubes with attached flat caps by Stellar Scientific (Baltimore, MD) also work.
4. Single-channel and multichannel pipettes: P2, P20, P200, and P1000.
5. Aerosol-resistant pipette tips with filter: 2, 10, 20, 200, and 1000 μL.
6. Microcentrifuge for 0.2-mL tubes.
7. Disposable reagent reservoir for multichannel pipetting.
8. Vortex mixer.
9. Flash drive.

### *2.3 Quantification of* Cryptosporidium *18S Molecules Using qPCR*

1. Luna® Universal qPCR Master Mix (New England Biolabs, Ipswich, MA).

2. The Global Enteric Multicenter Study (GEMS) modified forward primer: standard desalted purification, 5 μM,
   5′-GGGTTGTATTTATTAGATAAAGAACCA-3′.
   The Global Enteric Multicenter Study (GEMS) reverse primer: standard desalted purification, 5 μM,
   5′-AGGCCAATACCCTACCGTCT-3′.
3. Nuclease-free $H_2O$.
4. Centrifuge tube: 15–50 mL.
5. Single-channel and multichannel pipettes: P2, P20, P200, and P1000.
6. Aerosol-resistant pipette tips with filter: 2, 10, 20, 200, and 1000 μL.
7. Vortex mixer.
8. Microcentrifuge for 1.5-mL tubes.
9. Disposable reagent reservoirs for multichannel pipetting.
10. A 384-well PCR plate.
11. MicroAmp clear adhesive film.
12. Mini plate spinner.
13. Bio-Rad CFX384 Real-Time PCR Detection System.
14. Ultraviolet (UV) PCR workstation.
15. Flash drive.

### 2.4 iNextEra Library Preparation

1. Illumina DNA Prep Kit (Illumina, Inc., San Diego, CA).
2. Q5 High-Fidelity DNA Polymerase (New England Biolabs).
3. 1 M Tris–HCl (pH 8.0).
4. 1 M $MgCl_2$.
5. PEG-8000 (powder or 25% solution).
6. 10 mM dNTPs (Roche Diagnostic Corporation, Indianapolis, IN).
7. Nuclease-free $H_2O$.
8. iNextEra5 and iNextEra7 index primers, standard desalted purification (5 μM).
9. T100 thermal cycler (Bio-Rad Laboratories, Inc., Hercules, CA).
10. Centrifuge tube: 15–50 mL.
11. Single-channel and multichannel pipettes: P2, P20, P200, and P1000.
12. Aerosol-resistant pipette tips with filter: 2, 10, 20, 200, and 1000 μL.
13. Vortex mixer.

14. Microcentrifuge for 0.2-mL and 1.5-mL tubes.
15. 1.5-mL microcentrifuge tubes.
16. 96-well semi-skirted PCR plates.
17. Mini plate spinner.
18. 96-well plate sealing foil.
19. 0.2-mL eight-well strip tubes with attached flat caps.
20. UV PCR workstation.
21. Horizontal gel electrophoresis system.
22. Agarose.
23. Weigh boats.
24. Electronic scale.
25. Microwave.
26. Erlenmeyer flask.
27. GelRed® nucleic acid stain.
28. GeneRuler DNA ladder (Thermo Fisher Scientific).
29. DNA loading dye.
30. 0.5X Tris–borate–ethylenediaminetetraacetic acid (EDTA) (TBE) buffer.
31. Gel documentation system.

### 2.5 Library Quantification, Pooling, and Purification

1. Qubit™ 1X dsDNA HS Assay Kit (Thermo Fisher Scientific).
2. Qubit™ Flex Fluorometer (Thermo Fisher Scientific).
3. Single-channel and multichannel pipette: P2, P20, P200, and P1000.
4. Aerosol-resistant pipette tips with filter: 2, 10, 20, 200, and 1000 μL.
5. Tris-low EDTA (TLE): 10 mM Tris pH 8.0 and 0.1 mM EDTA.
6. Disposable reagent reservoir for multichannel pipetting.
7. 0.2-mL eight-well strip tubes with attached flat caps.
8. 1.5-mL microcentrifuge tubes.
9. Microcentrifuge for 0.2 mL- and 1.5-mL tubes.
10. SpeedBead Magnetic Carboxylate (Global Life Sciences Solutions Operations UK Ltd., Buckinghamshire, England) prepared according to [15].

### 2.6 Hybridization Bait Capture

1. T100 thermal cycler (Bio-Rad Laboratories, Inc.).
2. Water bath.
3. Single-channel and multichannel pipette: P2, P20, P200, and P1000.

4. Aerosol resistant pipette tips with filter: 2, 10, 20, 200, and 1000 μL.
5. 0.2-mL eight-well strip tubes with attached flat caps.
6. 1.5-mL microcentrifuge tubes.
7. Microcentrifuge for 0.2-mL and 1.5-mL tubes.
8. Heat block.
9. 50-mL centrifuge tube.
10. 0.2 mL PCR 8 Strip Magnetic Separator.
11. Nuclease-free $H_2O$.
12. Hyb N, Hyb S, Hyb D, and Hyb R (Daicel Arbor Biosciences, Ann Arbor, MI).
13. Block C, Block O, and Block X (Daicel Arbor Biosciences).
14. Streptavidin Beads (Daicel Arbor Biosciences).
15. Binding Buffer (Daicel Arbor Biosciences).
16. Wash Buffer (Daicel Arbor Biosciences).
17. Buffer E (Daicel Arbor Biosciences).
18. Baits (CryptoCap_100K bait set Daicel Arbor Biosciences catalog #308596.V5).
19. PCR primers to amplify sequencing libraries after capture, for example:

    Illumina P5 forward primer: standard desalted purification, 5 μM,

    **5′-AATGATACGGCGACCACCGA-3′**

    Illumina P5 forward primer: standard desalted purification, 5 μM,

    **5′-CAAGCAGAAGACGGCATACGA-3′**.
20. KAPA HiFi Fidelity Buffer (5X) (Roche Diagnostic Corporation).
21. 1 U/μL KAPA HiFi HotStart (Roche Diagnostic Corporation).
22. 10 mM dNTPs (Roche Diagnostic Corporation).
21. Horizontal gel electrophoresis system.
22. Agarose.
22. Weigh boats.
23. Electronic balance.
24. Microwave.
25. Erlenmeyer flask.
26. GelRed® nucleic acid stain.
27. GeneRuler DNA ladder (Thermo Fisher).
28. DNA loading dye.

29. 0.5X TBE buffer.
30. Gel documentation system.

### 2.7 Captured Pool Quantification, Purification, and Final Pooling for Sequencing

1. Qubit™ 1X dsDNA HS Assay Kit (Thermo Fisher Scientific).
2. Qubit™ Flex Fluorometer (Thermo Fisher Scientific).
3. Single-channel and multichannel pipette: P2, P20, P200, and P1000.
4. Aerosol resistant pipette tips with filter: 2, 10, 20, 200, and 1000 μL.
5. Tris-low EDTA (TLE): 10 mM Tris pH 8.0 and 0.1 mM EDTA.
6. Disposable reagent reservoir for multichannel pipetting.
7. 0.2-mL eight-well strip tubes with attached caps.
8. 1.5-mL microcentrifuge tubes.
9. Microcentrifuge for 0.2-mL and 1.5-mL tubes.
10. SpeedBead Magnetic Carboxylate (Global Life Sciences Solutions Operations UK Ltd.).

### 2.8 Sequence QC and Bioinformatics Analysis

1. The software program SeqKit v.2.5 [16].
2. The software program FastQC v0.11.9 [17].
3. The software program Trimmomatic v.0.39 [18].
4. The software program Samtools v.1.16 [19].
5. The software program Burrows–Wheeler Aligner alignment (BWA) tool v.0.7 [20].

## 3 Methods

### 3.1 DNA Isolation

1. The iNextEra library preparation protocol works on double-stranded DNA; therefore, the DNA should not be denatured during isolation.
2. Sample DNA can be obtained from various sources including environmental, fecal, or purified oocyst (*see* **Note 1**). Details for DNA isolation have been described previously [14]. As long as the sample DNA is double-stranded and is quantifiable, it can be used for iNextEra library preparation.
3. Protocols for whole-genome amplification (WGA) of DNA from single oocysts and for DNA concentration that is below detection limits are available but have additional constraints [13].

### *3.2 DNA Quantification by Qubit*

1. Remove DNA from the freezer and defrost at room temperature.
2. While DNA is defrosting, calibrate the Qubit™ Flex fluorometer using manufacturer's instruction.
3. Add 98 μL of Qubit™ 1X dsDNA HS Working Solution to Qubit™ Flex Tube Strips (we reduce the volume from the manufacturer's protocol to minimize costs).
4. Thoroughly vortex the DNA and spin using microcentrifuge. Then add 2 μL of DNA to the strip tubes containing 98 μL of 1X dsDNA HS Working Solution.
5. Vortex and spin the tubes and put them in a dark chamber for 2 min.
6. When the incubation is done, quantify the DNA using the Qubit™ Flex fluorometer.
7. Transfer the data using a flash drive, open the file using MS excel (or another equivalent program).
8. There will be multiple columns in the file but disregard everything except for the column "Sample Name" and "Original Sample Conc."
9. The values found there are 2X concentrated versus the manufacturer's protocol because of our reduction in working solution volume, so divide the observed value by 2 to get the actual sample concentration in ng/μL (*see* **Notes 3** and **4**).

### *3.3 Quantification of* Cryptosporidium *DNA via qPCR of 18S Ribosomal Gene*

1. Defrost DNA, primers, and LUNA universal qPCR master mix at room temperature. https://www.neb.com/en-us/luna/luna-universal-qpcr-and-rt-qpcr
2. Prepare the qPCR master mix by adding all the qPCR reagents in a 15-mL centrifuge tube (Table 1; *see* **Note 5**).
3. Mix by vortexing.
4. Transfer 15 μL of master mix to 384-well qPCR plate using a multichannel pipette.

**Table 1**
**qPCR master mix recipe for *Cryptosporidium* quantification**

| Organism | Target | Reagents | Volume (μL) |
|---|---|---|---|
| *Cryptosporidium* spp. | 18S | LUNA universal qPCR master mix | 10 |
| | | Nuclease-free $H_2O$ | 4 |
| | | GEMS modified forward primer (5 μM) | 0.5 |
| | | GEMS reverse primer (5 μM) | 0.5 |
| | | Total | 15 |

**Table 2**
***Cryptosporidium* qPCR conditions**

| qPCR cycling conditions: 40X | | |
|---|---|---|
| **Step** | **Temperature** | **Time (min)** |
| 1 | 95 °C | 1:00 |
| 2 | 95 °C | 0:15 |
| 3 | 60 °C | 0:30 |

**Table 3**
**2X TMP custom buffer**

| Reagents | Volume (μL) |
|---|---|
| 1 M Tris–HCl (pH 8.0) | 20 |
| 1 M $MgCl_2$ | 10 |
| 25% PEG-8000 | 640 |
| Nuclease-free $H_2O$ | 330 |
| Total | 1000 |

5. Add 5 μL of template DNA and mix by pipetting up and down ≥3 times.
6. Add 5 μL of nuclease free water to the negative control.
7. Seal the plate very well using microAmp clear adhesive film.
8. Gently vortex the plate to make sure DNA and master mix are properly mixed.
9. Spin down the plate using a mini plate spinner.
10. Use thermal cycling conditions in a CFX384 real-time PCR detection system. (Table 2).
11. When the run is complete, the $C_Q$ ($C_T$) values will be displayed on the screen. Click on the "Quantification Data" tab, copy the data, paste it into an Excel file, and then transfer the file using a flash drive.

### 3.4 iNextEra Library Preparation Protocol

1. Normalize all DNA samples to ~10 ng/μL (quantified on Qubit or plate reader; *see* **Notes 2**, **6**, and 7). Place ≥20 μL in plates or strip tubes (less can be used, but this amount is recommended for handling ease). Ensure these (and all other tubes/plates below) are labeled clearly.
2. Prepare 2x TMP buffer in a 1.5-mL tube (Table 3).

3. Set thermal cycler to incubate at 53 °C with the lid set to 80 °C. Allow it to preheat.
4. Set up tagmentation. Prepare a master mix of TMP buffer and transposome (BLT) following the per-sample recipe in Table 4. Transfer 3.5 μL of master mix in 8-well strip tube/96-well plate and then add 2.5 μL (total 25 ng input) of DNA. Gently mix by pipetting up and down ≥3 times to make sure DNA is mixed with the master mix. Try to avoid excessive foaming.
5. Incubate the mixture at 53 °C for 30 min in a preheated thermal cycler.
6. While the tagmentation is going on, set up PCRs.
7. Defrost PCR reagents and primers at room temperature.
8. Prepare the PCR master mix by adding all the PCR reagents to a 15-mL centrifuge tube (Table 5).
9. Mix well by vortexing or pipetting up and down. Then pour the solution in a reagent reservoir to facilitate the use of multichannel pipette.
10. Immediately after the tagmentation is finished, transfer 51.5 μL of master mix to 6 μL tagmented DNA from **step 5**.
11. For each reaction, add unique iNextEra primers (Table 6). This will make a total volume of 62.5 μL.
13. If a 96-well semi-skirted plate is used, make sure the plate is properly sealed with adhesive film/sealing foil. If an eight-well strip tube is used, make sure the lids are properly closed.
14. Vortex gently to ensure thorough mixing. Try to avoid excessive foaming.
15. Spin the plate/tube well using a microcentrifuge/mini plate spinner.
16. Program or select the cycling conditions for the iNextEra PCR reaction (Table 7) and start the thermal cycler (*see* **Note 12**).
17. After cycling is complete, run 5 μL of PCR product on 1.5% agarose gel to ensure each sample amplified, yielding a smear in the desired size range of 300–700 bp (*see* Fig. 1 and **Note 8**).
18. Quantify the libraries using Qubit.

*This is a safe stopping point. Libraries can be stored at −20 °C.*

### 3.5 Preparation of Pools for Hybrid Capture

1. Following Qubit, categorize the libraries based on low to high $C_T$ values.
2. Make a pooling plan with ~4 libraries in each pool (*see* **Notes 9–11**).
3. Pool equimolar quantities of each.

**Table 4**
**Tagmentation mix per sample**

| Reagents | Volume (μL) |
|---|---|
| 2x TMP custom buffer | 3.0 |
| Transposome (BLT) | 0.5 |
| Total | 3.5 |

**Table 5**
**iNextEra PCR master mix recipe per sample**

| Reagents | Volume (μL) |
|---|---|
| Nuclease-free $H_2O$ | 40.25 |
| Q5 reaction buffer | 9.50 |
| 10 mM dNTPs | 1.25 |
| Q5 HF DNA polymerase | 0.50 |
| Total | 51.5 |

**Table 6**
**iNextEra index primers (add unique primers to each reaction separately)**

| iNextEra index primers (5 μM) | Volume (μL) |
|---|---|
| iNextEra 5 | 2.5 |
| iNextEra 7 | 2.5 |

**Table 7**
**iNextEra PCR conditions**

| iNextEra PCR cycling conditions | | | |
|---|---|---|---|
| **Step** | **Temperature** | **Time (min)** | **Number of cycles** |
| Initial strand repair | 72 °C | 3:00 | 1 |
| Initial denaturation | 98 °C | 3:00 | 1 |
| Denaturation | 98 °C | 0:45 | 10 |
| Annealing | 62 °C | 0:30 | |
| Extension | 72 °C | 2:00 | |
| Final extension | 72 °C | 1:00 | 1 |
| Hold at 4 °C | | | |

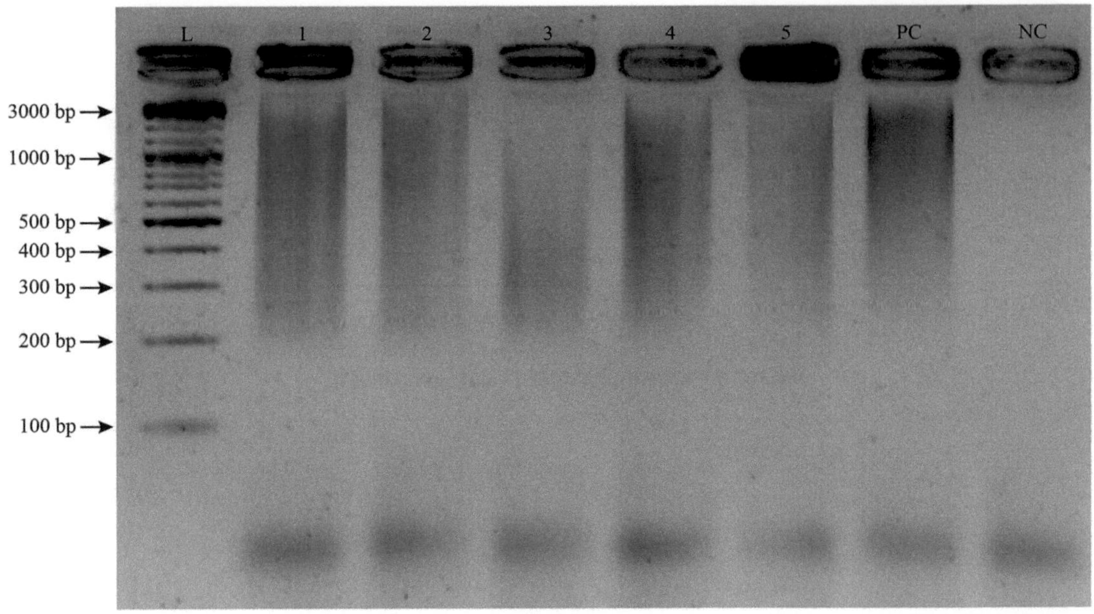


**Fig. 1** iNextEra libraries constructed from *Cryptosporidium* DNA. Lane L: DNA ladder, Lanes 1–5: iNextEra libraries resulting from 25 ng of input DNA, PC: positive control (25 ng Calf thymus) and NC: negative control

4. Purify the pools with SpeedBeads. Add a volume of SpeedBeads 1.25X the volume of pooled libraries, gently pipette or vortex to mix well, purify as normal, and resuspend in TLE buffer. Incubate 15 min. at room temperature, then place on magnet, and transfer the liquid to new tubes, leaving the beads behind.
5. Quantify the pool(s) with Qubit.
6. Ensure the pool concentration is $\geq$30 ng/μL. If achieving this concentration is challenging, hybrid capture is still feasible, but the P5/P7 PCR cycles must be adjusted accordingly (*see* **Notes** 7 and **12**).

*This is a safe stopping point. Libraries can be stored at −20 °C.*

### 3.6 Capture Hybridization Using myBaits

The entire capture hybridization protocol requires over 16 h to complete. For convenience, we describe the protocol in a two-day format.

*Day 1*

#### 3.6.1 Hybridization Master Mix Preparation

1. Defrost Hyb N, Hyb D, Hyb S, Hyb R, and Baits at room temperature.
2. While defrosting, set the heat block to 60 °C.
3. Vortex and spin all the tubes containing hybridization reagents and baits.

**Table 8**
**Hybridization master mix per sample**

| Reagents | Volume (μL) |
|---|---|
| Hyb N | 9.25 |
| Hyb D | 3.50 |
| Hyb S | 0.50 |
| Hyb R | 1.25 |
| Baits | 1.37 |
| Nuclease-free $H_2O$ | 4.13 |
| Total | 20.00 |

4. If there is any precipitation seen in the Hyb N and Hyb S tubes, let them sit on the heat block at 60 °C until the solutions are clear.
5. Prepare the hybridization master mix in a 1.5-mL microcentrifuge tube according to the recipe in Table 8.
6. Vortex and spin the microcentrifuge tube.
7. Incubate the hybridization master mix on the heat block at 60 °C for 10 min and then at room temperature for 5 min.
8. For each reaction, aliquot 18.5 μL of hybridization master mix into an eight-well strip tube labeled as "HYBs."

*3.6.2 Wash Buffer X Preparation*

1. Defrost and thoroughly homogenize wash buffer and Hyb S.
2. In a 50-mL microcentrifuge tube, prepare wash buffer X (Table 9). This step will generate enough wash buffer X to be used in multiple 0.2-mL tubes.
3. Mix the reagents very well by gently shaking the tube up and down, label it as "Wash Buffer X" with proper date, and keep in the refrigerator at 4 °C. It can be used up to one month after preparation.

*3.6.3 Blockers Mix Preparation*

1. Defrost Block O, Block C, and Block X at room temperature.
2. Vortex all the tubes containing hybridization reagents and baits and spin down using a microcentrifuge.
3. Prepare the Blockers mix in a 0.2-mL or 1.5-mL microcentrifuge tube (Table 10).
4. Mix by pipetting up & down ≥3 times.
5. For each reaction, aliquot 5 μL of blockers mix in an eight-well strip tube labeled as "LIBs."

**Table 9**
**Wash buffer X**

| Reagents | Unit volume | Desired volume |
|---|---|---|
| Hyb S | 400.0 μL | Unit volume × n/68 |
| Nuclease-free $H_2O$ | 39.6 mL | Here, "n" denotes the number of reactions. For example, with eight reactions, you will need 47.05 μL of Hyb S (400 × 8/68), 4.65 mL of nuclease-free $H_2O$ (39.6 × 8/68), and 1.17 mL of wash buffer (10 × 8/68). We also recommend a 10% allowance for measurement error. |
| Wash buffer | 10.0 mL | |

**Table 10**
**Blockers mix per reaction**

| Reagents | Volume (μL) |
|---|---|
| Block O | 2.5 |
| Block C | 2.5 |
| Block X | 0.5 |
| Total | 5.5 |

6. Add 7 μL of purified pool from Subheading 3.5 to the "LIBs," mix by pipetting up & down ≥3 times, and spin it down using a microcentrifuge.

*3.6.4 Setting Up the Overnight Hybridization Incubation*

1. Start a thermal cycler program as follows: 95 °C for 5 min, 63 °C for 5 min, and then hold at 63 °C. The lid should be heated to 105 °C.
2. Place the strip tube labeled as "LIBs" in the thermal cycler to incubate for 5 min at 95 °C. The double-stranded DNA libraries denature during this step, allowing blockers to hybridize to the DNA.
3. When the temperature reaches 63 °C in the second step of the thermal cycler program, place strip tubes labeled as "HYBs" with the "LIBs" to incubate for 5 min.
4. When the second step of the thermal cycle program finishes, leaving all the tubes in the thermal cycler, transfer 18 μL of "HYBs" to "LIBs" and mix the samples by pipetting up and down ≥3 times.
5. Discard the "HYBs" tubes. Briefly spin down the "LIBs," return to the thermal cycler, close the lid, and start the overnight incubation at 63 °C for 16–24 h. This step introduces baits and generates bait-target hybrids.

*Day 2*

Start at least 90 min before intended hybridization stop time (you can stop the hybridization anytime between 16–24 h).

*3.6.5 Bead Preparation*

1. Turn on the water bath and set it to 63 °C. It takes around 30 min for the water bath to reach the desired temperature.
2. Take the wash buffer X from the refrigerator and heat it in the water bath for at least 30 min before use.
3. While the wash buffer X is heating up, prepare the beads.
4. Take the beads and binding buffer from the refrigerator.
5. Vortex the tube containing beads, apply a quick spin using a microcentrifuge, and transfer 30 μL of beads to eight-well trip tubes for each reaction.
6. Place the strip tubes in the 0.2 mL PCR 8 Strip Magnetic Separator.
7. Wait until the solution is clear and discard the supernatant while the tubes are still on the magnetic stand.
8. Remove the strip tubes from the magnetic stand and add 200 μL binding buffer, vortex and spin (avoid precipitation), and then place it back to the magnetic stand.
9. Repeat **steps** 7 and **8** a total of three times.
10. Resuspend the washed beads in 70 μL of binding buffer.
11. Label the strip tube according to number of reactions.

*3.6.6 Binding Beads and Hybrids*

1. Place the strip tubes containing beads in the thermal cycler and incubate at 63 °C for 2 min.
2. Transfer each capture reaction in tubes labeled "LIBs" to the respective strip tube containing beads. Gently mix the solution by pipetting.
3. Incubate the libraries + beads on the thermal cycler at 63 °C for a total of 5 min. Make sure to gently vortex or pipette up and down at the 2.5-minute mark to keep the beads suspended in the solution, followed by brief spin down using microcentrifuge.

*3.6.7 Bead Washing to Remove Unbound DNA*

1. Place the strip tube containing beads in the magnetic stand, wait until the solution is clear and discard the supernatant.
2. Add 180 μL of preheated wash buffer X to each tube while the strip tubes are still on the magnetic stand.
3. Remove the strip tube from the magnetic stand, place it in the thermal cycler, and incubate at 63 °C for 15 s to keep the solution warm. Then gently mix by vortexing, taking extreme care to cover the lids with a finger so they do not pop open, spin

it down, and place back in the thermal cycler. If it is uncomfortable to mix by vortexing, it is also acceptable to mix by pipetting with the tubes still in the thermal cycler.

4. Incubate at 63 °C for 2.5 min, quickly vortex and spin or pipette to mix, and then incubate again for 2.5 min. An easy way to do this is to set a timer for 5 min and mix at the 2.5-minute mark.
5. Repeat **steps 1–4** for a total of four times. After the last wash and pelleting, remove as much liquid as possible without touching the bead pellet.
6. Add 30 μL of buffer E to the washed beads and thoroughly mix by pipetting.

#### *3.6.8 Library Amplification*

1. Defrost the PCR reagents at room temperature.
2. Prepare the PCR master mix by adding all PCR reagents in a 2-mL microcentrifuge tube (Table 11).
3. Add 10 μL of captured product with beads to the master mix, vortex the tube, and spin it down using a microcentrifuge. DO NOT separate the beads from the liquid at this stage, because the target DNA strands are still attached to the beads.
4. Select the cycling condition for the appropriate P5/P7 PCR reaction (Table 12) and start the thermal cycler (*see* **Note 12**).
5. Following PCR amplification, run 5 μL of captured products on 1.5% agarose gel to ensure each sample amplified, yielding a smear in the desired size range (300–700 bp; Fig. 2).
6. Quantify the pool(s) with Qubit.

*This is a safe stopping point. Libraries can be stored at −20 °C.*

### 3.7 Preparation of Final Pools for Sequencing

1. Following SpeedBead cleanup and quantification, normalize the pools using standard pooling guide and make the final pool with appropriate volume.

**Table 11**
**P5/P7 PCR master mix recipe for library amplification**

| Reagents | Volume(μL) |
|---|---|
| Nuclease-free $H_2O$ | 17.5 |
| KAPA HiFi Fidelity buffer | 10.0 |
| 10 mM dNTPs | 1.5 |
| KAPA HiFi HotStart | 1.0 |
| P5/P7 primer (5 μM) | 10.0 |
| Total | 40.0 |

**Table 12**
**P5/P7 PCR conditions**

| P5/P7 PCR cycling conditions | | | |
|---|---|---|---|
| **Step** | **Temperature** | **Time (min)** | **Number of cycles** |
| Initial denaturation | 98 °C | 2:00 | 1 |
| Denaturation | 98 °C | 0:20 | 8–18 |
| Annealing | 60 °C | 0:30 | |
| Extension | 72 °C | 1:00 | |
| Final extension | 72 °C | 5:00 | 1 |
| Hold at 4 °C | | | |

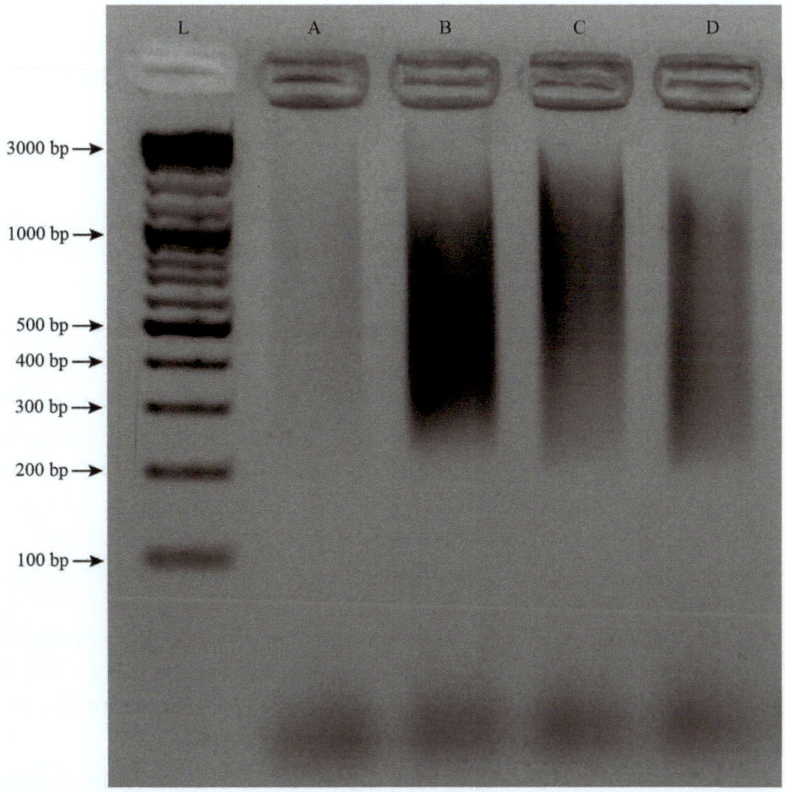


**Fig. 2** Libraries after capture and P5/P7 PCR amplification. Lane L: DNA Ladder and Lanes A–D: Captured libraries after P5/P7 PCR. The intensity of the smears varies depending on the input concentration of the pooled libraries, the proportion of target, and efficiency of capture

2. Purify the pool with SpeedBeads. Add a volume of SpeedBeads 1.25X the volume of pool, gently pipette or vortex to mix well, purify as normal, and resuspend in TE buffer. Incubate 15 min. at room temperature and then place on magnet and transfer the liquid to new tubes, leaving the beads behind.
3. Quantify the pool with Qubit (*see* **Notes 13-15**).
4. The pool is now ready to be sent off to sequencing (*see* **Notes 16** and **17**).

*This is a safe stopping point. Libraries can be stored at −20 °C.*

### 3.8 Sequence QC and Bioinformatics Analysis

1. Obtain demultiplexed Illumina reads from sequencing service provider as gzipped FASTQ files (*see* **Notes 18** and **19**).
2. Install any software needed (*see* Subheading 2.8).

#### 3.8.1 Count and Trim Sequence Reads

1. Assess read count and quality using SeqKit v.2.5. Use the (*stats -a*) option for tabular results (https://bioinf.shenwei.me/seqkit/usage/). Refer to Read_quality_check.sh (https://github.com/jkissing/HybridCapture-PublicScripts/blob/main/Read_quality_check.sh) [21] for an example of the script.
2. Check for the presence and type of adapters with FastQC v0.11.9 (https://www.bioinformatics.babraham.ac.uk/projects/fastqc/).
3. Remove adapters, low-quality reads (<Q20), and short-read length (<50 b) with Trimmomatic v.0.39 (https://github.com/usadellab/Trimmomatic) (*see* **Note 20**).
4. Assess read count and quality using SeqKit v.2.5 to make sure there are enough reads after trimming (*see* **Notes 21** and **22**). Refer to Read_trimming.sh (https://github.com/jkissing/HybridCapture-PublicScripts/blob/main/Read_trimming.sh) [21] for an example of the script.

#### 3.8.2 Map Sequence Reads to a Reference Genome

1. Download the *Cryptosporidium* spp. reference genome assembled sequence of your choice and appropriate species from National Center for Biotechnology Information (NCBI) or an accepted data resource.
2. Make an index of the assembled reference genome sequence with Samtools v.1.16 and Burrows–Wheeler Aligner (BWA) tool v.0.7.
3. Align paired trimmed reads to the reference sequence using the BWA tool v.0.7. Sort and mark duplicate reads with the following options (sort) and (markdup), respectively, in Samtools v.1.16.

4. Make an index of the output Bam files using Samtools v.1.16. Refer to Maping_trimmed_paired_reads.sh (https://github.com/jkissing/HybridCapture-PublicScripts/blob/main/Mapping_trimmed_paired_reads.sh) [21] for an example of the script.
5. Assess unique mapped reads with Samtools v.1.16.
6. Remove all contaminants (e.g., host DNA and bacterial DNA) prior to using these data for further analysis (*see* **Note 23**).

## 4 Notes

1. Oocyst DNA is often contaminated with bacteria.
2. DNA quantification can be performed using a Qubit, Qubit Flex, plate reader, NanoDrop, or other DNA quantification platform (including TapeStation or running on an agarose gel). Dye-based fluorometric quantification (e.g., Qubit, Qubit Flex, and plate reader) is superior to UV absorbance (e.g., NanoDrop), especially at low concentration. If concentration is determined using NanoDrop, all concentrations should be doubled for all protocols. Agarose gel electrophoresis or TapeStation (or similar) can be used to determine DNA length in addition to quantity, which is desirable, because fragments <200 bp are inefficiently converted to library molecules via iNextEra. Agarose gel electrophoresis is low cost but has limited sensitivity. TapeStation has higher sensitivity but has significant cost per sample. Thus, we describe DNA quantification using a Qubit/Qubit Flex but recommend additional QC of ~5% of samples via TapeStation.
3. This protocol saves 100 μL of 1X dsDNA HS Working Solution in each reaction.
4. If limited volumes of DNA are available, increase 1X dsDNA HS Working Solution to 99 μL and use 1 μL of sample DNA, and do not divide the observed value by 2.
5. If limited volumes of DNA are available, the volumes of Table 1 and Subheadings 3.3, **step 4**; 3.3, **step 5**; and 3.3, **step 6** can be divided by two (i.e., do half-volume reactions). You can also use this half-volume reaction to save reagents regardless the DNA volume.
6. iNextEra libraries can be constructed with starting DNA concentrations of <1 ng/μL to >60 ng/μL. The protocol works best with starting DNA concentrations of 2–20 ng/μL. We recommend ~10 ng/μL because typical measurement errors are well tolerated. The number of cycles used in Subheading 3.4, **step 16** will vary, depending on the amount of starting DNA (less starting DNA requires more cycles; *see* **Note 12**). Normalizing the DNA concentrations at Subheading 3.4, **step**

**1** facilitates using the same number of cycles for all samples within a batch.

7. *Critical consideration:* The number of target molecules available for capture is directly related to the amount of starting DNA. Significant dilution of templates for normalization is NOT recommended. If the concentrations among samples vary greatly, it is better to do multiple smaller batches than to significantly dilute samples to match the lowest concentration.
8. The desired size range depends on the specific sequencing instrument that will be used. For most Illumina instruments, it is acceptable for some fragments to be larger than 700 bp, but if a substantial portion of the library is larger than 1000 bp, double-sided size selection with SpeedBeads (or Pipin Prep or similar) should be used at Subheading 3.5, **step 4** to limit the size range.
9. There is a trade-off in the number of samples pooled together. The more samples that are pooled together, the lower the per-sample enrichment costs (i.e., pools of eight spread the enrichment costs among twice as many samples, thereby cutting the enrichment cost per sample by 50%). However, the more samples that are pooled together, the higher the odds of jackpotting (i.e., one sample being overrepresented in the pool following enrichment). Jackpotting is a higher risk when double enrichments (*see* **Note 14**) are used. The precautions in the note below reduce the extent of jackpotting, but in practice, it is generally best to limit pool sizes for hybridization capture enrichment of *Cryptosporidium* from fecal DNA libraries.
10. Capture hybridization works best with $C_T$ of 18 to 24 and can work on samples up to 30. If the $C_T$ value is below 18, it indicates a high abundance of *Cryptosporidium* DNA, allowing for direct sequencing without the need for capture hybridization (i.e., proceed from Subheading 3.4, **step 18** to Subheading 3.7). When the $C_T$ value exceeds ~24, the number of *Cryptosporidium* DNA molecules is limiting, and more variation in the number of duplicate reads and genome breadth are observed. When the $C_T$ value is above 30, few *Cryptosporidium* DNA molecules will be available in the starting pool of molecules for library preparation. Thus, the library will not represent the whole *Cryptosporidium* genome, and the capture enrichment process will yield a high percentage of duplicate reads covering only a small portion of the genome.
11. Libraries within a pool should have similar abundance of *Cryptosporidium* (thus similar $C_T$ values; *see* Subheading 3.3, **step 11**), size range (*see* Subheading 3.4, **step 17**), initial starting concentrations (thus similar numbers of cycles during Subheading 3.4, **step 16**), final concentrations (*see* Subheading

3.4, **step 18**), and similar species composition (i.e., all *C. parvum*, not a mix of different species). Ideally, small aliquots of the libraries would be used to obtain ≥100,000 reads to estimate better these parameters, but in practice, the available values through Subheading 3.4, **step 18** are typically used.

12. The optimal number of PCR cycles for both iNextEra and P5/P7 depends on the input concentration. For iNextEra libraries, with the following input concentrations, use the corresponding PCR cycles: >10 ng/μL, normalize to 10 ng/μL and use 10 cycles; 5–9.9 ng/μL, use 12 cycles; 1–4.99 ng/μL, use 14 cycles; and <1 ng/μL, use 16 cycles (libraries from <1 ng of starting material can be sequenced, but enrichment often yields jackpotting or limited genomic breadth of coverage). For P5/P7 PCR, try to minimize the cycles where possible. The use of unnecessary PCR cycles may result in excessive duplicates and jackpotting, thereby reducing the breadth of genome coverage and biasing read number per sample within a pool.
13. The efficiency of the hybrid capture can be checked by the same qPCR protocol that is used for the quantification of *Cryptosporidium* DNA via qPCR of 18S ribosomal gene (*see* Subheading 3.3 above).
14. Following SpeedBead cleanup and quantification of single captured products, these can be pooled again (two pools together; total 4 + 4 libraries) and undergo another round of capture as described above. This is known as "double enrichment," which will produce more on-target reads (i.e., increase depth of coverage), but it increases PCR duplicates, thus reducing library diversity and genome breadth of mapped reads. Double enrichment is particularly useful when single enrichments are variable and increasing the depth of sequencing coverage is needed.
15. Results from earlier variants of these protocols demonstrated enrichment factors of ~500X single enrichment and ~ 2000X double enrichment [4].
16. Here, we assume sequencing will be Illumina paired-end 150 base reads (PE150), which is currently most efficiently obtained on a NovaSeq X instrument. It is also possible to sequence iNextEra libraries on the AVITI instruments by Element Biosystems. It is also possible to use other read lengths, but PE150 is the maximum read length recommended for libraries resulting from Illumina bead tagmentation, including iNextEra.
17. It is generally recommended to obtain >10 million reads per sample. It is advisable to obtain a modest number of reads (e.g., 1–10 M reads per sample) from the pools to ensure the

enrichment went well, duplication rates are not too high, and jackpotting is minimal and then to obtain more reads from the best sample pools in a second run (e.g., up to 50 M reads per sample).

18. Sequences are demultiplexed into individual samples by the sequence provider, or appropriate demultiplexing tools (e.g., Stacks: process_radtags, [22]) should be used.
19. Read1 and Read2 fastq.gz files should be obtained for each sample. These files will need to be moved to the computer where analyses will be conducted and separately backed up.
20. Ensure software packages from Subheading 2.8 are installed at your institution's high-performance computing center (HPCC) or on your computer system. We generally recommend using local HPCCs, but national resources, such as the GalaxyProject, are available. Check recommendations for installing the software at their websites.
21. Trimmomatic outputs paired and unpaired reads with the former being used in subsequent analysis.
22. SeqKit (rmdup) can also be used to remove duplicate reads before mapping or Samtools markdup with the option -r to remove duplicates after mapping.
23. Sequences from human fecal samples should be screened for human DNA sequences, with human sequences removed prior to deposition into any public database.

## Acknowledgments

This work was supported in part by NIH R01AI148667. M.G. is supported by the Division of Intramural Research, National Institute of Allergy and Infectious Diseases (NIAID)/National Institute of Health (NIH). We thank the *Cryptosporidium* diagnostics and research community for sharing their precious samples for troubleshooting and analysis. We thank Asis Khan and Eliza Alves-Ferreira for useful discussions on their experience with the first generation of *Cryptosporidium* hybrid capture baits.

**Competing Interest Statement** The authors have no conflict of interests to declare that are relevant to the content of this chapter.

## References

1. Kotloff KL, Nataro JP, Blackwelder WC, Nasrin D, Farag TH, Panchalingam S, Wu Y, Sow SO, Sur D, Breiman RF, Faruque AS, Zaidi AK, Saha D, Alonso PL, Tamboura B, Sanogo D, Onwuchekwa U, Manna B, Ramamurthy T, Kanungo S, Ochieng JB,

Omore R, Oundo JO, Hossain A, Das SK, Ahmed S, Qureshi S, Quadri F, Adegbola RA, Antonio M, Hossain MJ, Akinsola A, Mandomando I, Nhampossa T, Acacio S, Biswas K, O'Reilly CE, Mintz ED, Berkeley LY, Muhsen K, Sommerfelt H, Robins-Browne RM, Levine MM (2013) Burden and aetiology of diarrhoeal disease in infants and young children in developing countries (the Global Enteric Multicenter Study, GEMS): a prospective, case-control study. Lancet 382(9888):209–222. https://doi.org/10.1016/S0140-6736(13)60844-2

2. Gilbert IH, Vinayak S, Striepen B, Manjunatha UH, Khalil IA, Van Voorhis WC (2023) Safe and effective treatments are needed for cryptosporidiosis, a truly neglected tropical disease. BMJ Glob Health 8(8). https://doi.org/10.1136/bmjgh-2023-012540
3. Hawash Y (2014) DNA extraction from protozoan oocysts/cysts in feces for diagnostic PCR. Korean J Parasitol 52(3):263–271. https://doi.org/10.3347/kjp.2014.52.3.263
4. Bayona-Vásquez NJ, Sullivan AH, Beaudry MS, Khan A, Baptista RP, Petersen KN, Bhuiyan M, Brunelle B, Robinson G, Chalmers RM, Alves-Ferreira E, Grigg ME, Kissinger JC, Glenn TC (2024) Whole genome targeted enrichment and sequencing of human-infecting *Cryptosporidium* spp. Res Sq. https://doi.org/10.21203/rs.3.rs-4294842/v1
5. Weber R, Bryan RT, Bishop HS, Wahlquist SP, Sullivan JJ, Juranek DD (1991) Threshold of detection of Cryptosporidium oocysts in human stool specimens: evidence for low sensitivity of current diagnostic methods. J Clin Microbiol 29(7):1323–1327. https://doi.org/10.1128/jcm.29.7.1323-1327.1991
6. Kaupke A, Gawor J, Rzezutka A, Gromadka R (2017) Identification of pig-specific *Cryptosporidium* species in mixed infections using Illumina sequencing technology. Exp Parasitol 182:22–25. https://doi.org/10.1016/j.exppara.2017.09.020
7. Hadfield SJ, Robinson G, Elwin K, Chalmers RM (2011) Detection and differentiation of Cryptosporidium spp. in human clinical samples by use of real-time PCR. J Clin Microbiol 49(3):918–924. https://doi.org/10.1128/jcm.01733-10
8. Glenn TC, Pierson TW, Bayona-Vásquez NJ, Kieran TJ, Hoffberg SL, Thomas Iv JC, Lefever DE, Finger JW, Gao B, Bian X, Louha S, Kolli RT, Bentley KE, Rushmore J, Wong K, Shaw TI, Rothrock MJ Jr, McKee AM, Guo TL, Mauricio R, Molina M, Cummings BS, Lash LH, Lu K, Gilbert GS, Hubbell SP, Faircloth BC (2019) Adapterama II: universal amplicon sequencing on Illumina platforms (TaggiMatrix). PeerJ 7:e7786. https://doi.org/10.7717/peerj.7786
9. Gnirke A, Melnikov A, Maguire J, Rogov P, LeProust EM, Brockman W, Fennell T, Giannoukos G, Fisher S, Russ C, Gabriel S, Jaffe DB, Lander ES, Nusbaum C (2009) Solution hybrid selection with ultra-long oligonucleotides for massively parallel targeted sequencing. Nat Biotechnol 27(2):182–189. https://doi.org/10.1038/nbt.1523
10. Glenn TC, Faircloth BC (2016) Capturing Darwin's dream. Mol Ecol Resour 16(5):1051–1058. https://doi.org/10.1111/1755-0998.12574
11. Jones A, Stanley D, Ferguson S, Schwessinger B, Borevitz,J, Warthmann N (2023) Cost-conscious generation of multiplexed short-read DNA libraries for whole-genome sequencing. PLoS ONE, 18(1), e0280004. https://www.protocols.io/view/cost-conscious-generation-of-multiplexed-short-rea-14egnx27zl5d/v2
12. Khan A, Alves-Ferreira EVC, Vogel H, Botchie S, Ayi I, Pawlowic MC, Robinson G, Chalmers RM, Lorenzi H, Grigg ME (2024) Phylogenomic reconstruction of *Cryptosporidium* spp. captured directly from clinical samples reveals extensive genetic diversity. bioRxiv. https://doi.org/10.1101/2024.04.17.589752
13. Dadzie FA, Beaudry MS, Deyanov A, Slanis H, Duong MQ, Turner R, Khan A, Arias CA, Kissinger JC, Glenn TC, de Paula Baptista R (2024) Evaluating the benefits and limits of multiple displacement amplification with whole-genome Oxford nanopore sequencing. bioRxiv. https://doi.org/10.1101/2024.02.09.579537
14. Mead JR, Arrowood MJ (eds) (2020) *Cryptosporidium*: methods and protocols, 1st edn. Springer, New York
15. Rohland N, Reich D (2012) Cost-effective, high-throughput DNA sequencing libraries for multiplexed target capture. Genome Res 22(5):939–946. https://doi.org/10.1101/gr.128124.111
16. Shen W, Sipos B, Zhao L (2024) SeqKit2: a Swiss army knife for sequence and alignment processing. iMeta 3(3):e191. https://doi.org/10.1002/imt2.191
17. Andrews S (2010) FastQC: A quality control tool for high throughput sequence data. https://www.bioinformatics.babraham.ac.uk/projects/fastqc/

18. Bolger AM, Lohse M, Usadel B (2014) Trimmomatic: a flexible trimmer for Illumina sequence data. Bioinformatics 30(15): 2114–2120. https://doi.org/10.1093/bioinformatics/btu170
19. Danecek P, Bonfield JK, Liddle J, Marshall J, Ohan V, Pollard MO, Whitwham A, Keane T, McCarthy SA, Davies RM, Li H (2021) Twelve years of SAMtools and BCFtools. GigaScience 10(2). https://doi.org/10.1093/gigascience/giab008
20. Li H, Durbin R (2009) Fast and accurate short read alignment with Burrows-Wheeler transform. Bioinformatics 25(14):1754–1760. https://doi.org/10.1093/bioinformatics/btp324
21. Kissinger JC (2025) HybridCapture-PublicScripts. https://github.com/jkissing/HybridCapture-PublicScripts
22. Catchen J, Hohenlohe PA, Bassham S, Amores A, Cresko WA (2013) Stacks: an analysis tool set for population genomics. Mol Ecol 22(11):3124–3140. https://doi.org/10.1111/mec.12354

# Chapter 5

# Preparation of *Cryptosporidium* for Whole-Genome Sequencing

**Guy Robinson and Rachel M. Chalmers**

## Abstract

The application of whole-genome sequencing (WGS) to *Cryptosporidium* has been challenged by the relatively small numbers of oocysts present in feces and lack of suitable culture systems to generate sufficient DNA that is required. Here, we provide a method for the preparation of sufficient, high-quality *Cryptosporidium* DNA for WGS from feces, incorporating parasite purification stages and an optional whole-genome amplification step.

**Key words** *Cryptosporidium*, DNA extraction, Whole-genome sequencing, WGS, Whole-genome amplification

## 1 Introduction

Whole-genome sequencing (WGS) has revolutionized our understanding of pathogens and is now being used routinely, for example, in diagnosis, antibiotic resistance monitoring, and genotyping many bacteria, as predicted by Didelot and colleagues in 2013 [1]. For *Cryptosporidium* and other hard-to-culture organisms, these technologies have remained firmly in the domain of research because of the demands of efficient DNA extraction and adequate genome coverage, paucity of reference genomes, and lack of quality assured and standardized bioinformatic pipelines. Whole-genome sequencing has been used to improve classification, genotyping, and molecular epidemiology; seek new drug and vaccine targets; and investigate host–pathogen interactions and virulence factors [2–4]. New *Cryptosporidium* genomes are made available on the National Center for Biotechnology Information (NCBI) website (www.ncbi.nlm.nih.gov) and can be analyzed using the CryptoDB resource (www.cryptoDB.org). Some of these developments were enabled by the use either of rigorous oocyst purification procedures [5, 6] or whole-genome amplification [7]; *Cryptosporidium*-

Jan R. Mead and Michael J. Arrowood (eds.), *Cryptosporidium: Methods and Protocols*, Methods in Molecular Biology, vol. 2978, https://doi.org/10.1007/978-1-0716-4824-7_5,

positive stools are often low in volume and may contain relatively small numbers of oocysts to generate sufficient DNA required for WGS. These protozoa are also outnumbered by the presence of non-*Cryptosporidium* DNA in feces (e.g., bacterial and human DNA). Recently, enrichment methods have been developed using RNA baits, which allow for the specific isolation of *Cryptosporidium* DNA from these problematic samples ([8, 9]; *see* also Chap. 4). However, despite these advances, the methods and technology are not yet widely available and more basic methods may still be preferred in certain settings.

Here, we provide a method for high-quality DNA preparation for WGS from feces (*see* Fig. 1). The method incorporates semi-purification of oocysts by salt flotation (which may also be used prior to regular DNA extraction and PCR, *see* Chap. 2), further purification and concentration by immunomagnetic separation (IMS) (*see* Fig. 2), surface sterilization using bleach, and finally spin-column DNA extraction [5]. An option to use whole-genome amplification (WGA) is also described.

## 2 Materials

### 2.1 Preparation of DNA from Next-Generation Sequencing (NGS)

1. Saturated salt solution (cold preparation method): NaCl solution, 1.2 SG. Approximately 1.5 kg of sodium chloride is required to saturate 4 liters of oocyst-free distilled water by adding the sodium chloride in small quantities, taking care to maintain continuous, brisk stirring (e.g., with a magnetic stirrer). Continue to add sodium chloride until 1.2 SG is achieved as measured by hydrometer. Store in the dark at ambient temperature.
2. Immunomagnetic separation (IMS) kit, adapted from [10], usually containing 10X reagent A (detergent solution), 10X reagent B (protein-blocking solution) and paramagnetic beads coated with anti-*Cryptosporidium* antibodies.
3. 1X reagent A: Dilute the supplied 10X reagent A 1:9 with oocyst-free distilled water on the day of processing. Prepare enough for 1 mL per sample, plus an additional wastage 1 mL.
4. Glass Leighton tubes appropriate for the commercial IMS kit (*see* **Note 1**).
5. Hydrochloric acid solution: 0.1 M HCl. Add 90 μL of concentrated hydrochloric acid (1.18 SG) to 90 mL of oocyst-free distilled water. Mix well and make up to 100 mL with oocyst-free distilled water. The solution may be stored in the range 20 ± 5 °C for up to one month. Commercial supplies are available.

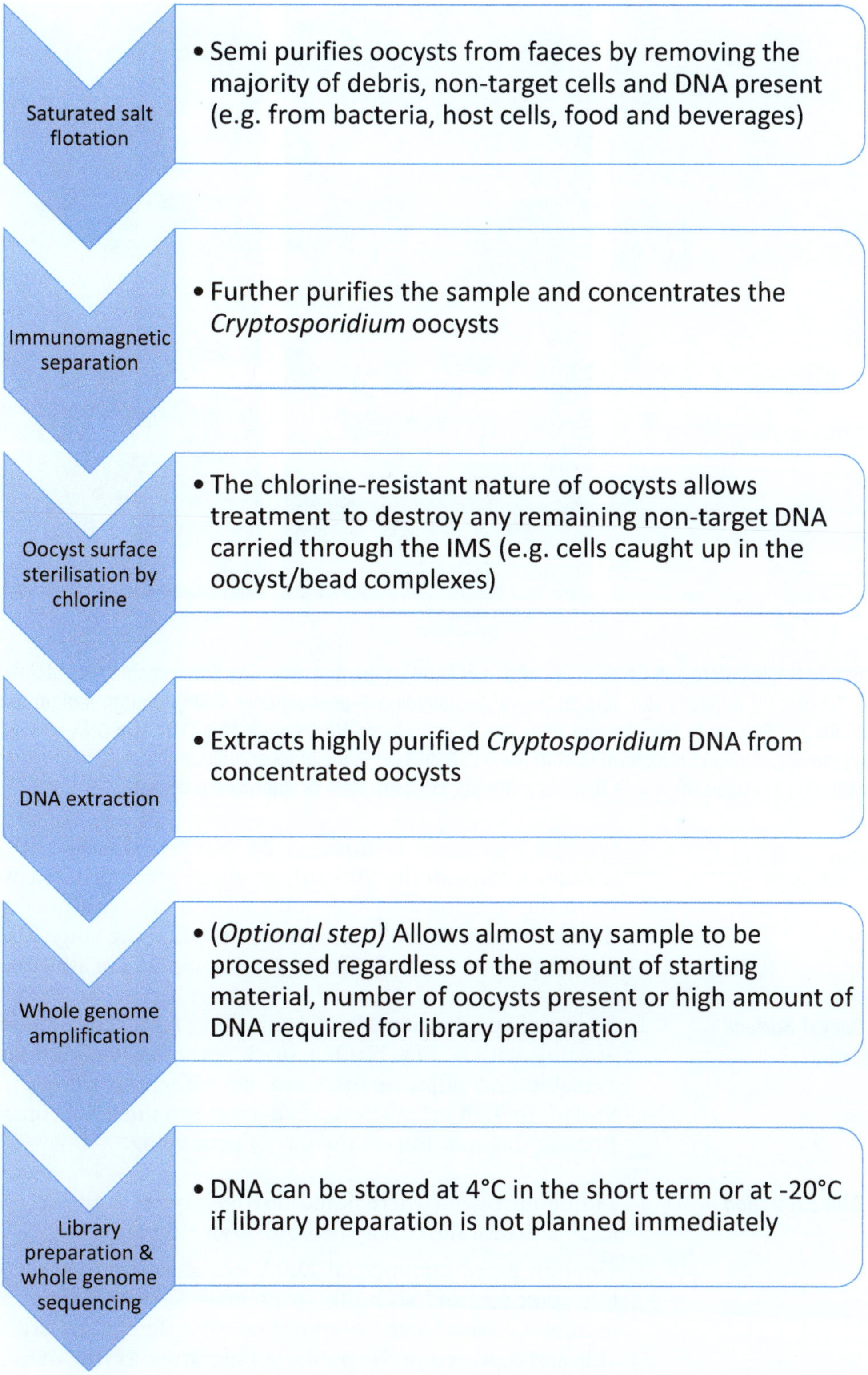


**Fig. 1** Workflow for *Cryptosporidium*-positive fecal sample preparation for whole-genome sequencing

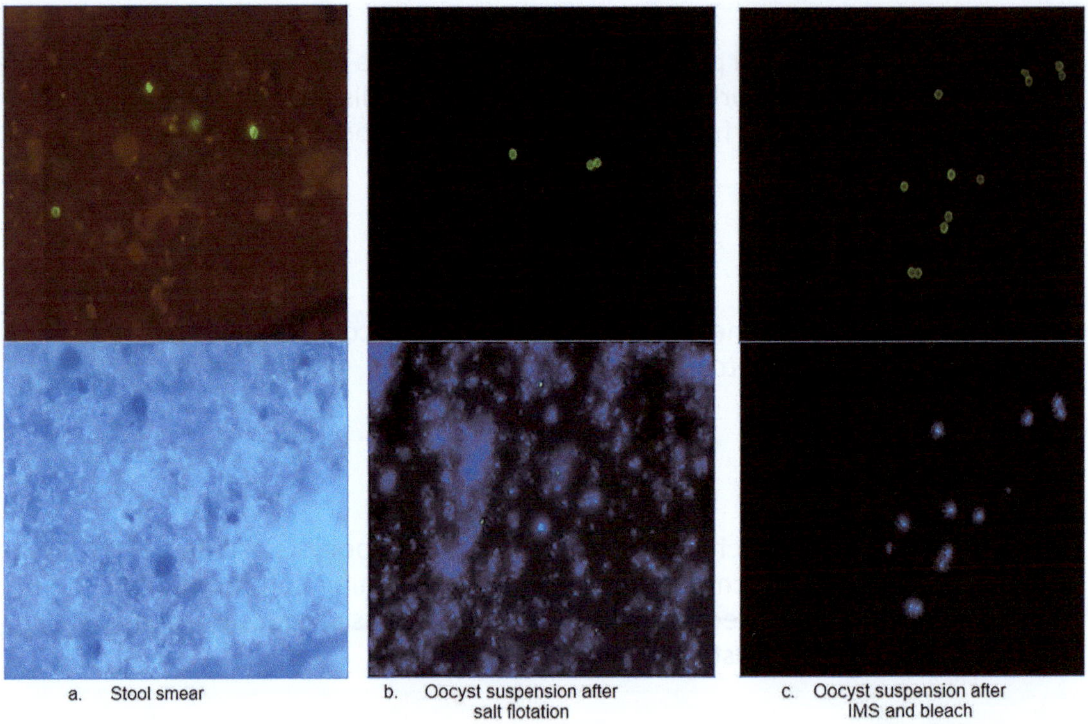


**Fig. 2** Stool sample before salt flotation (**a**), after salt flotation (**b**), and after IMS and bleaching (**c**), stained with immunofluorescent antibody (top image) and 4′,6-diamidino-2-phenylindole (DAPI) (bottom image) to demonstrate the purification of oocysts and reduction in nontarget DNA by each step DAPI is a DNA intercalating dye, and the bright blue in image (**a**) is from the bacterial load in the stool sample, much reduced in image (**b**) and eliminated in image (**c**) where there is a greater concentration of fluorescing oocysts

6. Sodium hydroxide solution: 1 M NaOH. Dissolve 40 g of sodium hydroxide in 900 mL of oocyst-free distilled water. Cool the solution and make up to 1000 mL with oocyst-free distilled water. The solution may be stored in the range $20 \pm 5$ ° C for up to one month. Commercial supplies are available.

### 2.2 Oocyst Surface Sterilization

1. Sodium hypochlorite: NaOCl solution (1.2% active chlorine). Starting with a 14% NaOCl stock (other concentrations are available and adjustments must be made accordingly), add 860 μL to 9.14 mL nuclease-free water in a universal container. Prepare this solution on the day of processing.

### 2.3 DNA Extraction

1. Liquid nitrogen (where liquid nitrogen is not available, a dry–ice/methanol slurry may be used as an alternative).
2. Column-based commercial DNA extraction kits usually contain some sort of lysis buffer, proteinase K, wash buffers, and a membrane-based spin column. Several different kits are available and reported in the published literature for the extraction of *Cryptosporidium* DNA. The example given below which is

used in our laboratory is the QIAamp DNA mini kit (Qiagen Inc., Germantown, MD).

3. Nuclease-free water.

#### 2.4 Optional Step: Whole-Genome Amplification (WGA) (See Note 2)

1. Commercial WGA kit. The use of WGA to boost *Cryptosporidium* DNA for WGS has been reported in the published literature [7]. Several different kits are available.

## 3 Methods

Wear appropriate personal protective equipment (PPE) during all sample processing steps.

#### 3.1 Preparation of DNA for WGS, Modified from [5]

1. Salt flotation procedure: Before use, ensure that the salt solution is mixed by inversion and allowed to settle for 5 min.
2. Transfer 10 mL of saturated salt solution in a 15-mL centrifuge tube (one per sample).
3. Add approximately 1 to 2 g of feces (*see* **Note 3**) with an applicator stick, or pipette 1 to 2 mL liquid feces, to the saturated salt solution, cap the tube, and mix thoroughly.
4. Gently overlay the salt/feces slurry with 3 mL of oocyst-free distilled water.
5. Place the centrifuge tube in a bench top centrifuge with swing out buckets, add a balance tube, if necessary, and centrifuge at 1100× g for 5 min (*see* **Note 4**).
6. Gently swirl the top 3 mL of fluid (the overlain fraction) and transfer to a clean 15-mL tube.
7. Wash the oocysts by topping up with oocyst-free distilled water, centrifuging at 1100× g for 5 min (*see* **Note 4**) and decanting (tipping off) the supernatant.
8. Resuspend the pellet in 1 mL of oocyst-free distilled water.

#### 3.2 Immunomagnetic Separation

1. Take the IMS reagents from the refrigerator and allow them to reach ambient temperature (*see* **Note 5**).
2. Add 1 mL of 10X reagent A and 1 mL of 10X reagent B to each clean labeled Leighton tube.
3. Transfer the entire 1 mL oocyst suspension to the Leighton tube and rinse out the sample tube with 1 mL of oocyst-free distilled water before also adding this to the Leighton tube.
4. Add 8 mL of oocyst-free distilled water to the Leighton tube.
5. Mix the antibody-coated paramagnetic beads thoroughly and add 100 μL to each suspension in the Leighton tubes.

6. Incubate the samples by placing the Leighton tubes into rotating mixer at ambient temperature for at least 60 min and rotate typically at approximately 20 rpm.
7. Remove from the rotor and place each Leighton tube into a magnetic particle concentrator with the flat side against the magnet.
8. Rock gently according to the manufacturer's instructions (typically either through a 90° motion or attached to a rotating mixer) (*see* **Note 6**).
9. One by one, with the remaining Leighton tubes still moving on the rocker, remove the magnet block containing first tube from the rocker and pour the non-captured particular matter carefully into a 15-mL centrifuge tube and retain in case further processing is required (*see* **Note 7**). Using a serological pipette, aspirate and dispose of any remaining supernatant from the Leighton tube without disturbing the beads (and captured oocysts) against the magnet.
10. Remove the Leighton tube from the magnet, add 0.8 mL of 1X reagent A, and rinse the beads (and captured oocysts) from the side of the tube by repeatedly pipetting the solution over them.
11. Transfer the beads (and captured oocysts) to a clean, labeled 1.5-mL flip-top microcentrifuge tube.
12. Add 0.2 mL of 1X reagent A to the Leighton tube (*see* **Note 8**), rinse out any remaining beads, and add this to the microcentrifuge tube.
13. Repeat the transfer of beads to 1.5-mL tubes with any remaining samples.
14. Gently resuspend the samples in the microcentrifuge tube by inversion several times and place the microcentrifuge tubes into a second particle concentrator (appropriate for the 1.5-mL tubes).
15. Rock according to the manufacturer's instructions (typically either through a 180° motion or attached to a rotating mixer).
16. One by one, with the remaining tubes still moving on the rocker, remove the first sample, still attached to the magnet, taking care not to disturb the pellet.
17. Immediately remove the supernatant from the inside of the cap and the tube (taking care not to disturb the pellet) using a clean fine-tipped disposable transfer pipette (Pastette), discard, and repeat with each of the remaining samples.
18. Remove all of the 1.5-mL tubes from the magnets and add 50 μL of 0.1 M HCl to each tube.

19. Vortex the tubes thoroughly for 10 s to separate the oocysts from the beads and stand the tubes at room temperature for 5 min (*see* **Note 9**).
20. Vortex each tube again for 10 s, immediately afterward placing the tubes into the dissociation block (*see* **Note 10**).
21. Lay the dissociation block flat on the workbench so that the tubes are horizontal with the magnets below them and leave for at least 30 s.
22. During this time, add 5 μL of 1.0 M NaOH to clean labeled 1.5-mL screw-capped microcentrifuge tubes (one for each test sample).
23. Return the dissociation block to an upright position and transfer the supernatant from each tube to the labeled NaOH-containing 1.5-mL screw-capped tube, without disturbing the beads at the back wall of the tube. Gently mix the sample using the same pipette tip.
24. Perform a second acid treatment on each sample by repeating the above steps and transfer this into the same tube.
25. Add 90 μL of oocyst-free distilled water to each of the tubes and mix well.
26. Proceed immediately to treat the suspension with NaOCl as described below. If this is not possible straight away, store the suspensions at 2–8 °C for no longer than 20 h.

### 3.3 Oocyst Surface Sterilization

1. Add 200 μL of 1.2% NaOCl solution to each 200 μL oocyst suspension.
2. Immediately vortex briefly and centrifuge at 1100× *g* for 5 min with soft acceleration–deceleration to minimize damage to oocysts (*see* **Note 4**).
3. Gently aspirate the supernatant into a suitable plastic waste container (*see* **Note 11**).
4. Wash the oocyst pellet by adding 200 μL of nuclease-free water to the 1.5-mL tube, vortex briefly to mix, centrifuge at 1100× g for 5 min with soft acceleration–deceleration profile to minimize damage to oocysts, and remove supernatant as previously to the waste container.
5. Repeat the nuclease-free water wash step twice more and finally resuspend the oocysts in 200 μL of nuclease-free water.
6. Proceed to DNA extraction immediately. If DNA extraction is not possible straight away, perform the freeze–thaw process, but following the final freeze, transfer the tubes to a chilled rack and store at −20 °C until ready for DNA extraction to be completed.

### 3.4 DNA Extraction

There are several commercial kits available to perform DNA extraction, but this example method using Qiagen's QIAamp DNA mini kit has been successfully used during the preparation of DNA for NGS [5, 6].

1. Disrupt the oocysts by eight freeze–thaw cycles of 1 min in liquid nitrogen and 1 min at 100 °C ± 5 °C, gently flicking the tubes between each cycle to resuspend any settled oocysts/debris.
2. Microcentrifuge briefly to spin any condensation back into the tube.
3. Add 200 μL of AL buffer and 20 μL of proteinase K (*see* **Note 12**) to each tube (*see* **Note 13**) before mixing by vortex for 15 s and incubating for 30 min at 54 to 58 °C.
4. After incubation at 54 to 58 °C for 30 min, remove the tubes from the dry block, cool in a rack for 5 min and briefly vortex, followed by brief centrifugation, and then place back in the tube rack.
5. Add 200 μL ethanol to each tube and vortex for 15 s before briefly centrifuging to remove any drops of sample from around the lid.
6. Carefully apply the mixture to the top of the appropriate QIAamp spin column sitting in a 2-mL collection tube, without wetting the rim and centrifuge at 6000× *g* for 1 min.
7. Discard the collection tube containing the filtrate and place the spin column in a second collection tube.
8. Add 500 μL of buffer AW1 to each tube (*see* **Note 14**) and centrifuge using at 6000× *g* for 1 min.
9. Discard the collection tube containing the filtrate into a waste disposal container (not containing bleach waste) and place the spin column in a third collection tube.
10. Add 500 μL of buffer AW2 to each tube (*see* **Note 14**) and centrifuge at 20,000× g for 4 min.
11. Discard the collection tube containing the filtrate into a waste disposal container and place the spin column into a clean, labeled 1.5-mL tube (*see* **Note 15**).
12. Add 50 μL of nuclease-free water to the membrane in each tube and incubate at room temperature for 5 min before centrifuging at 6000× g for 1 min.
13. Discard the spin columns and store the extracted DNA at −18 to −22 °C until use, either for library preparation if sufficient DNA is present or by WGA to increase the amount of available DNA.

### 3.5 Optional Step: Whole-Genome Amplification

1. Follow the manufacturer's instructions of the commercially available kit.
2. Store WGA DNA at 4 °C in the short term; for longer-term storage, keep at 18–22 °C.

## 4 Notes

1. Different manufacturers use different-sized Leighton tubes, so check the product insert before purchasing.
2. Several factors can influence whether WGA is required prior to WGS including the number of oocysts present in a sample, amount of sample available, and the amount of start DNA required for NGS library preparation (library kit requirements can vary between 1 ng and over 100 ng of starting DNA). We have used multiple displacement amplification (MDA) for WGA to enable WGS from any sample, but have not extensively validated it. However, several published studies use this technology prior to WGS [7].
3. All stools should be well mixed prior to transferring to the salt solution.
4. Centrifugation at speeds higher than 1100× g for longer periods of time (>5 min) is not advised as some parasites may deform or rupture and collapse.
5. Make sure that any crystals present in the reagents have dissolved.
6. It is important that the tubes are continuously moved at this stage to prevent additional unwanted particulate matter becoming stuck to the beads. If the motion stops for more than 10 s, samples will need removing from the magnet and resuspending, before being fully recaptured.
7. If only a few beads remain in the Leighton tube against the magnet, the majority still need to be recovered from the supernatant. Return the supernatant back to the Leighton tube and incubate for a further incubation as before. If this still returns few beads, then return the supernatant back to the tube and incubate again but reducing the speed of the rocking.
8. Take care not to disturb any fecal debris on the front side of the tube when rinsing.
9. Vortexing breaks the bonds between the oocysts and the beads. The acid stops the oocysts from rebinding with the beads.
10. Make sure all of the samples are at the bottom of the tube.

11. Do not discard waste into container with DNA extraction waste as some buffers contain guanidine thiocyanate that can react with the bleach to create toxic gases.
12. Aliquot sufficient reagents at the start to reduce the risk of contaminating stock reagents.
13. A new tip must be used for each sample tube, and each tube lid should be removed individually at the point of adding reagents.
14. Providing the tip does not touch any of the tubes, and there is no bridging; one tip can be used for all samples as the DNA is immobilized on the column and is no longer a backsplash risk.
15. This tube should remain closed at all times after labeling and during the extraction procedure to prevent cross-contamination and should only be opened to admit the spin column as described.

## Acknowledgments

The research leading to these methods has received funding from the European Union Seventh Framework Programme ([FP7/2007-2013] [FP7/2007-2011]) under Grant Agreement No: 311846.

**Competing Interest Statement** The authors have no conflict of interests to declare that are relevant to the content of this chapter.

### References

1. Didelot X, Bowden R, Wilson DJ (2012) Transforming clinical microbiology with bacterial genome sequencing. Nat Rev Genet 13: 601–612
2. Khan A, Shaik JS, Grigg ME (2018) Genomics and molecular epidemiology of *Cryptosporidium* species. Acta Trop 184:1–14
3. Ryan U, Hijjawi N (2015) New developments in *Cryptosporidium* research. Int J Parasitol 45: 367–373
4. Agyabeng-Dadzie F, Xiao R, Kissinger JC (2024) *Cryptosporidium* genomics—current understanding, advances, and applications. Curr Trop Med Rep 11:92–103
5. Hadfield SJ, Pachebat JA, Swain MT, Robinson G, Cameron SJS, Alexander JL, Hegarty MJ, Elwin K, Chalmers RM (2015) Generation of whole genome sequences of new *Cryptosporidium hominis* and *Cryptosporidium parvum* isolates directly from stool samples. BMC Genomics 16:650
6. Gilchrist CA, Cotton JA, Burkey C, Arju T, Gilmartin A, Lin Y, Ahmed E, Steiner K, Alam M, Ahmed S, Robinson G, Zaman SU, Kabir M, Sanders M, Chalmers RM, Ahmed T, Ma JZ, Haque R, Faruque ASG, Berriman M, Petri WA (2018) Genetic diversity of *Cryptosporidium hominis* in a Bangladeshi community as revealed by whole genome sequencing. J Infect Dis 218(2):259–264
7. Guo Y, Li N, Lysén C, Frace M, Tang K, Sammons S, Roellig DM, Feng Y, Xiao L (2015) Isolation and enrichment of *Cryptosporidium* DNA and verification of DNA purity for whole-genome sequencing. J Clin Microbiol 53:641–647
8. Bayona Vasquez NJ, Sullivan AH, Beaudry MS, Khan A, Baptista RP, Petersen KN, Bhuiyan MIU, Brunelle B, Robinson G, Chalmers RM, Alves-Ferreira EVC, Grigg ME, Kissinger JC, Glenn TC (2024) Whole genome targeted enrichment and sequencing of

human-infecting *Cryptosporidium* spp. bioRxiv. 2024.03.29.586458; also submitted to Nature Communications

9. Khan A, Alves-Ferreira EVC, Vogel H, Botchie S, Ayi I, Pawlowic M, Robinson G, Chalmers RM, Lorenzi H, Grigg ME (2024) Phylogenomic reconstruction of *Cryptosporidium* spp. captured directly from clinical samples reveals extensive genetic diversity. bioRxiv. 2024.04.17.589752; also submitted to Nature Communications
10. Anon (2010) The microbiology of drinking water (2010)—part 14—methods for the isolation, identification and enumeration of *Cryptosporidium* oocysts and *Giardia* cysts. Environment Agency, Bristol. Available at: http://www.environment-agency.gov.uk/static/documents/Research/Part_14-oct20-234.pdf

# Chapter 6

# A Real-Time PCR-Based Method for the Sequence Determination of *Cryptosporidium hominis* and *Cryptosporidium parvum gp60* Subtypes During Epidemiological Investigations

**Guy Robinson, Kristin Elwin, and Rachel M. Chalmers**

## Abstract

The *gp60* gene has been widely used to subtype *Cryptosporidium* for many years. Predominantly, this has involved nested polymerase chain reaction (PCR) amplification, gel electrophoresis, and Sanger sequencing of the amplicons. However, laboratories may no longer have gel electrophoresis capability, and during outbreak investigations a more timely approach may be required. Here, we describe a real-time PCR assay combined with high-resolution melt curve analysis that streamlines the PCR detection of *gp60* amplicons of *Cryptosporidium parvum* and *Cryptosporidium hominis* for sequencing during public health investigations.

**Key words** *Cryptosporidium*, *gp60*, Real-time PCR, Subtyping, Molecular epidemiology

## 1 Introduction

Primary diagnostic tests for *Cryptosporidium* only detect the presence of the genus, which limits the epidemiological usefulness. To determine the potential source of an infection, as a minimum, species determination is required, but this is only suggestive in broad terms with *Cryptosporidium hominis* predominantly indicating anthroponotic transmission and *Cryptosporidium parvum* potentially zoonotic [1, 2]. However, human-to-human *C. parvum* transmission also occurs, either as onward spread from a zoonotic source or with subtypes that are human adapted. To identify such infections, and to link any *C. parvum* or *C. hominis* cases with each other or with suspected sources, more discriminatory methods are required [2–4].

The ultimate resolution for molecular epidemiology is from whole genome sequencing data. However, the time and expense, as well as the lack of standardized bioinformatic pipelines for the

Jan R. Mead and Michael J. Arrowood (eds.), *Cryptosporidium: Methods and Protocols*, Methods in Molecular Biology, vol. 2978, https://doi.org/10.1007/978-1-0716-4824-7_6,

epidemiological investigation of such data, is currently a barrier to routine use in real time, particularly for public health purposes [5]. Although multilocus genotyping approaches are available, for example, see Chap. 8, the most commonly used *Cryptosporidium* spp. subtyping method, with over 20 years of use, is sequence analysis of the hypervariable region of the 60-kDa glycoprotein (*gp60*) gene [6–9]. This single gene method is particularly useful in settings where intraspecies diversity at this locus is high, such as *C. hominis* in low- and middle-income countries [2, 3, 10, 11].

Traditionally, the polymerase chain reaction (PCR) used to amplify the polymorphic region of the *gp60* gene is a nested assay requiring gel electrophoresis to detect amplicons prior to Sanger sequencing (*see* Chap. 7). However, by using real-time PCR to amplify and detect *gp60* amplicons prior to sequencing, this process is streamlined, allowing for improved workflow and reduced time to subtype result. Here, we describe a *gp60* real-time PCR that incorporates a high-resolution melt (HRM) to determine the presence of *C. hominis* and *C. parvum* amplicons that are suitable for sequencing. The assay can also be used to determine *Cryptosporidium cuniculus* subtypes. The primers used for this assay include the previously published forward AL3532 [7] and a newly described reverse primer that together amplify a 400–600 bp fragment, depending upon the subtype, which is sufficient to differentiate between different subtype families and their suffix variations [3]. The assay has been successfully used in the investigation of outbreaks and incidents in England and Wales since 2023 [12, 13].

## 2 Materials

### 2.1 *gp60* Real-Time PCR and HRM Analysis

1. Vortex mixer and microcentrifuge for 1.5-mL tubes, or a combination instrument (e.g., Grant Instruments Combi-spin PCV-2400).
2. PCR workstation with ultraviolet (UV) decontamination.
3. Variable volume pipettes.
4. Filter-protected pipette tips.
5. Real-time PCR thermocycler with HRM capability (e.g., Qiagen Rotor-Gene Q HRM) (*see* **Note 1**).
6. Qiagen Type-it HRM PCR Kit (*see* **Note 1**).
7. 10-μM forward primer (AL3532): 5′-TCC GCT GTA TTC TCA GCC-3′.
8. 10-μM reverse primer (gp60_500R): 5′-CCA GAG ATA TAT CTT GGT GCG GG-3′.
9. Nuclease-free water.

10. 1.5-mL microcentrifuge tubes.
11. 0.2-mL flat-capped PCR tubes.

## 3 Methods (*See* Note 1)

### 3.1 *gp60* Real-Time PCR and HRM Analysis

1. Allow the deoxyribonucleic acid (DNA) samples and PCR reagents (*see* Table 1) to thaw separately at room temperature.
2. Momentarily vortex the DNA samples and PCR reagents before centrifuging briefly to remove contents from the tube lids.
3. In a DNA-free PCR workstation, prepare sufficient PCR master mix in a 1.5-mL microcentrifuge tube for the required number of samples, plus a no-template control (nuclease-free water), a *Cryptosporidium*-positive DNA control, and 20% extra for wastage, based on the details for a single reaction ) (*see* Table 1 and **Note 2**).
4. Briefly vortex the master mix and spin down in a centrifuge for a few seconds to remove contents from the tube lid.
5. Pipette 22 μL of prepared master mix into each 0.2-mL PCR tube, one per sample.
6. Add 3 μL of nuclease-free water to the no-template control tube.
7. Move the tubes to a designated DNA template addition PCR workstation and UV decontaminate the workstation and tools used for reagent preparation.
8. Add 3 μL of each DNA sample to its corresponding tube.
9. Finally, to further reduce the risk of cross contamination, add 3 μL of *Cryptosporidium*-positive DNA to the last tube (*see* **Note 3**).

**Table 1**
**Recipe for *gp60* real-time HRM PCR master mix**

| Master mix components | 1× reaction (μL) |
|---|---|
| 2× type-it HRM PCR master mix | 12.5 |
| 10-μM forward primer (AL3532) | 1.75 |
| 10-μM reverse primer (gp60_500R) | 1.75 |
| Nuclease-free water | 6 |
| [DNA] | [3] |
| Total reaction volume | 25 |

**Table 2**
**Conditions for the *gp60* real-time HRM PCR**

| Stage | Temperature (°C) | Duration (s) |
|---|---|---|
| Hold (Hot start) | 95 °C | 300 |
| Cycling (45 cycles) | 95 °C<br>63 °C<br>72 °C | 10<br>30<br>15 |
| Pre-melt | 80 °C | 120 |
| HRM | 80 to 87 °C (increasing 0.1 per step) | 2 per step |

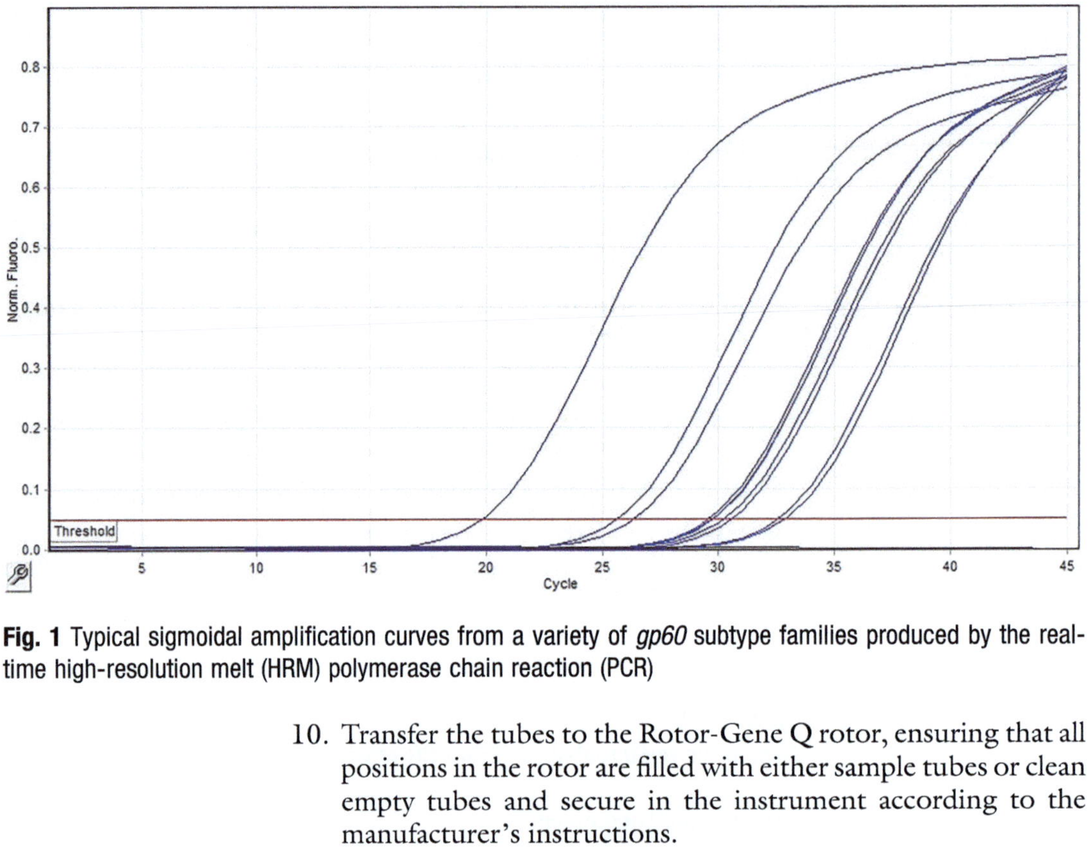


**Fig. 1** Typical sigmoidal amplification curves from a variety of *gp60* subtype families produced by the real-time high-resolution melt (HRM) polymerase chain reaction (PCR)

10. Transfer the tubes to the Rotor-Gene Q rotor, ensuring that all positions in the rotor are filled with either sample tubes or clean empty tubes and secure in the instrument according to the manufacturer's instructions.
11. Select and run the HRM PCR cycling program as described in Table 2.
12. Once the run has completed, the amplification curves should be analyzed in the Green channel, applying a fluorescence threshold of 0.05 and inspecting for typical sigmoidal curves (*see* Fig. 1 and **Note 4**).

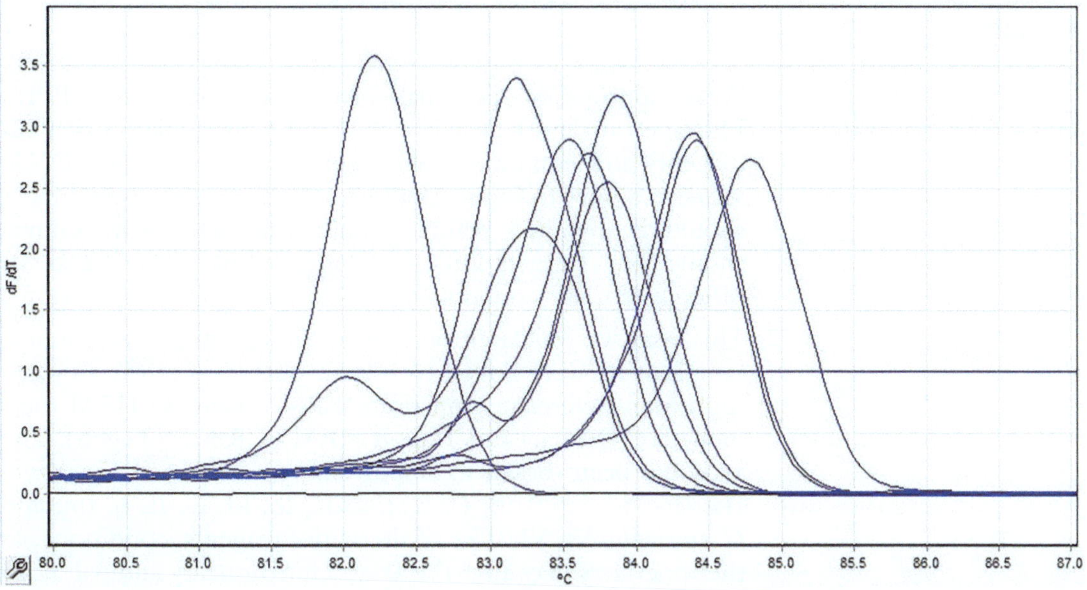


**Fig. 2** High-resolution melt curves from a variety of *gp60* subtype families which fall within the range 81.6–85.1 °C

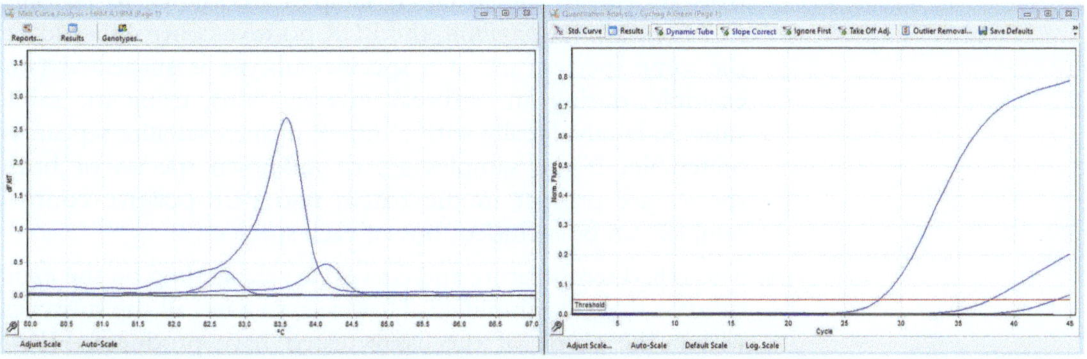


**Fig. 3** An example of two late amplification and non-sigmoidal curves, typical of weak amplification of *gp60* PCR-positive samples with insufficient template. The HRM curves are also well below the threshold, which can be adjusted to determine the melt temperature of the weak peaks. Samples with weak peaks may provide good-quality sequences

13. Positive amplification must be confirmed by checking the HRM curves by applying an initial threshold of 1.00 and manually adjusting to capture the peaks as required (*see* Figs. 2 and 3 and **Notes 5–8**).
14. Subtypes of positive samples are determined by Sanger sequencing the PCR amplicons. If this is not undertaken or sent for sequencing the same day, store samples at 2–8 °C for up to a week or −20 °C for longer periods (*see* **Note 9**).

## 4 Notes

1. Wear appropriate personal protective equipment (PPE). Extreme care must be taken to avoid contamination of PCR reagents and equipment with genomic and amplified DNA. Work in a low- to high-risk direction: reagent tubes and racks should be handled before sample tubes and racks. Gloves should be changed after touching sample tubes/racks or other potentially contaminated areas such as the discard box.
2. This assay was validated on human clinical samples and animal samples that were linked to human outbreaks at the Cryptosporidium Reference Unit using a Rotor-Gene Q HRM thermocycler (Qiagen) and Type-it HRM PCR Kit (Qiagen). The assay has been shown to amplify subtype families: *C. parvum* IIa, IIc, IId, IIe, IIm; *C. hominis* Ia, Ib, Id, Ie, If, Ij, Im; and *C. cuniculus* Va, Vb. Use of alternative reagents, conditions, or thermocyclers may alter the assay's functionality and will need to be optimized and validated to ensure satisfactory performance.
3. Consideration should be made as to which PCR-positive controls are run in the batch. If samples are known to be only *C. hominis* or *C. parvum*, the PCR-positive control should be the same species, and if a specific subtype is suspected (for example, during an outbreak investigation), using the same subtype family, ideally with a different microsatellite repeat, is preferable. If the samples are of unknown species or both species are present in the batch, two PCR-positive control samples can be included, one of each species.
4. The HRM temperature and curves vary depending on the *gp60* subtype that was amplified. However, to date, all the subtypes within the scope of the assay tested have produced HRM temperatures in the range 81.6–85.1 °C (*see* Fig. 2).
5. Occasionally, weak amplification can be seen as a late, low amplification curve that might not reach the threshold or have enough cycles to complete a sigmoidal shape (*see* Fig. 3).
6. As detection is based on a nonspecific DNA intercalating dye, any nonspecific double-stranded DNA artifacts (e.g., primer dimers, secondary structures) will also result in an increase in fluorescence that will be detected as late, low amplification. However, the HRM curves for these usually appear as broad, messy peaks and fall outside of the *gp60* HRM temperature range. This is the reason that it is important to confirm the presence of a typical peak within the correct temperature range (*see* Fig. 4). As the HRM program is set to increase from 80–87 °C, these nonspecific peaks are not generally seen in the assay analysis.

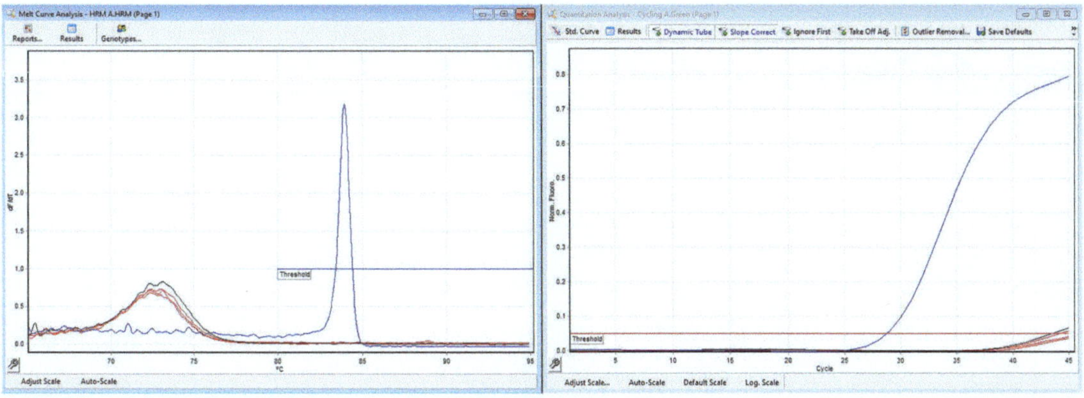


**Fig. 4** An example of four *gp60* PCR-negative samples showing weak amplification curves (including the black no-template control) and broad, messy nonspecific HRM curves between 70 °C and 75 °C. A *gp60*-positive amplification and HRM curve is also shown for reference

7. Strong peaks can be removed and the traces autoscaled if required, and the threshold adjusted to determine the melt temperature of weak positives.
8. While the HRM curves produced by most *gp60* families give single peaks, those of the family If give an unusual double peak due to the melting properties of the specific sequence in this subtype (*see* Fig. 5).
9. Various commercial providers are available to provide Sanger sequencing by mailing the PCR amplicons and an aliquot of the primers at ambient temperature, with the sequencing data usually returned via email the following day. However, if facilities are available and/or a rapid turnaround is required, in-house sequencing can be performed according to the specific instrument manufacturer's instructions.

## Competing Interest Statement

The authors have no conflict of interests to declare that are relevant to the content of this chapter.

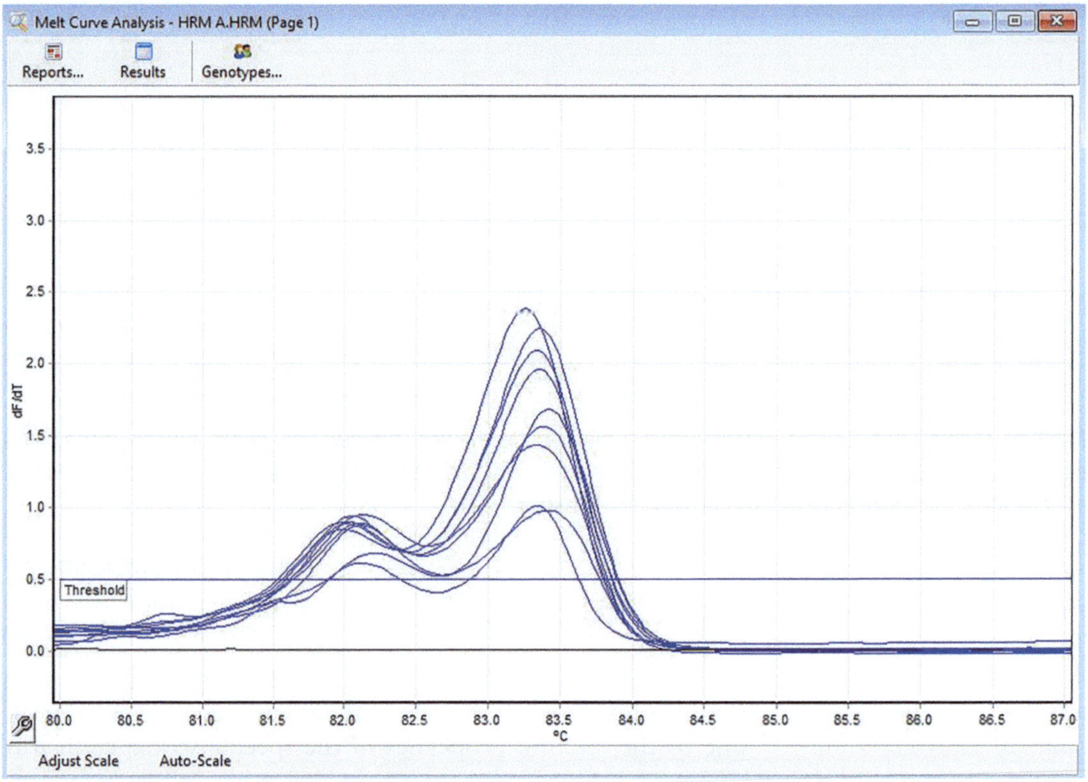


**Fig. 5** Nine examples of characteristic HRM curves generated from amplicons belonging to the *C. hominis* If *gp60* subtype family

## References


1. Xiao L (2010) Molecular epidemiology of cryptosporidiosis: an update. Exp Parasitol 124:80–99
2. Feng Y, Ryan UM, Xiao L (2018) Genetic diversity and population structure of *Cryptosporidium*. Trends Parasitol 34:997–1011
3. Xiao L, Feng Y (2017) Molecular epidemiologic tools for waterborne pathogens *Cryptosporidium* spp. and *Giardia duodenalis*. Food Waterborne Parasitol 8-9:14–32
4. Chalmers RM, Robinson G, Elwin K, Elson R (2019) Analysis of the *Cryptosporidium* spp. and *gp60* subtypes linked to human outbreaks of cryptosporidiosis in England and Wales, 2009 to 2017. Parasit Vectors 12:95
5. Morris A, Robinson G, Swain MT, Chalmers RM (2019) Direct sequencing of *Cryptosporidium* in stool samples for public health. Front Public Health 7:360
6. Strong WB, Gut J, Nelson RG (2000) Cloning and sequence analysis of a highly polymorphic *Cryptosporidium parvum* gene encoding a 60-kilodalton glycoprotein and characterization of its 15- and 45-kilodalton zoite surface antigen products. Infect Immun 68:4117–4134
7. Peng MM, Matos O, Gatei W, Das P, Stantic-Pavlinic M, Bern C, Sulaiman IM, Glaberman S, Lal AA, Xiao L (2001) A comparison of *Cryptosporidium* subgenotypes from several geographic regions. J Eukaryot Microbiol Suppl:28S–31S
8. Alves M, Xiao L, Sulaiman I, Lal AA, Matos O, Antunes F (2003) Subgenotype analysis of *Cryptosporidium* isolates from humans, cattle, and zoo ruminants in Portugal. J Clin Microbiol 41:2744–2747
9. Sulaiman IM, Hira PR, Zhou L, Al-Ali FM, Al-Shelahi FA, Shweiki HM, Iqbal J, Khalid N, Xiao L (2005) Unique endemicity of cryptosporidiosis in children in Kuwait. J Clin Microbiol 43:2805–2809
10. Gilchrist CA, Cotton JA, Burkey C, Arju T, Gilmartin A, Lin Y, Ahmed E, Steiner K,

Alam M, Ahmed S, Robinson G, Zaman SU, Kabir M, Sanders M, Chalmers RM, Ahmed T, Ma JZ, Haque R, Faruque ASG, Berriman M, Petri WA (2018) Genetic diversity of *Cryptosporidium hominis* in a Bangladeshi community as revealed by whole-genome sequencing. J Infect Dis 218:259–264

11. Tichkule S, Cacciò SM, Robinson G, Chalmers RM, Mueller I, Emery-Corbin SJ, Eibach D, Tyler KM, van Oosterhout C, Jex AR (2022) Global population genomics of two subspecies of *Cryptosporidium hominis* during 500 years of evolution. Mol Biol Evol 39:msac056
12. Peake L, Inns T, Jarvis C, King G, Rabie H, Henderson J, Wensley A, Jarratt R, Roberts C, Williams C, Orife O, Browning L, Neilson M, McCarthy C, Millar P, Love N, Elwin K, Robinson G, Mannes T, Young N, Chalmers R, Elson R, Vivancos R (2023) Preliminary investigation of a significant national *Cryptosporidium* exceedance in the United Kingdom, August 2023 and ongoing. Euro Surveill 28:pii=2300538
13. Peake L, Bardsley M, Bartram S, Bharuchi S, Howkins J, Robinson G, Charlett A, Chalmers R, Bird S, Young N (2024) A large cryptosporidiosis outbreak associated with an animal contact event in England: a retrospective cohort study, 2023. Epidemiol Infect 152: e91

# Chapter 7

## *Cryptosporidium* Genotyping for Epidemiology Tracking

**Matthew H. Seabolt and Dawn M. Roellig**

### Abstract

*Cryptosporidium* genotyping has made significant contributions to the species structure and population genetics of *Cryptosporidium* spp. In addition, the standardized method has shown utility in epidemiologic investigations identifying case linkages and contamination sources that could not be determined with traditional epidemiologic tools. The standardized method for *Cryptosporidium* genotyping from stool specimens described here comprises multiple individual protocols to amplify and sequence regions of the small subunit ribosomal ribonucleic acid (SSU rRNA) and *gp60* genes of *Cryptosporidium*.

**Key words** Genotyping, Subtyping, Molecular epidemiology, PCR, DNA sequencing

## 1 Introduction

Molecular methods have rewritten *Cryptosporidium* taxonomy; what was once considered a single species causing cryptosporidiosis in humans and other animals is now known to be a genus of nearly 50 species, many with multiple subtypes [1]. As all forms are morphologically indistinguishable by traditional diagnostics, molecular tools must be used to identify and distinguish the responsible etiologic agent in specimens.

The utility of these *Cryptosporidium* genotyping methods has expanded beyond population genetics and includes molecular epidemiology of cryptosporidiosis. In 2002, the previously identified *Cryptosporidium parvum* human genotype and bovine genotype were renamed *Cryptosporidium hominis* and *C. parvum* following biological distinction between the two species [2]. *Cryptosporidium* specimens have been systematically collected and characterized since then to reveal *C. hominis* and *C. parvum* account for >95% of human cryptosporidiosis [3]. In addition to cryptosporidiosis surveillance, genotyping has been used in outbreak investigations to differentiate outbreaks within the same geographic locale, link seemingly sporadic cases, confirm foodborne and waterborne

Jan R. Mead and Michael J. Arrowood (eds.), *Cryptosporidium: Methods and Protocols*, Methods in Molecular Biology, vol. 2978, https://doi.org/10.1007/978-1-0716-4824-7_7,

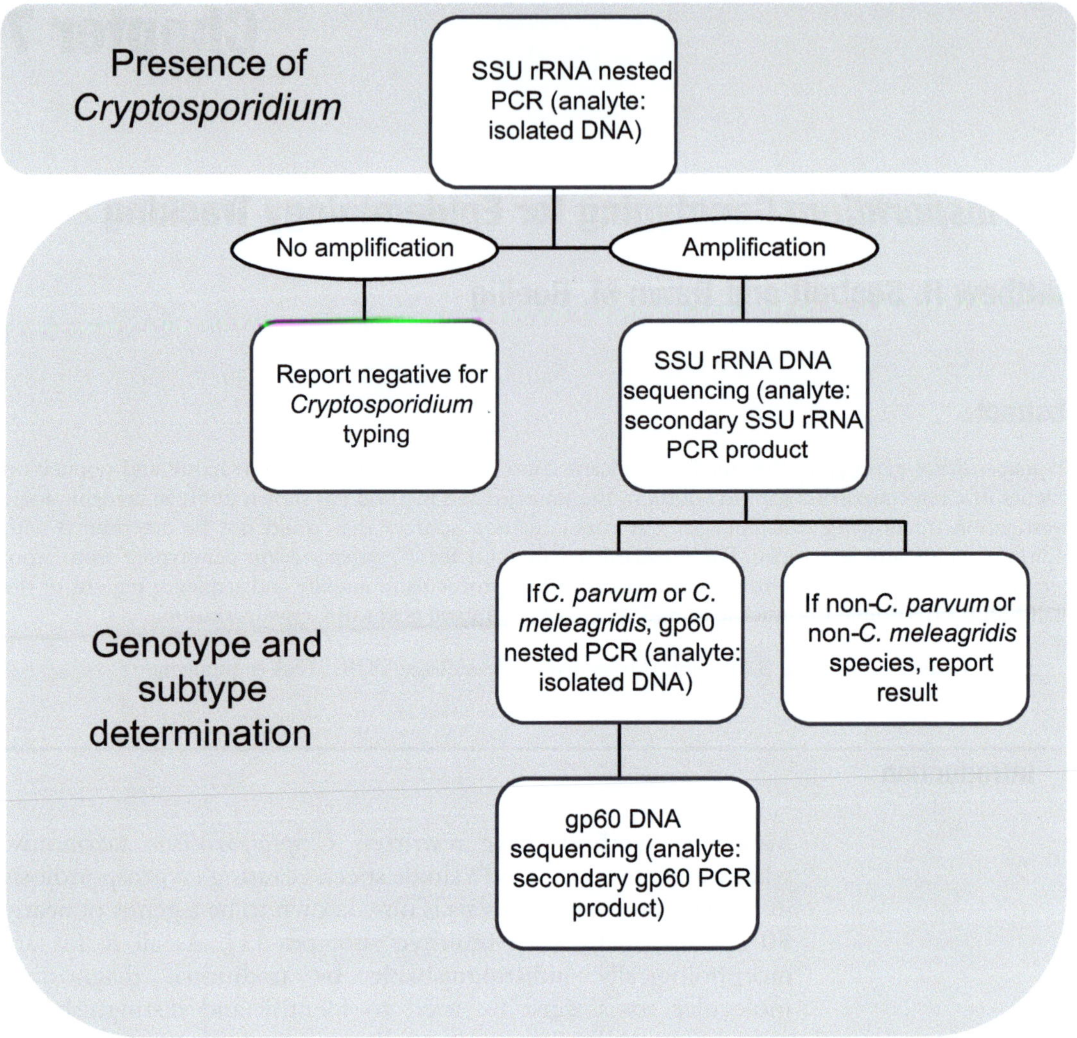


**Fig. 1** *Cryptosporidium* genotyping flowchart. Multiple protocols comprise the standardized *Cryptosporidium* genotyping method. Amplification of the small subunit ribosomal ribonucleic acid (SSU rRNA) gene by polymerase chain reaction (PCR) is completed first, where the resulting product is used in deoxyribonucleic acid (DNA) sequencing of the SSU rRNA gene. If *C. parvum* or *C. hominis* are identified, PCR amplification and subsequent DNA sequencing of the *gp60* gene are completed

contamination sources in outbreaks, and identify rare species and/or subtypes in sporadic and outbreak cases [4–8].

Multiple individual protocols comprise the standardized method for *Cryptosporidium* genotyping (*see* Fig. 1). A nested polymerase chain reaction (PCR) is used to amplify the small subunit ribosomal ribonucleic acid (SSU rRNA) gene, a high copy number and conserved gene interspersed with highly polymorphic regions, in all species and genotypes of *Cryptosporidium* [9, 10]. Next, deoxyribonucleic acid (DNA) sequencing of the SSU rRNA amplicon is used to determine and/or confirm the species of *Cryptosporidium* based on sequence identity to reference sequences. For specimens that are identified as either *C. parvum* or *C. hominis*, amplification and subsequent DNA (Sanger dideoxy) sequencing of the *gp60* gene is used to determine the subtype based on sequence identity to the reference sequences and the number of trinucleotide repeats [11, 12, 10].

In addition to classical PCR and sequencing methods, real-time PCR-based platforms, such as BioFire, have also become common and valuable tools for the rapid detection and species identification of *Cryptosporidium*. These systems can potentially offer high sensitivity and specificity, enabling the simultaneous detection of multiple pathogens, including *Cryptosporidium*.

## 2 Materials

### 2.1 PCR Analysis (SSU rRNA or gp60)

#### 2.1.1 Equipment

1. Vortex mixer.
2. PCR/ultraviolet (UV) workstation.
3. Microcentrifuges for 1.5-mL tubes and 0.2-mL tubes.
4. Thermal cycler (ABI GeneAmp® PCR System 9700; Applied Biosystems, Foster City, CA).
5. Horizontal gel electrophoresis system.
6. Electrophoresis power supply.
7. Camera photographic hood (documenting gels).
8. Balance.
9. Microwave.

#### 2.1.2 Consumables

1. 1.5-mL microcentrifuge tubes.
2. 0.2-mL PCR tubes.
3. Aerosol-resistant pipette tips with filter: 10 μL, 20 μL, 100 μL, 200 μL, and 1000 μL.

#### 2.1.3 Reagents

1. SSU rRNA gene primers:

   Primary SSU rRNA gene PCR primers:

   (a) Forward (F1): 5′-TTC TAG AGC TAA TAC ATG CG-3′.

   (b) Reverse (R1): 5′-CCC ATT TCC TTC GAA ACA GGA-3′.

   Secondary SSU rRNA gene PCR primers:

   (a) Forward (F2): 5′-GGA AGG GTT GTA TTT ATT AGA TAA AG-3′.

   (b) Reverse (R2): 5′-CTC ATA AGG TGC TGA AGG AGT A-3′.

2. *gp60* gene primers:

   Primary PCR *gp60* gene primers:

   (a) Forward (F1): 5′-ATG AGA TTG TCG CTC ATT ATC G-3′.

   (b) Reverse (R1): 5′-TTA CAA CAC GAA TAA GGC TGC-3′.

   Secondary PCR *gp60* gene primers:

   (a) Forward (F2): 5′′-TCC GCT GTA TTC TCA GCC-3′.

   (b) Reverse (R2): 5′-GGA AGG AAC GAT GTA TCT-3′.

3. GeneAmp® 10× PCR Buffer I with 15 mM $MgCl_2$ (Applied Biosystems).
4. GeneAmp® 10× PCR Buffer II (Applied Biosystems).
5. 100 mM deoxynucleoside triphosphate (dNTP).
6. 25 mM $MgCl_2$.
7. 100 mg bovine serum albumin (BSA).
8. GoTaq® DNA polymerase, 5 U/μL (Promega, Madison, WI).
9. Agarose low electroendosmosis (LE) (molecular biology grade).
10. 10× Tris-Borate-EDTA (TBE).
11. 100-bp DNA ladder (e.g., TriDye™, New England Biolabs, Ipswich, MA).
12. Nucleic acid gel stain (e.g., GelRed® 10,000× in water, Biotium, Hayward, CA).
13. Gel loading dye (bromophenol blue).
14. Distilled and deionized water.

## 2.2 Sequence Analysis

### 2.2.1 Equipment

1. Vortex mixer.
2. PCR/UV workstation.
3. Microcentrifuges for 1.5-mL tubes and 0.2-mL tubes.
4. ABI GeneAmp® PCR System 9700 (Applied Biosystems).
5. Plate mixer.
6. Vacuum centrifuge (Eppendorf Vacufuge).
7. ABI 3500 Genetic Analyzer (Applied Biosystems).

### 2.2.2 Consumables

1. 1.5-mL microcentrifuge tubes.
2. 0.2-mL PCR tubes.
3. Aerosol-resistant pipette tips with filter: 10 μL, 20 μL, 100 μL, 200 μL, and 1000 μL.

4. Millipore, MultiScreen high-throughput screening (HTS) Vacuum Manifold (Thermo Fisher Scientific Inc., Waltham, MA).
5. Millipore, MultiScreen HTS PCR plate (Thermo Fisher Scientific).
6. Centri-Sep spin column (Princeton Separations, Adelphia, NJ).
7. MicroAmp Optical 96-Well Reaction Plate (Applied Biosystems).
8. 3100 Genetic Analyzer Plate Septa 96 well (Applied Biosystems).

#### 2.2.3 Reagents

1. Secondary SSU rRNA gene PCR primers:
   (a) Forward (F2): 5′-GGA AGG GTT GTA TTT ATT AGA TAA AG-3′.
   (b) Reverse (R2): 5′-CTC ATA AGG TGC TGA AGG AGT A-3′.
2. Secondary PCR *gp60* gene primers:
   (a) Forward (F2): 5′-TCC GCT GTA TTC TCA GCC-3′.
   (b) Reverse (R2): 5′-GGA AGG AAC GAT GTA TCT-3′.
3. BigDye® Terminator v3.1 Cycle Sequencing Kit (Applied Biosystems).
4. Distilled and deionized water.

## 3 Methods

### 3.1 PCR Analysis (SSU rRNA or gp60)

1. Defrost PCR reagents at room temperature.
2. Prepare the PCR master mixture solution by adding all the PCR reagents in a 1.5-mL microcentrifuge tube as suggested in appropriate PCR target reaction (*see* Tables 1 and 2 and **Notes 1** and **2**).
3. Mix well by vortexing briefly and centrifuge the tube for 5 s to remove any solution trapped in the tube cap.
4. Add 49 μL (for primary PCR) or 48 μL (for secondary PCR) of the PCR master mixture prepared above to each 0.2-mL PCR tube.
5. Add 1 μL UV-treated distilled water to the tube designated as the blank negative control, or add 2 μL PCR product of Primary PCR-negative control to the secondary PCR tube designated as Secondary PCR-negative control.
6. Add 1 μL of the unknown DNA samples to the tubes with the master mixture for primary PCR, or add 2 μL of the primary PCR products to the tubes with the master mixture for secondary PCR (*see* **Notes 3** and **4**).

**Table 1**
**PCR master mixture recipe for SSU rRNA gene of *Cryptosporidium***

| SSU rRNA gene | | | | | |
|---|---|---|---|---|---|
| **Primary PCR master mixture** | | | **Secondary PCR master mixture** | | |
| **Reagent** | **Per 50 μL reaction (μL)** | **Final concentration** | **Reagent** | **Per 50 μL reaction (μL)** | **Final concentration** |
| Distilled water | 31.35 | | Distilled water | 29.85 | |
| dNTP (1.25 mM) | 4 | 100 μM | dNTP (1.25 mM) | 4 | 100 μM |
| 10× PCR buffer (no $MgCl_2$) | 5 | 1× | 10× PCR buffer (no $MgCl_2$) | 5 | 1× |
| $MgCl_2$ (25 mM) | 4 | 2 mM | $MgCl_2$ (25 mM) | 4 | 2 mM |
| Primer F1 (10 μM) | 1.25 | 250 nM | Primer F2 (10 μM) | 2.5 | 500 nM |
| Primer R1 (10 μM) | 1.25 | 250 nM | Primer R2 (10 μM) | 2.5 | 500 nM |
| BSA (10 mg/mL) | 2 | 400 ng/μL | GoTaq (5 units/μL) | 0.15 | 0.75 unit |
| GoTaq (5 units/μL) | 0.15 | 0.75 unit | | | |

**Table 2**
**PCR master mixture recipe for *gp60* gene of *Cryptosporidium***

| *gp60* gene | | | | | |
|---|---|---|---|---|---|
| **Primary PCR master mixture** | | | **Secondary PCR master mixture** | | |
| **Reagent** | **Per 50 μL reaction (μL)** | **Final concentration** | **Reagent** | **Per 50 μL reaction (μL)** | **Final concentration** |
| Distilled water | 31.35 | | Distilled water | 29.85 | |
| dNTP (1.25 mM) | 8 | 200 μM | dNTP (1.25 mM) | 8 | 200 μM |
| 10× PCR buffer (w/15 mM $MgCl_2$) | 5 | 1× | 10× PCR buffer (w/15 mM $MgCl_2$) | 5 | 1× |
| Primer F1 (10 μM) | 1.25 | 250 nM | Primer F2 (10 μM) | 2.5 | 500 nM |
| Primer R1 (10 μM) | 1.25 | 250 nM | Primer R2 (10 μM) | 2.5 | 500 nM |
| BSA (10 mg/mL) | 2 | 400 ng/μL | GoTaq (5 units/μL) | 0.15 | 0.75 unit |
| GoTaq (5 units/μL) | 0.15 | 0.75 unit | | | |

**Table 3**
***Cryptosporidium* PCR conditions for genotyping**

| Gene | Bp | Primers (10 μM) | 1′ program | 2′ program |
|---|---|---|---|---|
| SSU rRNA (18S) | ~800 | Forward (F1): 5′-TTC TAG AGC TAA TAC ATG CG-3′<br>Reverse (R1): 5′-CCC ATT TCC TTC GAA ACA GGA-3′<br>Forward (F2): 5′-GGA AGG GTT GTA TTT ATT AGA TAA AG-3′<br>Reverse (R2): 5′-CTC ATA AGG TGC TGA AGG AGT A-3′ | 55<br>94 °C<br>3 min<br>35 cycles:<br>94 °C<br>45 s<br>55 °C<br>45 s<br>72 °C<br>60 s<br>72 °C<br>5 min | 58<br>94 °C<br>3 min<br>35 cycles:<br>94 °C<br>45 s<br>58 °C<br>45 s<br>72 °C<br>60 s<br>72 °C<br>5 min |
| *gp60* | ~800–850 | Forward (F1): 5′-ATG AGA TTG TCG CTC ATT ATC G-3′<br>Reverse (R1): 5′-TTA CAA CAC GAA TAA GGC TGC-3′<br>Forward (F2): 5′-TCC GCT GTA TTC TCA GCC-3′<br>Reverse (R2): 5′-GGA AGG AAC GAT GTA TCT-3′ | 52<br>94 °C<br>3 min<br>35 cycles:<br>94 °C<br>45 s<br>52 °C<br>45 s<br>72 °C<br>60 s<br>72 °C<br>5 min | 52<br>94 °C<br>3 min<br>35 cycles:<br>94 °C<br>45 s<br>52 °C<br>45 s<br>72 °C<br>60 s<br>72 °C<br>5 min |

7. Add 1 μL positive control DNA to the tube designated as Primary PCR-positive control or add 2 μL PCR product of Primary PCR-positive control to the secondary PCR tube designated as Secondary PCR-positive control. The positive controls should not be handled until this step to avoid cross contamination.
8. Mix sample tubes by centrifuging briefly.
9. Select the cycling program for the appropriate PCR reaction (*see* Table 3). Start the thermal cycler.
10. Run amplified products on a 1.5% agarose gel. For SSU rRNA PCR, a band of ~800 bp should be present to be considered positive for *Cryptosporidium* DNA. For *gp60* PCR, samples with a band of 800–850 bp on the agarose gel are considered positive, indicating presence of *Cryptosporidium parvum* or *C. hominis* DNA in the sample analyzed. If no band is visible, or if the band is the incorrect size, the sample is considered negative. The negative controls must be negative. The positive controls must be positive. The test should be repeated if any of these quality control samples fail.

## 3.2 Sequence Analysis (SSU rRNA Gene or gp60 Gene)

### 3.2.1 Cleaning of Secondary PCR Products

1. Use positive secondary PCR products of SSU rRNA or *gp60* nested PCR in DNA sequencing.
2. Load each approximately 40-μL PCR reaction solution (after running the agarose gel) into an individual well in the MultiScreen HTS PCR plate.
3. Place the MultiScreen HTS PCR plate on top of the MultiScreen HTS Vacuum Manifold.
4. Apply vacuum at 25 inches Hg for 10 min or until wells have emptied (5 12 min; wells will need more time to dry if pressure is <25 inches Hg).
5. Allow 30 extra seconds under vacuum after the wells appear empty to be sure all liquid has filtered. The filters will appear shiny after they are dry.
6. Remove the plate from manifold, blot from underneath with paper towels, and add 50 μL distilled water into each well.
7. Mix samples on a plate mixer/rocker for 10 min.
8. Transfer the sample from each well into its own individual 1.5-mL tube.

### 3.2.2 DNA Sequencing Reaction

1. Thaw sequencing reaction reagents at room temperature.
2. The cleaned PCR products will be sequenced using two sequencing primers (the respective forward and reverse secondary PCR); thus, there are two sequencing reactions for each PCR product. For each sequencing reaction of 20 μL, prepare the following:
   (a) Distilled water: 12.5 μL.
   (b) BigDye Terminator Buffer: 4 μL.
   (c) Primer (10 μM): 2 μL.
   (d) BigDye Terminator: 1 μL.
   (e) Total: 19.5 μL.
3. Add 19.5 μL of the master mixture to each 0.2-mL PCR tube.
4. In the PCR template area, add 0.5 μL of cleaned PCR product to each PCR tube with master mixture solution.
5. Mix by centrifuging briefly.
6. Select and start the cycling program for DNA sequencing as follows: 25 cycles of denaturation at 96 °C, 10 s; annealing at 50 °C, 5 s; extension at 60 °C, 4 min; hold 4 °C until removed from the thermal cycler.

### 3.2.3 Cleaning of Sequencing Reaction Products

1. Rehydrate the Centri-Sep spin columns to be used: Tap dehydrated contents down, add 0.8 mL distilled water, cap the top, mix well by vortexing, remove bubbles by inverting and sharply tapping the column, and allow hydration for at least 30 min.

2. Remove the top cap and end stopper and place the column into a wash tube (2 mL).
3. Centrifuge at 750× *g* for 2 min. Keep track of position/orientation of column with respect to G forces, so next time it is placed in the same orientation.
4. Transfer 20 μL of amplified sequencing reaction products on top of the gel of each column. Do not disturb the top of the gel and do not touch the sides of the column.
5. Place loaded column in sample collection tube, with proper column orientation. Centrifuge for 2 min at 750× *g*.
6. Place the sample collection tubes in the vacuum centrifuge to dry the DNA products. DNA products are ready to be run on an ABI 3500 Genetic Analyzer or similar instrument.

#### 3.2.4 Sequence Analysis

1. Read the chromatograms generated by the DNA sequencer using the ChromasPro software (www.technelysium.com.au/ChromasPro.html) or similar program, assemble the two sequences generated by two sequencing primers for each sample, check the assembled contigs for sequence uncertainty, and trim the ambiguities on the 5′ and 3′ ends.
2. Align the generated SSU rRNA or *gp60* sequences with each other and reference sequences using the software MEGA (http://megasoftware.net/) or similar software. Check the sequence alignment for sequencing accuracy.
3. Determine sequence designation based on sequence identity to reference sequences. Determine subtype designation based on sequence identity to reference sequences and the number of trinucleotide repeats.
   (a) Subtypes of *C. parvum* and *C. hominis* at the *gp60* locus are named based on their subtype family designations and the number of each type of trinucleotide repeats. For each subtype, the name starts with the subtype family designation (Ia, Ib, Id, Ie, If, Ig, etc., for *C. hominis* and IIa, IIb, IIc, IId, IIe, IIf, IIg, IIh, IIi, etc., for *C. parvum*), followed by the number of TCA (represented by the letter A), TCG (represented by the letter G), and TCT (represented by the letter T) repeats found (*see* Table 4 and Fig. 2). In the IIa subtype family, a few subtypes have two copies of the ACATCA sequence right after the trinucleotide repeats, which are represented by "R2" (R1 for most subtypes). The subtype family Ia has different copies of a 15-bp repetitive sequence 5′-AAA/G ACG GTG GTA AGG-3′ (the last copy is 13 bp) shortly after the trinucleotide repeats, which is represented by the letter "R." The subtype family IIc was previously known as Ic. It differs from other subtype families by having no

**Table 4**
**Trinucleotide repeats present in the *Cryptosporidium gp60* gene and associated designation for subtype nomenclature**

| Trinucleotide repeat (5′ → 3′) | Subtype designation |
|---|---|
| TCA | A |
| TCG | G |
| TCT | T |
| ACATCA | R[a] |
| AAA/G ACG GTG GTA AGG | R[b] |
| C/AAG AA/G GGC A | R[c] |

[a]Only within *C. parvum* subtype family IIa
[b]Only within *C. hominis* subtype family Ia
[c]Only within *C. hominis* subtype family If

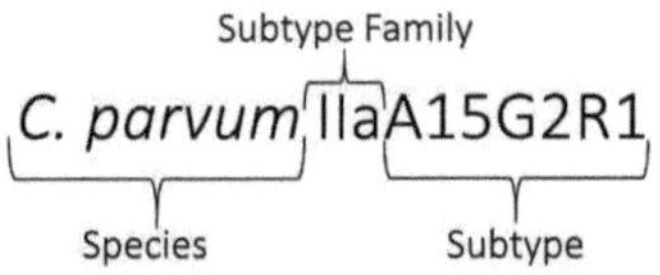


**Fig. 2** *Cryptosporidium* nomenclature. In the example, the species is *C. parvum* determined by sequencing. The subtype family is IIa determined by sequence alignment against reference sequences. The subtype designates 15 copies of the TCA repeat, two copies of TCG, and one copy of ACATCA

variation in the number of trinucleotide repeats (all A5G3) and by having extensive sequence polymorphism at the 3′ end of the gene. Thus, subtypes in the family are differentiated from each other by the additional letters "a" to "g," such as IIcA5G3a (reference sequence AF164491), IIcA5G3b (reference sequence AF164501), IIcA5G3c, IIcA5G3d (reference sequence AF440621), IIcA5G3e, IIcA5G3f, and IIcA5G3g. Occasionally, a few subtypes in other families have the same trinucleotide repeat sequence but differ from each other by one or two nucleotides in the sequence after the repeat region. They are differentiated from each other by the addition of lowercase letters, such as IIdA18G1a, IIdA18G1b, IIdA18G1c, and IIdA18G1d.

## 4 Notes

1. All master mixtures should be made in preprocessing spaces that are free of contaminants and specimens; the template should be added in a clean PCR workstation. Both precautions will reduce contamination.

2. All materials used in making the master mixture (all pipettes, the tray with 0.2-mL PCR tubes, the tubes labeled for primers and master mixture, and the tube with distilled water) should be exposed to UV light inside a UV Crosslinker for automatic crosslink. Do not expose the tubes containing reagents to the UV light.
3. PCR of all stool specimens should be run in duplicate.
4. Water or environmental specimens can be tested with the method, but success varies widely. A minimum of 5 replicates should be tested. The template volume can be adjusted up to 5 μL in primary PCR reactions. The volume of water added to the master mixture should be adjusted to accommodate the additional template to ensure that the final reaction volume remains 50 μL.

## Acknowledgments

This publication's contents are solely the responsibility of the authors and do not necessarily represent the official views of the Centers for Disease Control and Prevention nor does mention of trade names imply endorsement by the U.S. Government.

**Competing Interest Statement** The authors have no conflicts of interest to declare that are relevant to the content of this chapter.

## References

1. Ryan U, Fayer R, Xiao L (2014) *Cryptosporidium* species in humans and animals: current understanding and research needs. Parasitology 141(13):1667–1685. https://doi.org/10.1017/S0031182014001085
2. Morgan-Ryan UM, Fall A, Ward LA, Hijjawi N, Sulaiman I, Fayer R, Thompson RC, Olson M, Lal A, Xiao L (2002) *Cryptosporidium hominis* n. sp. (Apicomplexa: Cryptosporidiidae) from *Homo sapiens*. J Eukaryot Microbiol 49(6):433–440
3. Chalmers RM, Smith R, Elwin K, Clifton-Hadley FA, Giles M (2011) Epidemiology of anthroponotic and zoonotic human cryptosporidiosis in England and Wales, 2004-2006. Epidemiol Infect 139(5):700–712. https://doi.org/10.1017/S0950268810001688
4. Blackburn BG, Mazurek JM, Hlavsa M, Park J, Tillapaw M, Parrish M, Salehi E, Franks W, Koch E, Smith F, Xiao L, Arrowood M, Hill V, da Silva A, Johnston S, Jones JL (2006) Cryptosporidiosis associated with ozonated apple cider. Emerg Infect Dis 12(4):684–686
5. Fill MA, Lloyd J, Chakraverty T, Sweat D, Manners J, Garman K, Hlavsa MC, Roellig DM, Dunn JR, Schaffner W, Jones TF (2017) Cryptosporidiosis outbreak associated with a single hotel. J Environ Health 79(9):16–22
6. Hlavsa MC, Roellig DM, Seabolt MH, Kahler AM, Murphy JL, McKitt TK, Geeter EF, Dawsey R, Davidson SL, Kim TN, Tucker TH, Iverson SA, Garrett B, Fowle N, Collins J, Epperson G, Zusy S, Weiss JR, Komatsu K, Rodriguez E, Patterson JG, Sunenshine R, Taylor B, Cibulskas K, Denny L, Omura K, Tsorin B, Fullerton KE, Xiao L (2017) Using molecular characterization to support investigations of aquatic facility-associated outbreaks of cryptosporidiosis—Alabama, Arizona, and Ohio, 2016. MMWR 66(19):493–497. https://doi.org/10.15585/mmwr.mm6619a2

7. Feng Y, Tiao N, Li N, Hlavsa M, Xiao L (2014) Multilocus sequence typing of an emerging *Cryptosporidium hominis* subtype in the United States. J Clin Microbiol 52(2):524–530. https://doi.org/10.1128/JCM.02973-13
8. Xiao L, Hlavsa MC, Yoder J, Ewers C, Dearen T, Yang W, Nett R, Harris S, Brend SM, Harris M, Onischuk L, Valderrama AL, Cosgrove S, Xavier K, Hall N, Romero S, Young S, Johnston SP, Arrowood M, Roy S, Beach MJ (2009) Subtype analysis of *Cryptosporidium* specimens from sporadic cases in Colorado, Idaho, New Mexico, and Iowa in 2007: widespread occurrence of one *Cryptosporidium hominis* subtype and case history of an infection with the *Cryptosporidium* horse genotype. J Clin Microbiol 47(9):3017–3020. https://doi.org/10.1128/JCM.00226-09
9. Xiao L, Alderisio K, Limor J, Royer M, Lal AA (2000) Identification of species and sources of *Cryptosporidium* oocysts in storm waters with a small-subunit rRNA-based diagnostic and genotyping tool. Appl Environ Microbiol 66(12): 5492–5498
10. Xiao L, Fayer R, Ryan U, Upton SJ (2004) *Cryptosporidium* taxonomy: recent advances and implications for public health. Clin Microbiol Rev 17(1):72–97
11. Alves M, Xiao L, Sulaiman I, Lal AA, Matos O, Antunes F (2003) Subgenotype analysis of *Cryptosporidium* isolates from humans, cattle, and zoo ruminants in Portugal. J Clin Microbiol 41(6):2744–2747
12. Sulaiman IM, Hira PR, Zhou L, Al-Ali FM, Al-Shelahi FA, Shweiki HM, Iqbal J, Khalid N, Xiao L (2005) Unique endemicity of cryptosporidiosis in children in Kuwait. J Clin Microbiol 43(6):2805–2809

# Chapter 8

# A Multilocus Variable-Number Tandem-Repeat Analysis (MLVA) Scheme for Epidemiological Investigations of *Cryptosporidium parvum*

**Harriet Risby, Guy Robinson, and Rachel M. Chalmers**

## Abstract

A multi-locus variable-number tandem-repeat analysis (MLVA) scheme has been validated and used for subtyping *Cryptosporidium parvum*, to infer linkage between cases, and suspected sources, in outbreaks and for epidemiological surveillance. Here, we describe the multiplex polymerase chain reaction (PCR) assays, fragment sizing, and bioinformatics to identify alleles and create a seven-locus MLVA profile for each specimen. The PCRs to produce amplicons for Sanger sequencing to confirm new alleles are also described.

**Key words** *Cryptosporidium*, *C. parvum*, Multi-locus variable-number tandem-repeat analysis, MLVA, Subtyping

## 1 Introduction

Subtyping *Cryptosporidium* spp. in patient or animal specimens is important for outbreak investigations and epidemiological surveillance. *Cryptosporidium* spp. subtyping is most commonly done by sequence analysis of the 60-kDa glycoprotein gene (*gp60*), which is useful for allele discrimination and inferring some level of linkage between cases and to sources of infection ( [1]; *see* Chaps. 6 and 7). However, due to the recombinant nature of this parasite, a multi-locus scheme is required for more discriminatory genotyping [2].

The ultimate resolution for molecular epidemiology would be provided by whole genome sequencing. However, the time and expense to extract sufficient and suitable deoxyribonucleic acid (DNA) from stools (*see* Chap. 5), as well as the lack of standardized, quality-assured bioinformatic pipelines for epidemiological analyses, have been a barrier to routine use for public health purposes [3]. Multi-locus genotyping schemes such as polymerase chain reaction (PCR)-based multi-locus sequence typing (MLST) or multi-locus fragment typing (MLFT) have a long history of

Jan R. Mead and Michael J. Arrowood (eds.), *Cryptosporidium: Methods and Protocols*, Methods in Molecular Biology, vol. 2978, https://doi.org/10.1007/978-1-0716-4824-7_8,

application to gastrointestinal pathogens, and a few combinations of loci have been used to investigate *Cryptosporidium parvum* and *Cryptosporidium hominis* but without standardization [4–6].

Not all loci show the same variation or have the same primer sites across *Cryptosporidium* spp., and previous analysis has suggested that separate schemes are required for each species [5]. At a multinational and multidisciplinary meeting, a consensus agreement was reached, due to its multiple hosts and genetic diversity, to prioritize initially a standardized MLFT scheme for *C. parvum* [6]. The multi locus variable-number tandem-repeat (VNTR) analysis (MLVA), based on seven loci located on five *C. parvum* chromosomes, described here was developed and validated by the Cryptosporidium Reference Unit (CRU) [7]. The loci in chromosomal order are: cgd1_470_1429, cgd4_2350_796, cgd5_10_310, cgd5_4490_2935, cgd6_4290_9811, cgd8_4440_NC_506, and cgd8_4840_6355. The alleles (number of tandem repeats inferred from the fragment sizes) are then expressed in chromosomal order to provide an MLVA profile, e.g., 4-14-5-7-18-36-16.

Since 2021, this subtyping scheme has been applied during the investigation of *C. parvum* outbreaks in England and Wales, providing improved microbiological evidence to support decision-making by the outbreak control team [8–10]. Here, we described the wet laboratory processes of DNA amplification by multiplex (4-plex and 3-plex) PCR, fragment sizing by capillary electrophoresis, bioinformatic steps, and the reference PCRs for Sanger sequencing to confirm potentially new or unusual alleles.

## 2 Materials

### 2.1 Multiplex PCRs for Cryptosporidium parvum Loci

1. Vortex mixer and microcentrifuge for 1.5-mL tubes, or a combination instrument (e.g., Grant Instruments Combi-spin PCV-2400, Grant Instruments, Royston, United Kingdom).
2. PCR workstation with ultraviolet (UV) decontamination.
3. Variable volume pipettes.
4. Filter-protected pipette tips.
5. PCR thermocycler.
6. QIAGEN 2x Type-it Microsatellite PCR Master Mix (QIAGEN Inc., Germantown, MD).
7. 100 μM stock labeled forward and reverse primers (*see* Table 1).
8. Nuclease-free water.
9. 1.5-mL microcentrifuge tubes.
10. 0.2-mL PCR tubes or 96-well plates.
11. Adhesive foil seal for 96-well plates.

**Table 1**
**PCR primers used to amplify a panel of seven VNTR loci suitable for MLVA analysis in *Cryptosporidium parvum***

| | Locus | Direction | Primer sequence |
|---|---|---|---|
| 4-plex | cgd1_470_1429 (GRH) | Forward | **PET**-5′-CTC AGG AAG AGG AAG ATA CGG-3′ |
| | | Reverse | 5′-GGA AGG TAT GGC AGC AAA AG-3′ |
| | cgd4_2350_796 | Forward | **VIC**-5′-GGG TCA ATC AGG CAT GAG C-3′ |
| | | Reverse | 5′-TTG CAT GTT TAT CAT ATT ATT TCC CAT-3′ |
| | cgd8_4440_NC_506 | Forward | **NED**-5′-CTC AAT ATT TTT TCC ACA CCT GAA C-3′ |
| | | Reverse | 5′-ACT GCC TGA GAA AGG AAC CA-3′ |
| | cgd8_4840_6355 (MM19) | Forward | **6-FAM**-5′-GTT CCA GGA ATA TTT GAT TCT GC-3′ |
| | | Reverse | 5′-CTC CTA CGC CAA CTC CTA-3′ |
| 3-plex | cgd5_10_310 (MSF) | Forward | **VIC**-5′-AAG GTG AAG GAA TCA AAG GC-3′ |
| | | Reverse | 5′-TTT GTC CTT CTT GCC CTC GG-3′ |
| | cgd5_4490_2935 | Forward | **6-FAM**-5′-CAG TGA ATA ACT CTG AAC GGA AC-3′ |
| | | Reverse | 5′-TTG ATT TTG GGT TCG GTA TTG-3′ |
| | cgd6_4290_9811 (MSC6–5) | Forward | **NED**-5′-CAT TGG AAC GTA AAC AAA ACC A-3′ |
| | | Reverse | 5′-CTA GCC GAA TCT GGC GGT AT-3′ |

12. Vacuum desiccator (or other means of concentrating DNA) if weak results are obtained.

#### *2.2 Fragment Sizing by Capillary Electrophoresis (See Note 1)*

1. Vortex mixer and microcentrifuge for 1.5-mL tubes, or a combination instrument (e.g., Grant Instruments Combi-spin PCV-2400).
2. 96-well plate vortex mixer (e.g., Eppendorf MixMate) and microcentrifuge.
3. Applied Biosystems™ SeqStudio™ Genetic Analyzer and computer (Applied Biosystems, Foster City, CA).
4. Applied Biosystems™ SeqStudio™ Genetic Analyzer Cartridge v2.
5. Applied Biosystems™ SeqStudio™ Genetic Analyzer Cathode Buffer Container (CBC).
6. Variable volume pipettes and multichannel pipettes.
7. Filter-protected pipette tips.
8. Ice block reagent rack.
9. Reagent reservoirs.
10. Hi-Di Formamide.

11. GeneScan™ 600 LIZ™ dye Size Standard v2.0 (Applied Biosystems).
12. 1.5-mL microcentrifuge tubes.
13. 0.2-mL 8-tube strips and caps or 96-well plates.
14. Adhesive foil seal for 96-well plates.
15. Strip or plate septa (Applied Biosystems).

### 2.3 Fragment Analysis

1. Fragment Analysis software (e.g., BioNumerics or Geneious [*see* **Note 2**]).

### 2.4 VNTR Sequencing PCR

1. Vortex mixer and microcentrifuge for 1.5-mL tubes, or a combination instrument (e.g., Grant Instruments Combi-spin PCV-2400).
2. PCR workstation with UV decontamination.
3. Variable volume pipettes.
4. Filter-protected pipette tips.
5. PCR thermocycler.
6. HotStarTaq (QIAGEN).
7. 10× reaction buffer (QIAGEN).
8. 2 mM deoxynucleoside triphosphates (dNTPs) (BioGene, Kimbolton, Huntingdon, United Kingdom).
9. 5 mg/mL bovine serum albumin (BSA) (nonacetylated).
10. 10 μM stock labeled forward and reverse primers (*see* Table 2).
11. Nuclease-free water.
12. 1.5-mL microcentrifuge tubes.
13. 0.2-mL PCR tubes or 96-well plates.
14. Adhesive foil seal for 96-well plates.

## 3 Methods (*See* Note 3)

### 3.1 Primer Mix Preparation

1. Prepare 500 μL of working concentration (2 μM) of each primer mix (*see* Table 1 and **Note 4**):
   (i) 4-plex: Pipette 10 μL of each 100 μM stock primer into a 1.5 mL-tube and add 420 μL of nuclease-free water.
   (ii) 3-plex: Pipette 10 μL of each 100 μM stock primer into a 1.5-mL tube and add 440 μL of nuclease-free water.

### 3.2 Multiplex PCRs for Cryptosporidium parvum Loci

1. Thaw the DNA samples and PCR reagents (*see* Table 3) separately at room temperature.
2. Briefly vortex and centrifuge the DNA samples and reagents to mix and remove contents from the tube lids.

**Table 2**
**VNTR sequencing primers to amplify VNTR loci for the identification of new alleles in *Cryptosporidium parvum***

| Primer name | Sequence |
|---|---|
| cgd1SeqF | AGGAGGAAGTGGCGAATCAGG |
| cgd1SeqR | GATAGGGTAGTTTTACCGGGGTTG |
| cgd4SeqF | CAATGGATGCCAGACAAGCT |
| cgd4SeqR | AAGGCTACGAGCAGATTAACG |
| MSFSeqF | TGACCCTTCTATTGAGCCAC |
| MSFSeqR | ACTTCTTCCTCATCAGTAGC |
| cgd5SeqF | ATCCAGTAATTTCTGACATTTCTGAG |
| cgd5SeqR | TAATTAACATTTTCAGGGTCTCTGG |
| cgd6SeqF | GACTTGGATTTGGACTTACACC |
| cgd6SeqR | TCCAAATGATGTAAATACTCCAGC |
| cgd8SeqF | CCATATTGAATGCAATGCCAAATC |
| cgd8SeqR | GCATGGATTAACGACCAGTTG |
| MM19SeqF | GCGGAGAAGGAGGATTTAATTC |
| MM19SeqR | TTACAACTCCAACTCCACCAC |

**Table 3**
**MLVA multiplex PCR reaction mix components and volumes**

| 4-plex PCR mix<br>Reagent | Volume/reaction (μL) | 3-plex PCR mix<br>Reagent | Volume/reaction (μL) |
|---|---|---|---|
| 2x type-it kit | 12.5 | 2x type-it kit | 12.5 |
| 4-plex primer mix (2 μM) | 2.5 | 3-plex primer mix (2 μM) | 2.5 |
| nf$H_2O$ | 8 | nf$H_2O$ | 8 |
| Template DNA | 2 | Template DNA | 2 |
| Total volume | 25 | Total volume | 25 |

3. Two PCR reaction mixes are required, one for the 4-plex primers and one for the 3-plex primers as prepared in Subheading 3.1. See Table 3 for the volumes for a single reaction.
4. In a DNA-free PCR workstation, prepare sufficient PCR master mix of each type in 1.5 mL microcentrifuge tubes for the required number of samples (including PCR controls) plus 20% extra for wastage..

**Table 4**
**PCR cycling conditions for fragment analysis**

| Step | Time | Temperature |
|---|---|---|
| Initial heat inactivation | 5 min | 95 °C |
| *3-step cycling (30 cycles)* | | |
| Denaturation | 30 s | 95 °C |
| Annealing | 90 s | 63 °C |
| Extension | 30 s | 72 °C |
| Final extension | 30 min | 60 °C |
| Hold | ∞ | 4 °C |

5. Briefly vortex and centrifuge the master mixes to thoroughly mix and remove contents from the tube lids.
6. Pipette 23 μL of each PCR reaction mix into one 0.2-mL PCR tube or the well of a PCR plate per sample/control.
7. Pipette 2 μL of nuclease-free water to each of the two no-template control tubes/wells.
8. Move the tubes/PCR plate to a designated DNA template addition PCR workstation.
9. Aliquot 2 μL template DNA into each of two corresponding tubes/wells per sample.
10. Pipette 2 μL *C. parvum*-positive DNA to each of the two positive control tubes/wells.
11. Cap the 0.2-mL tubes or seal the plate with an adhesive foil seal.
12. Insert the 0.2-mL tubes/plate into a thermal cycler and run the PCR program with the cycle settings in Table 4.

### *3.3 PCR Products Preparation*

1. Thaw the Hi-Di Formamide (*see* **Note 3**) and collect the PCR products and place them on separate ice block racks.
2. Vortex and briefly centrifuge the reagents and PCR products to mix and remove the contents from the lid.
3. On a workbench, dilute the PCR products 10× with Hi-Di Formamide by pipetting 9 μL Hi-Di Formamide and 1 μL PCR products into 0.2-mL tubes/plate.
4. Cap the tubes or seal the 96-well plate with an adhesive foil seal.
5. Vortex and briefly centrifuge the diluted amplicons to mix and remove contents from the lid.

**Table 5**
**MLVA fragment sizing reaction mix for capillary electrophoresis**

| Reagent | Volume/reaction (μL) |
|---|---|
| Hi-Di Formamide | 12 |
| GeneScan 600 LIZ | 0.5 |
| Diluted PCR amplicons | 2 |
| Total volume | 14.5 |

### 3.4 Fragment Sizing by Capillary Electrophoresis

1. Ensure there are sufficient runs left on the installed in date cartridge and cathode buffer container before performing the capillary electrophoresis on the SeqStudio.
2. Vortex and briefly centrifuge the GeneScan 600 LIZ size standard to thoroughly mix and remove the contents from the lid.
3. Place the GeneScan 600 LIZ size standard on the reagent ice block rack.
4. On a workbench, prepare a capillary electrophoresis master mix according to the volumes in Table 5. Sufficient master mix is required for both the 3-plex and 4-plex amplicons.
5. Briefly vortex and centrifuge the master mix to thoroughly mix and remove the contents from the lid.
6. Pipette 12.5 μL master mix per well in 0.2 mL-tubes/PCR plate; two tubes/wells are required for each sample, one for the 3-plex and another for the 4-plex amplicons.
7. Pipette 2 μL of diluted PCR amplicons prepared in Subheading 3.2 into the corresponding tubes/wells.
8. Apply strip septa to the tube strips, or apply plate septa to the PCR plate.
9. Briefly vortex and centrifuge the tube strips/plate to ensure there are no bubbles in the wells.
10. Insert the tube strips/PCR plate into the SeqStudio Genetic Analyzer.
11. Prepare a plate plan for fragment analysis in Plate Manager using the Fragment Analysis application type:
    (a) Create a name for the plate and select default fragment analysis settings.
    (b) On the next page, highlight all the wells in use for fragment analysis and select the settings: 1) Size Standard: GS-600 LIZ (60–400), 2) Dye Set: G5 (DS-33), and 3) Run Module: FragAnalysis.

(c) Provide a label for each specimen and include the suffix "_3" or "_4" to denote the 3-plex or 4-plex amplicons, to assist in the linking of results from each sample later during data analysis.

(d) Save the plate setup in the computer drive accessible by the SeqStudio.

12. Using the SeqStudio touch screen, select "Setup run" and locate the plate setup created in **step 11**.
13. Use the "eject plate" button to bring the autosampler to the front and insert the plate before clicking "retract plate."
14. Click "Start Run."

#### 3.5 Fragment Size Analysis

1. Here, BioNumerics (bioMérieux) is used to illustrate the process (*see* **Note 2**).
2. Follow the software manufacturer's instructions to install the MLVA plug-in and then create an MLVA schema and database (*see* **Note 5**).
3. Import the .fsa fingerprint curve files generated by the SeqStudio into BioNumerics (bioMérieux), following the manufacturer's instructions for filename parsing to assemble the 3-plex and 4-plex fingerprint files together, and selecting the reference bands (assigning LIZ as the reference dye).
4. Select all the imported files in the database, and select "Assign VNTR" and select the schema created in **step 2** and described in **Note 5**.
5. VNTR copy numbers are automatically assigned based on the MLVA schema; however, each copy number should be manually checked in the Fingerprint curve processing window (*see* **Note 6**).
6. Any samples producing weak or negative alleles should be concentrated by desiccating the DNA using an appropriate instrument and resuspended in a lower volume of nuclease-free water and repeated.

#### 3.6 Recording and Expression of MLVA Results

1. Record the number of repeats for each locus separated by a hyphen in the order they appear in the genome (cgd1_470_1429, cgd4_2350_796, cgd5_10_310, cgd5_4490_2935, cgd6_4290_9811, cgd8_4440_NC_506, cgd8_4840_6355) to generate an MLVA profile (e.g., 4–14–5-7-18-36-16) (*see* **Notes 6–8**).

#### 3.7 VNTR Sequencing PCRs for Confirmation of New Alleles

1. Thaw the DNA samples and PCR reagents (*see* Table 6) separately at room temperature.
2. Briefly vortex and centrifuge the DNA samples and reagents to mix and remove contents from the tube lids.

**Table 6**
**VNTR sequencing master mix components and volumes**

| Component | Volume/reaction (μL) |
|---|---|
| 10× reaction buffer | 2 |
| 2 mM dNTP | 2 |
| BSA (5 mg/mL) | 2 |
| 10 mM forward primer | 0.4 |
| 10 mM forward primer | 0.4 |
| 5 units/μL HotStarTaq | 0.5 |
| Template DNA | 2 |
| nf$H_2O$ | 10.7 |
| Total | 20 |

3. In a DNA-free PCR workstation, prepare sufficient PCR master mixes in 1.5-mL microcentrifuge tubes for the required number of samples plus 20% extra for wastage, based on the details for a single reaction (*see* Table 6). One master mix should be prepared per VNTR locus.
4. Briefly vortex and centrifuge the master mix to thoroughly mix and remove contents from the tube lids.
5. Pipette 18 μL of PCR reaction mix into 0.2-mL PCR tubes or the wells of a PCR plate per sample.
6. Move the tubes/PCR plate to a designated DNA template addition PCR workstation.
7. Aliquot 2 μL DNA per well for each sample into the corresponding tubes/wells.
8. Cap the 0.2-mL tubes or seal the plate with an adhesive foil seal.
9. Insert the 0.2-mL tubes/plate into a thermal cycler and run the PCR program with the following cycle settings in Table 7.
10. Various commercial providers are available to provide Sanger sequencing by mailing the PCR amplicons and an aliquot of the primers at ambient temperature, with the sequencing data usually returned via email the following day. However, if facilities are available and/or a rapid turnaround is required, in-house sequencing can be performed according to the specific instrument manufacturer's instructions. For tips on analyzing sequencing files, refer to **Note 9**.

**Table 7**
**VNTR sequencing PCR cycling conditions**

| Step | Time | Temperature |
|---|---|---|
| Initial heat inactivation | 15 min | 95 °C |
| *3-step cycling (34 cycles)* | | |
| Denaturation | 30 s | 94 °C |
| Annealing | 90 s | 58 °C |
| Extension | 60 s | 72 °C |
| Final extension: NO | | |
| Hold | ∞ | 4 °C |

## 4 Notes

1. This protocol for fragment sizing uses an Applied Biosystems™ SeqStudio™ Genetic Analyzer. Appropriate alternative fragment sizing instruments can be used if they provide the required accuracy and precision, and their use is validated.
2. The BioNumerics software is being phased out by bioMérieux and will no longer be supported from December 2024. Alternative fragment sizing data analysis software (such as Geneious) may be used. BioNumerics is included in this chapter as it still functions where users already have a license and was used to validate the MLVA method at the CRU. Many of the examples and tips in these Notes were generated using BioNumerics but are applicable to alternative software.
3. Wear appropriate personal protective equipment (PPE) during all sample processing steps. Extreme care must be taken to avoid contamination of PCR reagents and equipment with genomic and amplified DNA. Work in a low- to high-risk direction: reagent tubes and racks should be handled before sample tubes and racks. Gloves should be changed after touching sample tubes/racks or other potentially contaminated areas such as the discard box. Particular caution should be taken when using Hi-Di Formamide as it is a reproductive toxin (teratogen), and pregnant women and those trying to conceive should avoid working with it.
4. The MLVA primers are supplied at 100 μM and are prepared at 2 μM working solution in 4-plex and 3-plex combinations. The 4-plex is set to amplify markers cgd1_470_1429, cgd4_2350_796, cgd8_4440_NC_506, and cgd8_4840_6355 using labeled forward primers with dyes PET, VIC, NED, and 6-FAM, respectively, and the 3-plex

markers cgd5_10_310, cgd5_4490_2935, and cgd6_4290_9811 using dyes VIC, 6-FAM, and NED, respectively (*see* Table 1).

5. Allele binning settings within an MLVA schema can be set up and modified in the MLVA management window in BioNumerics. Some initial bins for the MLVA schema can be created by running fragment sizing for samples with known alleles and repeat sizes, then applying those sizes to the corresponding bins. Any new alleles that are subsequently found in further samples can be verified by DNA sequencing (*see* **Note 6**), and bins can be created using their repeat sizes.
6. Tips for identifying true peaks and artifactual peaks during fragment analysis:
   - Assess the multicolored overview of all the channels to identify true peaks (tall and with a narrow base) and artifactual peaks described in the points below.
   - Alleles with peaks that fall into bins (set up previously by the MLVA schema) can be identified by the bin copy number. A profile can then be produced by displaying the seven alleles in chromosomal order such as in the following example: 4-14-5-7-27-32-15 (*see* Fig. 1).
   - Some artifactual peaks tend to appear wider at the base and can be observed in multiple samples in the same place (*see* Fig. 2).
   - Strong peaks may cause bleed-through from one channel to another. These are not true alleles and must be identified prior to calling allele copy numbers (*see* Fig. 3). This is particularly important to note when the bleed-through falls within the bin of another channel.
   - Stutter peaks do occasionally occur. They are usually visible as lower peaks in the next repeat bins lower than the main peak and get smaller the further away from the main peak they get (*see* Fig. 4).
   - Peaks that do not fall into bins but have an approximate size that may indicate potential new alleles can be confirmed with Sanger sequencing (*see* Fig. 5).
   - Weak or missing peaks can often be identified following concentration by desiccation of the DNA and the 3-plex and/or 4-plex PCRs repeated as required.
   - If bands are very strong, they can appear as a double peak with the correct size in the middle of the trough (*see* Fig. 6).
   - Occasionally, an artifact peak may fall within or next to the position of a bin that can mask the presence of a weak band at this location, or the band may appear as a "shoulder" to

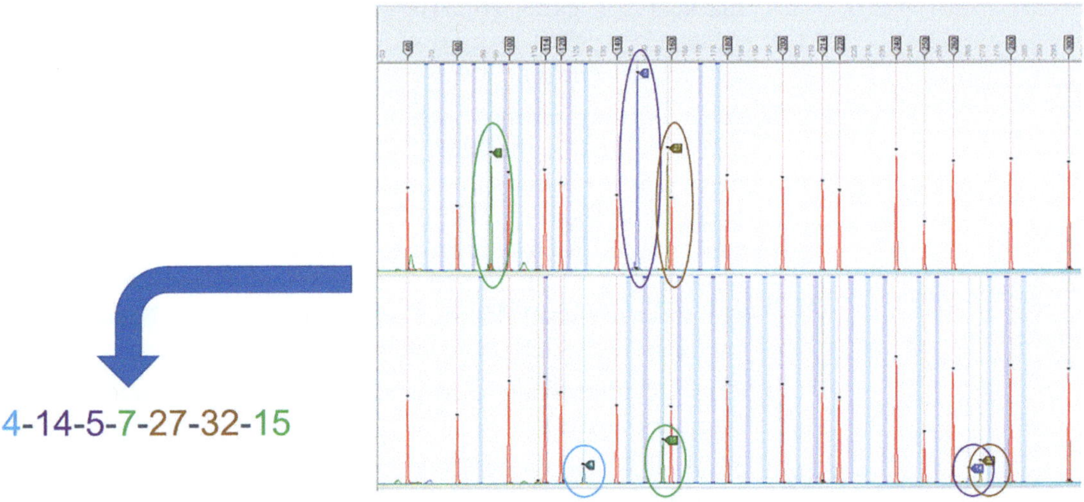


**Fig. 1** A multicolored overview showing the peaks of all alleles that are then combined in chromosomal order to produce a MLVA profile. The top half shows the results from the 3-plex PCR and the bottom half from the 4-plex PCR. The colors represent products labeled with the following fluorophores: Blue = PET, Purple = VIC, Green = 6-FAM, Brown = NED, and Red = LIZ (size marker). The flags show the bins that the alleles fall into under their respective dyes. The blue and purple stripes show the bin positions for loci labeled with the 6-FAM fluorophore

the artifactual peak (*see* Fig. 7). This is mainly a problem when weak bands are present, as stronger bands generally overshadow the artifact peak.

7. Sometimes, more than one allele may be present at one or more of the loci, indicating a mixed *C. parvum* infection (*see* Fig. 8). Multiple peaks can occur in samples with a population containing more than one subtype. Additional peaks are usually accepted if they are >25% of the largest peak. This is an example of a mixed MLVA profile with two alleles at each of the last two loci: 4-14-5-7-20-21/24-15/17.
8. Occasionally, there is no PCR amplification or the allele cannot be determined, so a "null" record is made. This is an example of a profile with null alleles at two loci: Ø-14-5-Ø-18-9-34.
9. The MLVA primers were not used for sequencing due to their short distance to the repeat region; therefore, the VNTR sequencing primers were designed to cover a larger area around this region. Upon assembly of the forward and reverse sequences, count the number of tandem repeats to determine the VNTR allele.

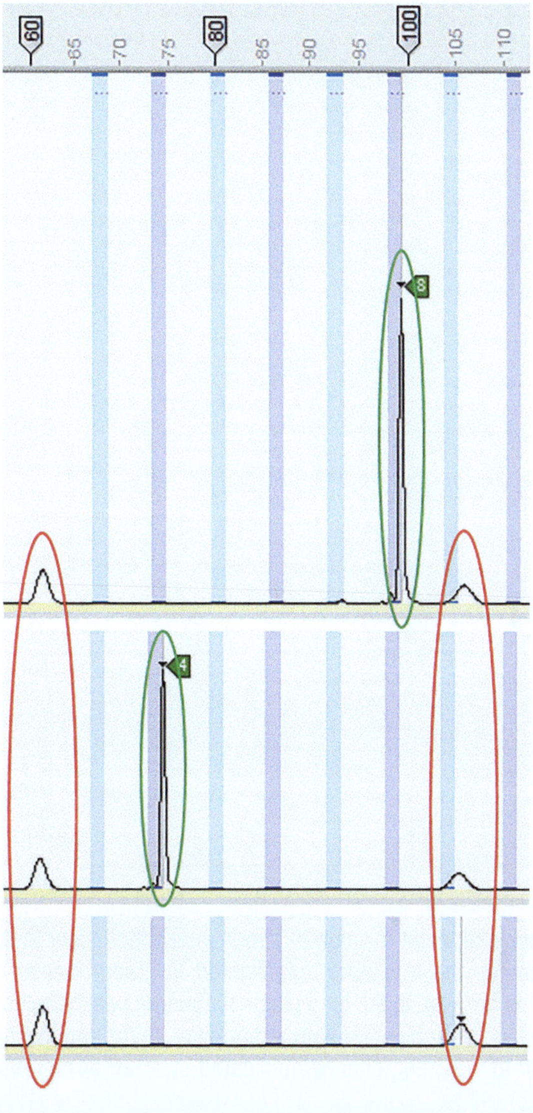


**Fig. 2** Examples of true peaks (circled in green) for alleles of 8 and 4 repeats, as well as short and wide artifactual peaks (circled in red), which can be observed at the same number of base pairs across three different samples, including the no-template control (NTC) (displayed in the bottom image) [9]

## Competing Interest Statement

The authors have no conflicts of interest to declare that are relevant to the content of this chapter.

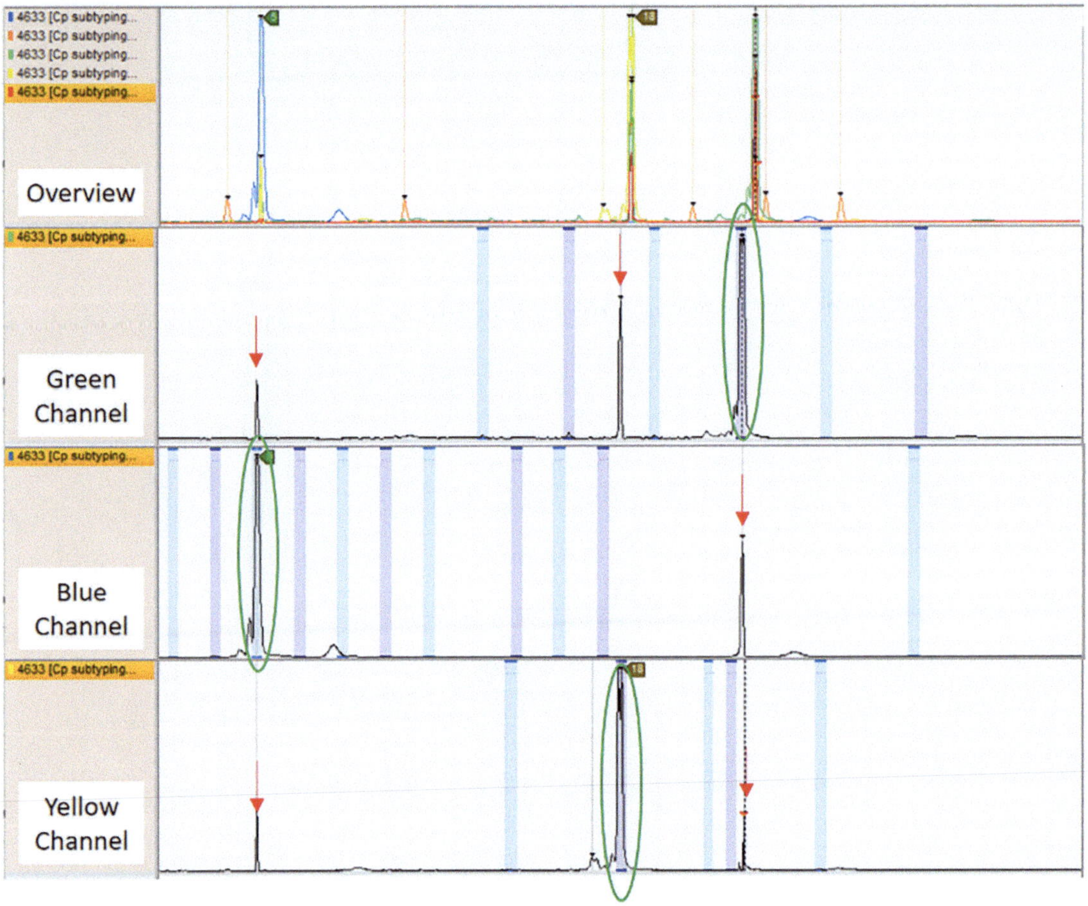


**Fig. 3** An example of bleed-through of strong peaks that occur in the same position across other channels. In the green channel, the true peak (highlighted in green) bleeds through to the blue and yellow channels (indicated by red arrows). In the blue channel, the true peak (circled in green) also bleeds through to the green and yellow channels (highlighted with red arrows). In the yellow channel, the true peak (circled in green) also bleeds through to the green channel (indicated by a red arrow) [9]

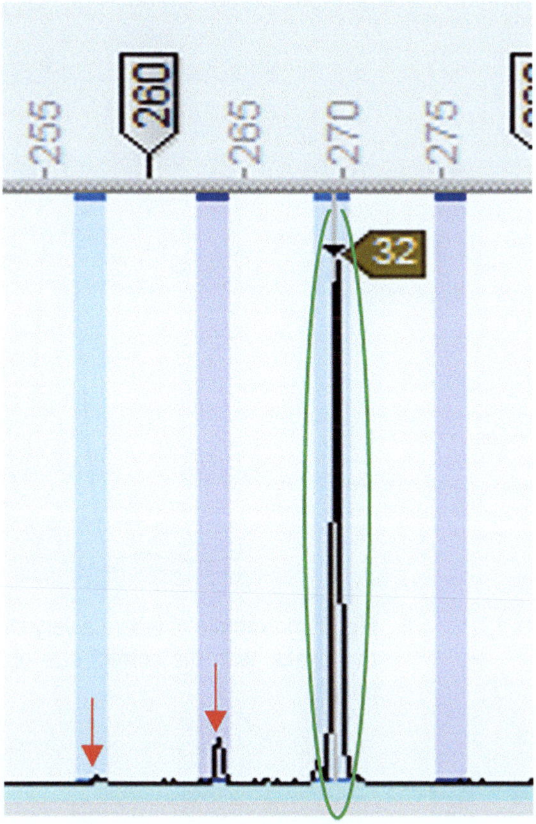


**Fig. 4** An example of a true peak (circled in green) as well as short and narrow stutter peaks (indicated by red arrows) [9]

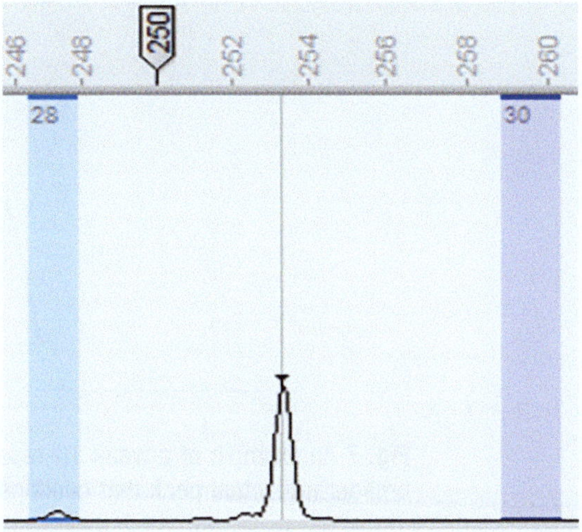


**Fig. 5** An example of a potential new allele that has an approximate size between two known allele bins (28 and 30 repeats) that may indicate an allele of 29 repeats

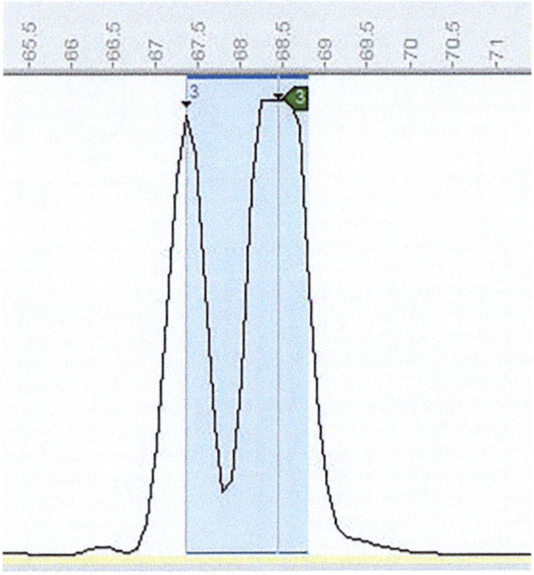


**Fig. 6** An example of where a very strong peak has inverted on itself generating two peaks, with the correct size of the amplicon falling in the middle of the trough

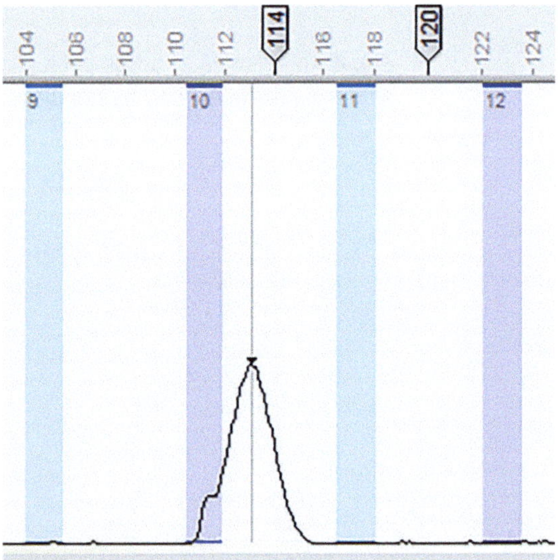


**Fig. 7** An example of a weak 10-repeat allele which is seen as a shoulder to a broader artifactual peak that overlaps the bin

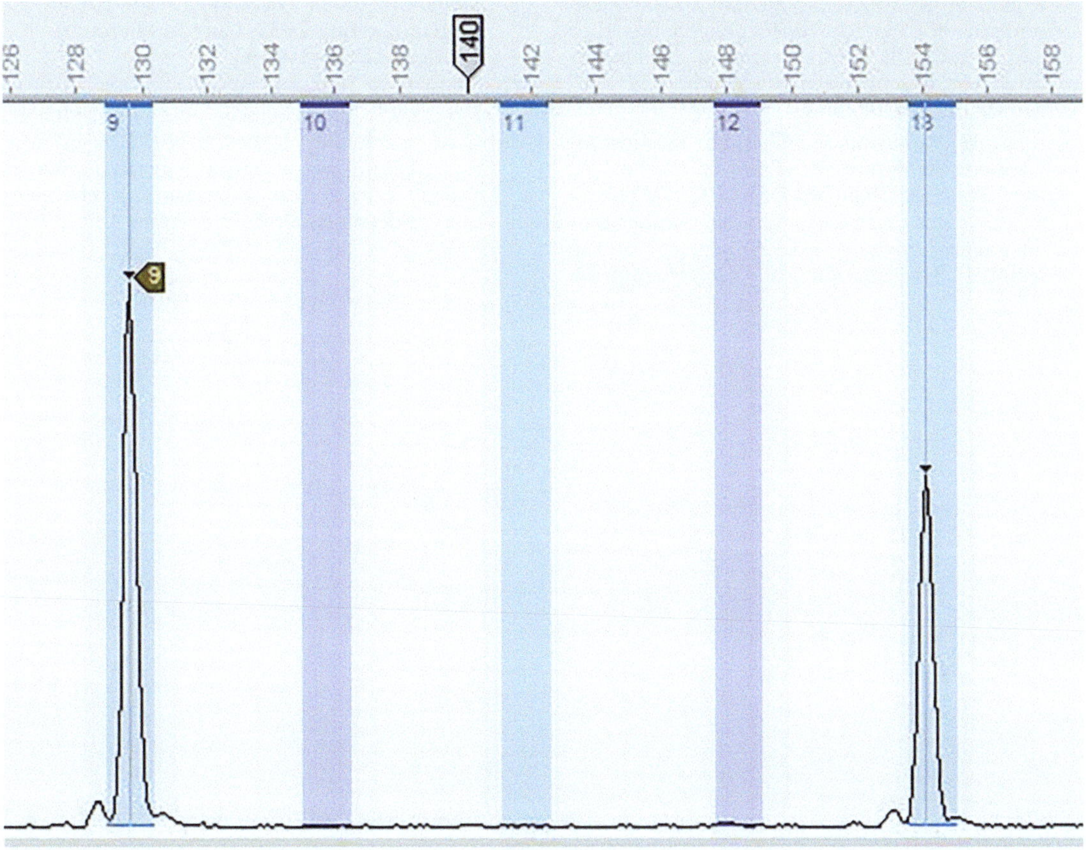


**Fig. 8** An example of a mixed infection where two separate alleles (9 and 13 repeats) can be seen at the same locus

## References


1. Feng Y, Tiao N, Li N, Hlavsa M, Xiao L (2014) Multilocus sequence typing of an emerging *Cryptosporidium hominis* subtype in the United States. J Clin Microbiol 52(2):524–530
2. Widmer G, Feng X, Tanriverdi S (2004) Genotyping of *Cryptosporidium parvum* with microsatellite markers. In: Spencer JFT, Ragout de Spencer AL (eds) Public health microbiology: methods and protocols. Humana Press, Totowa, pp 177–187
3. Morris A, Robinson G, Swain MT, Chalmers RM (2019) Direct sequencing of *Cryptosporidium* in stool samples for public health. Front Public Health 7:360
4. Xiao L, Feng Y (2017) Molecular epidemiologic tools for waterborne pathogens *Cryptosporidium* spp. and *giardia duodenalis*. Food Waterborne Parasitol 8-9:14–32
5. Robinson G, Chalmers RM (2012) Assessment of polymorphic genetic markers for multi-locus typing of *Cryptosporidium parvum* and *Cryptosporidium hominis*. Exp Parasitol 132:200–215
6. Chalmers RM, Pérez-Cordón G, Cacció SM, Klotz C, Robertson LJ, Participants of the Cryptosporidium genotyping workshop (EURO-FBP) (2018) *Cryptosporidium* genotyping in Europe: the current status and processes for a harmonised multi-locus genotyping scheme. Exp Parasitol 191:25–30
7. Robinson G, Pérez-Cordón G, Hamilton C, Katzer F, Connelly L, Alexander CL, Chalmers RM (2022) Validation of a multilocus genotyping scheme for subtyping *Cryptosporidium parvum* for epidemiological purposes. Food Waterborne Parasitol 27:e00151
8. Gopfert A, Chalmers RM, Whittingham S, Wilson L, van Hove M, Ferraro CF,

Robinson G, Young N, Nozad B (2022) An outbreak of *Cryptosporidium parvum* linked to pasteurised milk from a vending machine in England: a descriptive study, March 2021. Epidemiol Infect 150:e185

9. Risby H, Robinson G, Chandra N, King G, Vivancos R, Smith R, Thomas D, Fox A, McCarthy N, Chalmers RM (2023) Application of a new multi-locus variable number tandem repeat analysis (MLVA) scheme for the seasonal investigation of *Cryptosporidium parvum* cases in Wales and the northwest of England, spring 2022. Curr Res Parasitol Vector Borne Dis 4:100151
10. Peake L, Bardsley M, Bartram S, Bharuchi S, Howkins J, Robinson G, Charlett A, Chalmers R, Bird S, Young N (2024) A large cryptosporidiosis outbreak associated with an animal contact event in England: a retrospective cohort study, 2023. Epidemiol Infect 152: e91

# Chapter 9

# Optimized Genome Assembly Protocol for *Cryptosporidium*

**Wanyi Huang and Lihua Xiao**

## Abstract

Whole-genome sequence data have made significant contributions to studies of the biology and transmission of *Cryptosporidium* spp. Well-assembled genomes, free of sequence ambiguity and structural errors, are critical for comparative genomic analysis at the species and subtype levels. The optimized method for *Cryptosporidium* genome assembly described here comprises several individual protocols to generate chromosome- and contig-level genomes.

**Key words** PacBio, Illumina, Genome assembly, Hybrid assembly, Self-assembly

## 1 Introduction

High-quality genomes have revolutionized *Cryptosporidium* research by providing more accurate references for studying the parasite biology, transmission, and evolution [1]. Fully assembled and annotated genomes are critical for comparing gene content and characterizing copy number variation among species or subtypes [2].

The use of these genome assembly methods has expanded the number of valuable *Cryptosporidium* genomes available for study. *Cryptosporidium* spp. have genomes of approximately 9 Mb in eight chromosomes, encoding approximately 4000 protein-coding genes [3, 4]. With the development of next-generation sequencing technologies, whole-genome sequence data from *Cryptosporidium* spp. have increased rapidly in recent years. To date, 71 genomes of 17 *Cryptosporidium* spp. have been assembled in varying numbers of contigs using Illumina sequencing technology. More importantly, five *Cryptosporidium parvum* and one *Cryptosporidium hominis* isolates have been fully assembled into eight-chromosome genomes [2–6]. Four of them were generated using a combination of short-read (Illumina) and long-read (PacBio or Nanopore) sequencing.

Jan R. Mead and Michael J. Arrowood (eds.), *Cryptosporidium: Methods and Protocols*, Methods in Molecular Biology, vol. 2978, https://doi.org/10.1007/978-1-0716-4824-7_9,

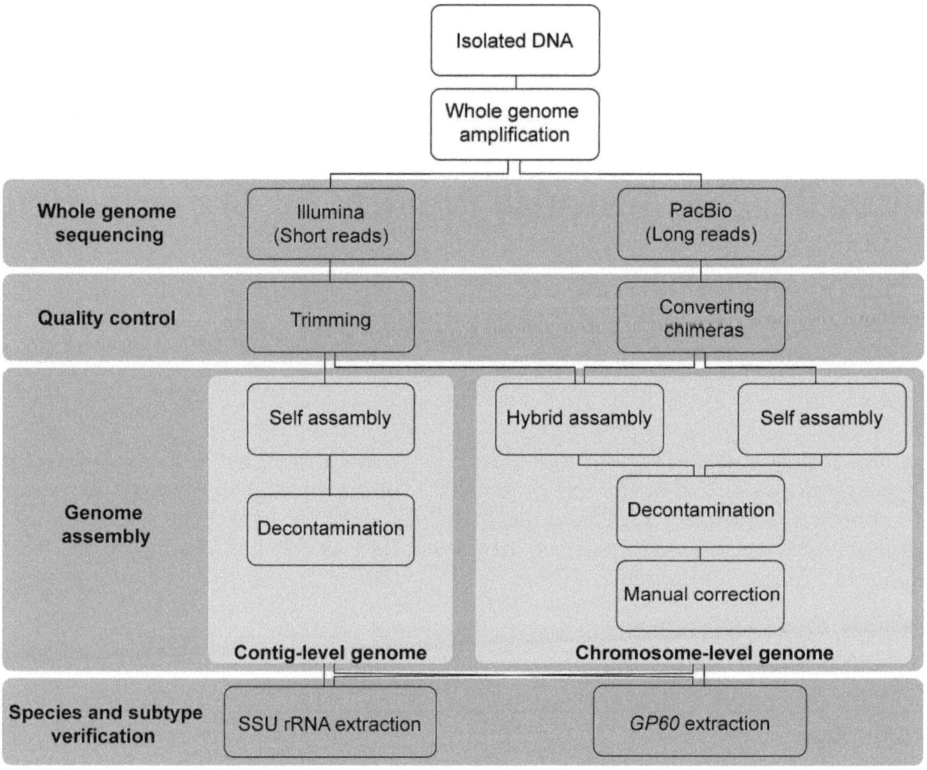


**Fig. 1** Flow diagram illustrating the procedure for *Cryptosporidium* genome assembly

The standardized method for *Cryptosporidium* genome assembly consists of several steps (*see* Fig. 1). After isolation of genomic deoxyribonucleic acid (DNA) and whole-genome amplification (described in other sections), whole-genome sequencing (WGS) is performed using PacBio/Nanopore and/or Illumina technologies. Next, Illumina sequence reads are trimmed for adapter sequences and poor sequence quality using fastp v0.23.4, while PacBio or Nanopore sequence reads are filtered for chimeras using 3rd-ChimeraMiner [7]. To generate contig-level genomes, all trimmed Illumina reads are assembled de novo using CLC Genomics Workbench or SPAdes v3.1. Contaminants are removed from the assemblies based on alignment and basic local alignment search tool (BLAST) data. To generate chromosome-level genomes, the PacBio/Nanopore and Illumina reads are assembled de novo using wengan v0.2 (hybrid assembly), and the PacBio/Nanopore reads are also assembled using hifiasm v0.19.8 (self-assembly). The hybrid assembly is polished using racon v1.5.0 and manually inspected against the mapped-back PacBio/Nanopore reads and the self-assembly. The sequences of small subunit ribosomal ribonucleic acid (SSU rRNA) and the 60-kDa glycoprotein encoding gene (*GP60*) are extracted to verify the species and the subtype based on sequence identity to the reference sequences and the standard subtype nomenclature [8]. Here, we have illustrated the assembly of *Cryptosporidium* genomes using data from both short-read (Illumina) and long-read (PacBio) sequencing.

## 2 Materials

### 2.1 Quality Control of Sequencing Data

1. Computer with Linux system.
2. Samtools v1.21 (http://www.htslib.org/).
3. 3rd-ChimeraMiner (https://github.com/dulunar/3rdChimeraMiner).
4. fastp v0.23.1 (https://github.com/OpenGene/fastp).
5. seqtk v1.4 (https://github.com/lh3/seqtk).

### 2.2 Genome Assembly

1. Computer with Linux system.
2. SPAdes v3.1 (http://cab.spbu.ru/software/spades/).
3. BLAST+ v2.16.0 (https://ftp.ncbi.nlm.nih.gov/blast/executables/blast+/LATEST/).
4. seqtk v1.4 (https://github.com/lh3/seqtk).
5. wengan v0.2 (https://github.com/adigenova/wengan).
6. hifiasm v0.19.8 (https://github.com/chhylp123/hifiasm).
7. racon v1.5.0 (https://github.com/lbcb-sci/racon).
8. Mauve v2.3.1 (https://darlinglab.org/mauve).
9. minimap2 v2.28 (https://github.com/lh3/minimap2).
10. IGV v2.17.1 (https://igv.org/doc/desktop/).

### 2.3 Species and Subtype Verification

1. Computer with Linux system.
2. BLAST+ v2.16.0 (https://ftp.ncbi.nlm.nih.gov/blast/executables/blast+/LATEST/).
3. seqtk v1.4 (https://github.com/lh3/seqtk).
4. MAFFT v7 (https://mafft.cbrc.jp/alignment/server/).

## 3 Methods

### 3.1 Quality Control of Sequencing Data

#### 3.1.1 Processing Short Reads from Short-Read Sequencing

1. Obtain sequence reads from Illumina.
2. Trim sequence reads for adapter sequences and poor sequence quality (Phred score < 30 and trimmed length < 60) using fastp v0.23.4. The high-quality reads are saved in *$samp.out*.fq.gz*. *Code:*

   For single-end data (gzip compressed)

*fastp -i $dir/$samp.fq.gz -o $dir/$samp.out.fq.gz -q 30 -u 50 -n 5 -l 60*

For paired-end data (gzip compressed)

*fastp -i $dir/$samp.R1.fq.gz -I $dir/$samp.R2.fq.gz -o $dir/$samp.out.R1.fq.gz -O $dir/$samp.out.R2.fq.gz -q 30 -u 50 -n 5 -l 60*

#### 3.1.2 Recognizing and Converting Chimeras in Long-Read Sequencing Data

1. Obtain sequence reads from PacBio.
2. Map PacBio reads to a reference genome using minimap2 and align the *.bam* file. *Code:*

   *minimap2 -Y -ax map-hifi $dir/RefGenome.fa --cs --MD $dir/$samp.fastq.gz | samtools view -Sb -o $dir/RefGenome.$samp.bam*
3. Split the *.bam* file obtained above and generate a shell script. This step was performed by using the script ExtIDsGenShell.pl downloaded from the 3rd-ChimeraMiner package. By performing this step, the *.bam* file is split into multiple *.bam* files saved in *$dir/result/example*; each *.bam* file contains "$number" reads ID (default was set to 100,000). Then, a shell script (example.runFind.sh) is generated and saved in *$dir/result/example* to run the 3rd-ChimeraMiner. *Code:*

   *perl $dir/ExtIDsGenShell.pl -i $dir/RefGenome.$samp.bam -d $dir/result -m $samp -n 100000 -r $dir/RefGenome.fa -sn 1*
4. Split the shell script example.runFind.sh into three according to the feature and run the chimera detection script. The chimeric reads are saved in *.chimera* files. *Code:*

   *split -l 3 RefGenome.$samp.runFind.sh split.Find.*

   *ls split.Find.* |grep -v log | perl -ne 'chomp;`nohup sh $_ &> $_.log &`;'*
5. Merge chimeras of each subchimera file to one file. Extract the direct chimera and transform the raw format to a better format for viewing. *Code:*

   *perl $dir/ChimerasDownstream.pl -d $dir/result/RefGenome.$samp -od $dir/result/RefGenome.$samp/chimeras.2 -n RefGenome.$samp -p 2*
6. Extract non-chimeric reads and save them in *$samp.nochimera.fq.gz*. *Code:*

   *awk 'NR==FNR{a[$0]++};NR!=FNR && ! a[$0]' $dir/result/RefGenome.$samp/chimeras.2/RefGenome.$samp.chimera.id $dir/$samp.reads.id.list.txt > $dir/samp.nochimera.id.txt*

   *seqtk subseq $dir/$samp.fastq.gz $dir/samp.nochimera.id.txt > $dir/$samp.nochimera.fq*

   gzip *$dir/$samp.nochimera.fq*
7. Split chimeric reads into normal segments. *Code:*

*perl $dir/SplitChimericRead.pl -i $dir/RefGenome.$samp.bam -f $dir/result/RefGenome.$samp/chimeras.2/RefGenome.$samp.chimera.id -c $dir/result/$samp/chimeras.2/RefGenome.$samp.normal.txt -o $dir/result/$samp/chimeras.2/$samp.chimera.split.fq*

### 3.2 Genome Assembly

#### 3.2.1 Generating a Contig-Level Genome

1. Generate a de novo assembly with the high-quality Illumina reads using CLC Genomics Workbench with a word size of 63 and bubble size of 400 (other parameters are default settings in the assembly workflow in CLC Genomics Workbench). Alternatively, generate a de novo assembly using SPAdes v3.1 with a k-mer of 63 and the careful mode. *Code:*

   For single-end data (gzip compressed)

   *spades.py --phred-offset 33 --pe1 $dir/$samp.out.R1.fq.gz -k 63 --careful -o $dir/$samp.assembley.fasta*

   For paired-end data (gzip compressed)

   *spades.py --phred-offset 33 --pe1–1 $dir/$samp.out.R1.fq.gz --pe1-2 $dir/$samp.out.R2.fq.gz -k 63 --careful -o $dir/$samp.assembley.fasta*

2. Compare the assembled DNA sequences with the public nucleotide (nt) database downloaded from National Center for Biotechnology Information (NCBI) using nucleotide-against-nucleotide (blastn). Alternatively, go to the NCBI website for BLAST (http://www.ncbi.nlm.nih.gov/blast/) and use the blastn tab. *Code:*

   *blastn -query $dir/$samp.assembley.fasta -db nt -outfmt 11 -out $dir/$samp.raw@nt.asn*

   *blast.formatter -archive $dir/$samp.raw@nt.asn -outfmt 18 > $dir/$samp.raw@nt.asn.report.tab*

3. Extract *Cryptosporidium* sequences from the original assembly. The remainder is the sequences from other organisms (*see* **Note 1**). *Code:*

   *grep -m 1 "Cryptosporidium" $dir/$samp.raw@nt.asn.report.tab >> $dir/$samp.api.IDlist.txt*

   *seqtk subseq $dir/$samp.assembley.fasta $dir/$samp.api.IDlist.txt > $dir/$samp.assembley.api.fasta*

4. Align and sort the assembly obtained from above against a reference genome of *Cryptosporidium* spp. using the "move contig" tool of Mauve v2.3.1 (*see* **Note 2**). Keep the aligned sequences and discard the unmapped ones (*see* **Note 1**). *Code:*

   *java -Xmx500m -cp $dir/Mauve.jar org.gel.mauve.contigs.ContigOrderer -output RefGenome.$samp.mauve $dir/RefGenome.fa -draft $dir/$samp.assembley.api.fasta*

#### 3.2.2 Generating a Chromosome-Level Genome

1. Generate a hybrid de novo assembly with the PacBio and Illumina reads using wengan v 0.2 with M model, a genome size of 9.5 Mb, haploid, and a k-mer of 63. The assembly is saved as a file ending with *.asm.wengan.fasta*. *Code:*

*wengan.pl -x pacraw -a M -s $dir/$samp.out.R1.fq.gz,$dir/$samp.out.R2.fq.gz -l $dir/$samp.nochimera.fq.gz -p asm2 -g 9.5 -D 1 -K 63*

2. Map PacBio long reads back to the new assembly (renamed as *$samp.wengan.asm*) using minimap2 v 2.28 with the map-hifi mode. *Code:*

   *minimap2 -ax map-hifi $dir/$samp.wengan.asm $dir/$samp.nochimera.fq.gz > $dir/$samp.wengan.sam*

3. Polish assembly using racon v1.5.0 with default settings. *Code:*

   *racon $dir/$samp.nochimera.fq.gz $dir/$samp.wengan.sam $dir/$samp.wengan.asm > $dir/$samp.wengan.racon.asm*

4. Search for telomeric repeats ("AAACCTAAACCT" or "TTTGGATTTGGA") throughout the polished assembly. Ensure that these sequences occur only at the beginning or end of contigs (*see* **Note 3**). Manually cut the contigs at the telomeric repeats if they occur in the middle contigs. The outcome is saved in *samp.wengan.racon.cut.asm.*

   *makeblastdb -in telomeric.repeats.fasta -dbtype nucl -title telomeric.repeats -parse.seqids -out telomeric.repeats*

   *blastn -query $dir/$samp.wengan.racon.asm -db telomeric.repeats -outfmt 6 -out $dir/$samp.wengan.racon.asm@telomeric.repeats.tab*

5. Generate a de novo assembly (self-assembly) with the PacBio reads using the hifiasm v0.19.8 with a k-mer of 63. *Code:*

   *hifiasm -o $dir/$samp.hifiasm.asm $dir/$samp.nochimera.fq.gz -k 63*

6. Compare the assembled DNA sequence with the public nucleotide (nt) database and extract *Cryptosporidium* sequences from the original assembly. *Code:*

   *blastn -query $dir/$samp.wengan.racon.cut.asm -db nt -outfmt 11 -out $dir/$samp.wengan.racon.cut.asm@nt.asn*

   *blast.formatter -archive $dir/$samp.wengan.racon.cut.asm@nt.asn -outfmt 18 > $dir/$samp.wengan.racon.cut.asm@nt.report.tab*

   *grep -m 1 "Cryptosporidium" $dir/$samp.wengan.racon.cut.asm@nt.report.tab*

   *>> $dir/$samp.wengan.racon.cut.asm.api.IDlist.txt*

   *seqtk subseq $dir/$samp.wengan.racon.cut.asm $dir/$samp.wengan.racon.cut.asm.api.IDlist.txt > $dir/$samp.wengan.racon.cut.api.asm*

7. Align the self-assembly against polished hybrid assembly using the "move contig" tool of Mauve v2.3.1. Manually extend the telomeric regions by comparing those assemblies. Name the extend assembly as *$samp.wengan.racon.cut.api.extend.asm.*

   *java -Xmx500m -cp $dir/Mauve.jar org.gel.mauve.contigs.ContigOrderer -output $dir/$samp.wengan.racon.cut.api.asm.*

*mauve.samp.hifiasm.api.asm $dir/$samp.wengan.racon.cut.api.asm -draft $dir/$samp.hifiasm.api.asm*

8. Align the extended assembly from the last step against published chromosome-level genomes, such as *C. parvum* IOWA-CDC [2], IOWA-BGF [6], IOWA-ATCC, and *C. hominis* IbA12G3_46287 (*see* **Note 4**). Manually sew up the broken chromosomes with 10 Ns according to the reference genome. Save the outcome as *samp.wengan.racon.cut.api.extend.sew.asm.*

   *java -Xmx500m -cp $dir/Mauve.jar org.gel.mauve.contigs.ContigOrderer -output $dir/$RefGenome.fa.mauve.samp.wengan.racon.cut.api.extend.asm $dir/$RefGenome.fa -draft $dir/$samp.wengan.racon.cut.api.extend.asm*

   *The sewed regions look like below:*

   *cacgtattatttacttattattttaaatttcctttatcttgttaattgacataNNNNNNNNNNNatttagccttgaataatcttttgcatgcagtcggcgccaattgtttggcgcctgacgag*

9. Map PacBio long reads back to the assembly from the last step using minimap2 v 2.28 with the map-hifi mode. Visualize the .*bam* file in IGV v2.17.1. Manually extend the telomeric regions and fulfill the Ns added in the last step based on the mapped reads. *Code:*

   *minimap2 -ax map-hifi $dir/$samp.wengan.racon.cut.api.extend.sew.asm $dir/$samp.nochimera.fq.gz > $dir/$samp.wengan.racon.cut.api.extend.sew.asm.sam*

   *samtools sort -O bam $dir/$samp.wengan.racon.cut.api.extend.sew.asm.sam -o $dir/$samp.wengan.racon.cut.api.extend.sew.asm.bam.*

### 3.3 Species and Subtype Verification

1. Download reference SSU rRNA and *GP60* sequences from NCBI to build up databases using makeblastdb in BLAST+ v2.16.0. *Code:*

   *makeblastdb -in $dir/Reference_SSU_GP60.fasta -dbtype nucl -title Reference_SSU_GP60 -parse_seqids -out $dir/Reference_SSU_GP60*

2. Compare genomes with the databases obtained above using blastn to locate the SSU rRNA and the *GP60.* Generate a list (*$samp.seq.id.txt*) of contig IDs and gene locations (start and end positions) based on the BLAST data. *Code:*

   blastn -query *$dir/$samp.genome.fasta* -db *$dir/Reference_SSU_GP60* -outfmt 6 -out *$dir/$samp@Reference_SSU_GP60.report.tab*

3. Extract the 18S rRNA and *GP60* sequences from genomes. *Code:*

   *seqtk subseq $dir/$samp.genome.fasta $dir/$samp.seq.id.txt > $dir/$samp.seq.fq*

4. Upload the extracted SSU rRNA or *GP60* sequences and the reference sequences to the MAFFT v7 website for sequence alignment.
5. Determine sequence designation based on sequence identity to reference sequences. Determine subtype designation based on sequence identity to reference sequences and the number of trinucleotide repeats (see *Cryptosporidium* genotyping).

## 4 Notes

1. Contaminants in WGS data include sequences from non-*Cryptosporidium* organisms, other *Cryptosporidium* spp., and closely related subtypes. The non-*Cryptosporidium* sequences would be discarded in Subheading 3.2.1, **step 3**. For samples with mixed *Cryptosporidium* species but with divergent genomes, the contaminants would be discarded in Subheading 3.2.1, **step 4**. For samples with mixed *Cryptosporidium* strains with similar genomes (final genomes larger than 9.1 Mb), we strongly recommend that these genomes be excluded from further analysis.
2. The genome with the highest genomic similarity in public databases could be selected as a reference genome.
3. Two or more chromosomes may be misassembled due to long telomeric repeats.
4. This step could be applied to *C. parvum*-related *Cryptosporidium* spp., including *C. hominis*, *C. tyzzeri*, *C. meleagridis*, and *C. cuniculus*. Use caution with other species that have divergent genomes. For these species, you may sew up the large contigs based on the self-assembly or PacBio reads.

## Competing Interest Statement

The authors have no conflicts of interest to declare that are relevant to the content of this chapter.

### References

1. Agyabeng-Dadzie F, Xiao R, Kissinger JC (2024) *Cryptosporidium* genomics - current understanding, advances, and applications. Curr Trop Med Rep 11(2):92–103. https://doi.org/10.1007/s40475-024-00318-y
2. Huang W, Tang K, Chen C, Arrowood MJ, Chen M, Guo Y, Li N, Roellig DM, Feng Y, Xiao L (2024) Sequence introgression from exogenous lineages underlies genomic and biological differences among *Cryptosporidium parvum* IOWA lines. Water Res 254:121333. https://doi.org/10.1016/j.watres.2024.121333
3. Abrahamsen MS, Templeton TJ, Enomoto S, Abrahante JE, Zhu G, Lancto CA, Deng M, Liu C, Widmer G, Tzipori S, Buck GA, Xu P, Bankier AT, Dear PH, Konfortov BA, Spriggs HF, Iyer L, Anantharaman V, Aravind L, Kapur

V (2004) Complete genome sequence of the apicomplexan, *Cryptosporidium parvum*. Science 304(5669):441–445. https://doi.org/10.1126/science.1094786

4. Baptista RP, Li Y, Sateriale A, Brooks KL, Tracey A, Sanders MJ, Ansell BRE, Jex AR, Cooper GW, Smith ED, Xiao R, Dumaine JE, Georgeson P, Pope B, Berriman M, Striepen B, Cotton JA, Kissinger JC (2021) Long-read assembly and comparative evidence-based reanalysis of *Cryptosporidium* genome sequences reveals expanded transporter repertoire and duplication of entire chromosome ends including subtelomeric regions. Genome Res 32(1): 203–213. https://doi.org/10.1101/gr.275325.121
5. Menon VK, Okhuysen PC, Chappell CL, Mahmoud M, Mahmoud M, Meng Q, Doddapaneni H, Vee V, Han Y, Salvi S, Bhamidipati S, Kottapalli K, Weissenberger G, Shen H, Ross MC, Hoffman KL, Cregeen SJ, Muzny DM, Metcalf GA, Gibbs RA, Petrosino JF, Sedlazeck FJ (2022) Fully resolved assembly of *Cryptosporidium parvum*. GigaScience 11:1–8. https://doi.org/10.1093/gigascience/giac010
6. Baptista RP, Xiao R, Li Y, Glenn TC, Kissinger JC (2023) New T2T assembly of *Cryptosporidium parvum* IOWA annotated with reference genome gene identifiers. bioRxiv. https://doi.org/10.1101/2023.06.13.544219
7. Lu N, Qiao Y, An P, Luo J, Bi C, Li M, Lu Z, Tu J (2023) Exploration of whole genome amplification generated chimeric sequences in long-read sequencing data. Brief Bioinform 24(5). https://doi.org/10.1093/bib/bbad275
8. Xiao L, Fayer R, Ryan U, Upton SJ (2004) *Cryptosporidium* taxonomy: recent advances and implications for public health. Clin Infect Rev 17(1):72–97. https://doi.org/10.1128/CMR.17.1.72-97.2004

# Chapter 10

# Evolutionary Genomic Analyses of *Cryptosporidium*

**Tianpeng Wang and Lihua Xiao**

## Abstract

The increasing amount of whole-genome sequencing (WGS) data allows us to study the evolution and transmission of *Cryptosporidium* spp. at the population level. Population genomics is helping us to understand the genetic structure and adaptive evolution of *Cryptosporidium* and the mechanism and determinants of emergence of hypertransmissible subtypes and variants. The integrated pipeline described here provides an example of the use of WGS data for evolutionary genomic analysis of *Cryptosporidium*.

**Key words** Population genomics, Population structure, Recombination, Evolution, Adaptation

## 1 Introduction

Over the past decade, there have been significant advances in comparative genomic analysis of *Cryptosporidium* spp. [1]. Whole-genome sequencing (WGS) data are rapidly accumulating, particularly for the major human-infecting species, *Cryptosporidium parvum* and *Cryptosporidium hominis*.

Using WGS data from different hosts, comparative genomics and population genetics have provided new insights into the evolution of *Cryptosporidium* species [2–7]. These results have revealed discordant phylogenetic relationships between *gp60* and WGS data [6, 7] and the presence of geographically segregated populations of *C. parvum* and *C. hominis* [2, 4, 5]. These studies have also highlighted the role of genetic recombination and selective sweeps in the emergence of evolution of hypertransmissible *Cryptosporidium* subtypes.

Here, we have summarized the procedures for evolutionary genomic analyses of *Cryptosporidium* spp. (*see* Fig. 1). The approaches and their utility are described in recent publications [2, 5]. In addition to the WGS data obtained by the researchers themselves, published WGS data with the consent of the original data submitters could be downloaded from NCBI (https://ncbi.

Jan R. Mead and Michael J. Arrowood (eds.), *Cryptosporidium: Methods and Protocols*, Methods in Molecular Biology, vol. 2978, https://doi.org/10.1007/978-1-0716-4824-7_10,

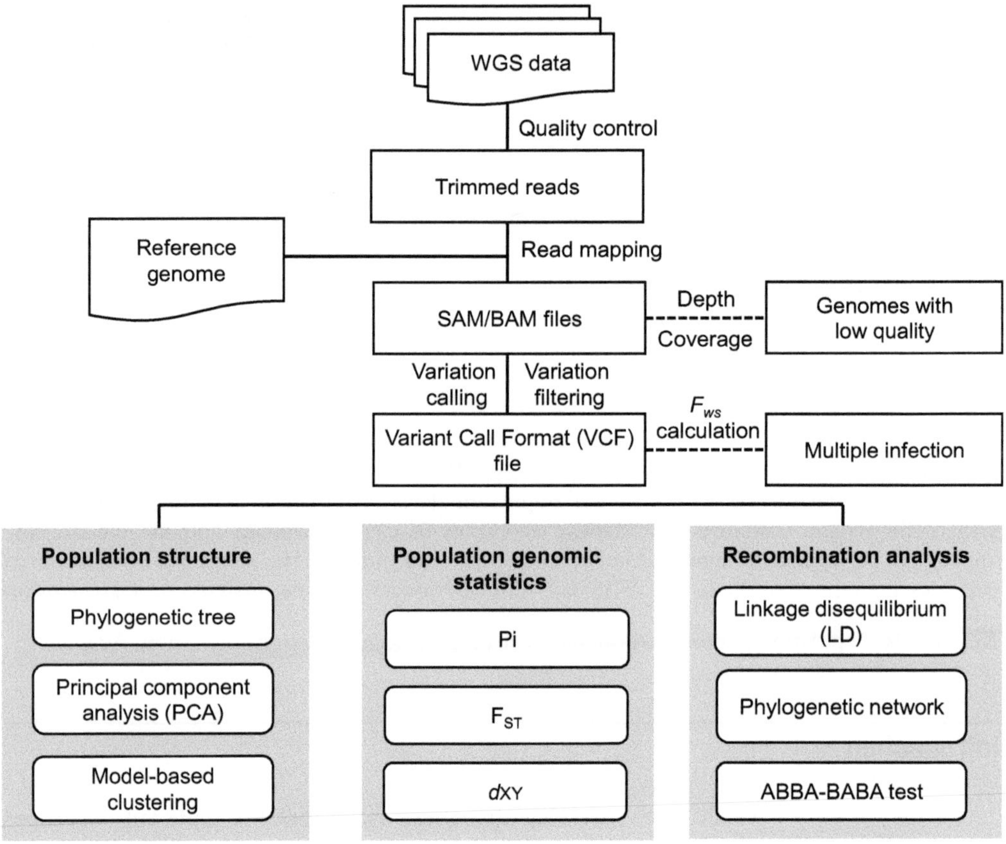


**Fig. 1** Flow diagram illustrating procedures for evolutionary genomic analyses of *Cryptosporidium*

nlm.nih.gov/) to obtain a comprehensive picture of the population structure and evolutionary history. First, quality control of the WGS data is performed by adapter trimming and removal of low-quality reads in the raw data using fastp v0.24.0. The trimmed reads are then mapped to the reference genomes using bwa-mem2 v2.2.1. After viewing and sorting the read alignments with samtools v1.21, polymerase chain reaction (PCR) duplications are removed with sambamba v1.0.1. Genomes with low average mapping depth (< 10×) and poor genome coverage (<95%) should be removed. Second, variants are detected using bcftools v1.21 or Genome Analysis Toolkit (GATK) v4.3. After a series of variant filtering steps, $F_{ws}$ values are calculated for each genome and potential multiple infections ($F_{ws} < 0.95$) are removed. Third, population structure is inferred using phylogenetic trees, principal component analysis (PCA), and model-based clustering. Fourth, population genomic statistics are calculated for each group or between groups, including nucleotide divergence (Pi), fixation index ($F_{ST}$), and genetic divergence (*d*XY). Finally, linkage disequilibrium (LD), phylogenetic network, and ABBA-BABA analyses are performed to investigate recombination events between different populations and isolates.

In addition to the pipeline described in this section, other alternative approaches can also be used for evolutionary genomic analysis of *Cryptosporidium* (*see* **Note 1**).

## 2 Software

### *2.1 Quality Control and Variant Calling*

1. Computer or server with Linux system.
2. R v4.3.3 (https://www.r-project.org/).
3. Python v2.7 and v3.10 (https://www.python.org/).
4. fastp v0.24.0 (https://github.com/OpenGene/fastp),
5. bwa-mem2 v2.2.1 (https://github.com/bwa-mem2/bwa-mem2),
6. samtools v1.21 (https://github.com/samtools/samtools),
7. sambamba v1.0.1 (https://github.com/biod/sambamba),
8. bcftools v1.21 (https://github.com/samtools/bcftools),
9. GATK v4.3 (https://github.com/broadinstitute/gatk).
10. VCFtools v0.1.16 (https://vcftools.github.io/index.html).
11. moimix (https://github.com/bahlolab/moimix),
12. SeqArray (https://github.com/zhengxwen/SeqArray).

### *2.2 Population Structure Analysis*

1. vcf2phylip (https://github.com/edgardomortiz/vcf2phylip),
2. IQ-TREE v2.3.6 (http://www.iqtree.org/).
3. Interactive Tree Of Life (iTOL) v7 (http://itol.embl.de/),
4. SNPRelate (https://github.com/zhengxwen/SNPRelate).
5. gdsfmt (https://github.com/cran/gdsfmt),
6. Plink v1.9 (https://www.cog-genomics.org/plink/1.9/).
7. fastStructure (https://rajanil.github.io/fastStructure/),
8. Cluster Markov Packager Across K (CLUMPAK) (https://tau.evolseq.net/clumpak/).

### *2.3 Population Genomic Statistics Calculation*

1. genomics_general (https://github.com/simonhmartin/genomics_general).

### *2.4 Recombination Analysis*

1. PopLDdecay v3.43 (https://github.com/hewm2008/PopLDdecay).
2. SplitsTree v6 (https://github.com/husonlab/splitstree6).
3. Dsuite (https://github.com/millanek/Dsuite).

## 3 Methods (*See* Note 2)

### *3.1 Quality Control and Variant Calling*

1. Obtain the sequencing reads in the FASTQ file format (*$sample*.fq* for not compressed or *$sample*.fq.gz* for gzip compressed).
2. Perform quality control on the FASTQ data using fastp, including trimming adapter contamination and filtering out low-quality reads (phred score <30 and trimmed length <60).

   For single-end data (not compressed):

```
'fastp -i $sample.fq -o $sample.out.fq -q 30 -u 50 -n 5 -l 60'
```

   For paired-end data (gzip compressed):

```
'fastp -i $sample.R1.fq.gz -I $sample.R2.fq.gz -o $sample.
out.R1.fq.gz -O $sample.out.R2.fq.gz -q 30 -u 50 -n 5 -l 60'
```

3. Index the reference genome (*$ref.fasta*) and map the trimmed reads to the reference genome using bwa-mem2. The mapping results are saved in the sequence alignment/map (SAM) file format (*$sample.sam*). The alignments in the SAM file should be sorted and written to the compressed binary version of a SAM file (BAM) using samtools to save space.

   First, the reference genome should be indexed:

```
'bwa-mem2 index $ref.fasta'
```

   Then, map the trimmed reads to the indexed reference genome. For single-end data:

```
'bwa-mem2 mem $ref.fasta $sample.out.fq.gz > $sample.sam'
```

   For paired-end data:

```
'bwa-mem2 mem $ref.fasta $sample.out.R1.fq.gz $sample.out.R2.
fq.gz > $sample.sam'
```

   Sort the alignments in the SAM file and write the sorted output to the BAM file. The unmapped reads can be filtered with the *-q* parameter. *Code:*

```
'samtools view -q 1 -b $sample.sam | samtools sort - -o
$sample.sorted.bam'
```

   The above mapping and sorting commands can be combined into one using the Linux pipe:

```
'bwa-mem2 mem $ref.fasta $sample.out.R1.fq.gz $sample.out.R2.
fq.gz | samtools view -q 1 -b | samtools sort - -o $sample.
sorted.bam'
```

Remove the duplicate reads in the BAM file using sambamba:

```
'sambamba markdup -t 10 -p -r $sample.sorted.bam $sample.
sorted.RM.bam'
```

4. Based on the mapping results, remove samples with low mapping depth (<10×) and insufficient genome coverage (<95%).
5. There are two commonly used methods for calling variations into the variant call format (VCF) file and filtering out the low-quality sites, including samtools/bcftools and the Genome Analysis Toolkit (GATK) (*see* **Note 3**).

   For samtools/bcftools, assuming that the location of each BAM file has been written to the $BAM.paths. *The codes are as below:*

```
# Generate the gzip-compressed VCF file containing genotype
likelihoods for one or more BAM file(s).
bcftools mpileup -Ou -A -a FORMAT/AD,FORMAT/DP -f $ref.fasta
-b $BAM.paths | bcftools call -mv -O z -o $all.vcf.gz --threads
8
# Sites with a phred-scaled quality score (QUAL) of less than
30, an average read depth (DP) of less than 25 and a mapping
quality of less than 30 are marked as FILTERED. Single
nucleotide polymorphisms (SNPs) within three base pairs of an
indel or other variant type are marked with SnpGap.
bcftools filter -s FILTERED --SnpGap 3 --set-GTs . -O v --
include "%QUAL>=30 && AVG(FMT/DP)>=25 && MQ>=30" $all.vcf.gz >
$all.Marked.vcf
# Set the genotype of the failed site (DP less than 3) to a
missing value (.).
bcftools filter -S . -e 'FMT/DP<3' $all.Marked.vcf > $all.
DPmarked.vcf
# Keep the filtered biallelic sites in $all.final.vcf.
bcftools view -m2 -M2 -v snps -O v -o $all.Biallelic.vcf $all.
DPmarked.vcf && grep -v "SnpGap" $all.Biallelic.vcf > $all.
final.vcf
```

For the GATK pipeline, a genomic VCF (GVCF) is created for each sample to record each site in the file for joint analysis of a cohort in subsequent steps. *The codes are as below:*

```
# Generate the gzip-compressed GVCF file for each sample.
gatk --java-options "-Xms8G -Xmx16G" HaplotypeCaller -R $ref.
fasta -I $sample.sorted.RM.bam -O $sample.g.vcf.gz -ERC GVCF
# Write all the GVCF file locations to the $GVCF.paths file.
Combine the GVCF files into one gzip-compressed GVCF file.
gatk --java-options "-Xms8G -Xmx16G" CombineGVCFs -R $ref.
fasta -V $GVCF.paths -O $all.g.vcf.gz
# Perform joint genotyping on the combined GVCF file and write
the genotyped information to a gzip-compressed VCF file.
gatk --java-options "-Xms8G -Xmx16G" GenotypeGVCFs -R $ref.
fasta -V $all.g.vcf.gz -O $all.genotype.vcf.gz
# Select SNPs from the original VCF file and perform the hard-
filtering.
gatk --java-options "-Xms8G -Xmx16G" SelectVariants -select-
type SNP -V $all.genotype.vcf.gz -O $all.snp.vcf.gz
gatk --java-options "-Xms8G -Xmx16G" VariantFiltration -R
$ref.fasta -V $all.snp.vcf.gz --filter-expression "QUAL <
30.0 || QD < 2.0 || MQ < 40.0 || FS > 60.0 || SOR > 3.0 ||
MQRankSum < -12.5 || ReadPosRankSum < -8.0" --filter-name
SNP_filter -O $all.snp.filter.vcf.gz
gatk --java-options "-Xms8G -Xmx16G" SelectVariants -R $ref.
fasta -V $all.snp.filter.vcf.gz --exclude-filtered --restrict-
alleles-to BIALLELIC -O $all.snp.pass.vcf.gz
```

6. Filter sites with minor allele frequency (MAF) less than 0.005 to remove the less informative sites including singletons, and also remove isolates with more than 20% missing sites using VCFtools. The passed records are written to the *$all.maf.miss.recode.vcf* file. *Code:*

```
'vcftools --gzvcf $all.snp.pass.vcf.gz --max-missing 0.8 --
maf 0.005 --out $all.maf.miss --recode --recode-INFO-all'
```

7. Evaluate the multiplicity of infection (MOI) of each genome using the moimix and SeqArray R packages by running *commands below:*

```
# Load the required packages.
library(moimix)
library(SeqArray)
# Load the VCF file.
vcf_file <- "$all.maf.miss.recode.vcf"
# Convert the VCF file to Graphic Data System (GDS) format for
Fws calculation.
seqVCF2GDS(vcf_file, "$Input.gds")
my_vcf <-seqOpen("$Input.gds")
# Calculate the Fws values for each sample and write the output
to the specified file.
```

```
fws_all <- getFws(my_vcf)
write.table(fws_all, file = "$output.Fws.txt", quote = F, row.
names = F, sep = "\t")
```

Finally, remove the potential multiple infections ($F_{ws} < 0.95$) using VCFtools. *Code:*

```
`cat $output.Fws.txt | awk '{if ($1<0.95) print $2}' >
Multi_Infec.ids && vcftools --vcf $all.maf.miss.recode.vcf --
remove Multi_Infec.ids --out $final.vcf --recode --recode-
INFO-all`
```

### 3.2 Population Structure Analysis

#### 3.2.1 Phylogenetic Analysis

1. Create alignments by concatenating SNPs in the VCF file using the vcf2phylip.py script. By default, the output would be in PHYLogeny Inference Package (PHYLIP) format named *$SNPaln.min4.phy*. We can use the *-f* argument to write a FASTA format named *$SNPaln.min4.fasta. Code:*

   ```
   'python vcf2phylip.py -i $final.vcf --output-prefix
   $SNPaln -f'
   ```

2. Build phylogenetic tree using IQ-TREE with 1000 bootstraps. Substitution models would be automatically selected using ModelFinder Plus (MFP) and optimized with the option *-bnni. Code:*

   ```
   'iqtree2 -s $SNPaln.min4.fasta -m MFP+ASC -B 1000
   -bnni -T AUTO --prefix $tree'
   ```

3. Submit the output file *$tree.contree* to the iTOL web server for tree visualization and annotation.

#### 3.2.2 Principle Component Analysis (PCA)

Perform the PCA with unlinked SNPs (linkage disequilibrium <0.4 in 10-kb window) using the SNPRelate R package by running *commands below:*

```
# Load the required packages.
library(gdsfmt)
library(SNPRelate)
# Load the VCF file and convert it to the GDS file format.
vcf.fn <- "$final.vcf "
snpgdsVCF2GDS(vcf.fn, "$input.gds", method="biallelic.only")
genofile <- snpgdsOpen("$input.gds")
# Keep the unlinked sites for analysis.
snpset <- snpgdsLDpruning(genofile, ld.threshold=0.4, slide.
max.bp = 10000)
snpset.id <- unlist(snpset)
pca <- snpgdsPCA(genofile, snp.id=snpset.id, num.thread=2)
# Write the PCA results to files, including the eigenvectors,
eigenvalues, and SNPs used. The data frames can be plotted
using the ggplot2 package.
```

```
tab <- data.frame(sample.id = pca$sample.id,
EV1 = pca$eigenvect[,1], EV2 = pca$eigenvect[,2],
EV3 = pca$eigenvect[,3], stringsAsFactors = FALSE)
write.table(tab, file = "$pca.features", quote = F, sep = "\t")
write.table(pca$eigenval, file = "$pca.eigenval", quote = F,
sep = "\t")
write.table(pca$snp.id, file = "$pca.usedSNPs", quote = F, sep
= "\t")
```

#### 3.2.3 Model-Based Clustering

1. Extract the same subset of SNPs used in PCA (*see* **Note 4**).
2. Transform the VCF file to the PLINK PED format using VCFtools and create the Browser Extensible Data (BED) file for model-based clustering:

```
'vcftools --vcf $pca.used.vcf --plink --out $structure'
'plink --noweb --file $structure --make-bed --out $structure'
```

Infer the population structure using fastStructure with K values from 2 to 15. Then choose the appropriate number of model components to explain the structure (*see* **Note 5**). *The commands are as below:*

```
for i in 'seq 2 15'
do
python structure.py --input=$structure --full --seed=100 --
output=$fastStructure -K ${i}
done
python chooseK.py --input=$fastStructure
```

3. The Q-matrix results (*$fastStructure.k.Q*) can be imported into R for plotting using the *barplot()* function. The results can also be submitted to the CLUMPAK web server.

### 3.3 Population Genomic Statistics Calculation

1. Convert the VCF to the needed file format *$input.geno* using the Python script parseVCF.py in the VCF_processsing directory. *Code:*

```
'python parseVCF.py -i $final.vcf -o $input.geno'
```

2. Calculate the population genomic statistics between the groups using the Python script *popgenWindows.py*. Assuming we have two groups A and B, a file *$popfile* is needed with two columns: The first is the sample name, and the second is the population name. The statistics are written to the *$output.csv* file. *Code:*

```
'python popgenWindows.py --windType coordinate -w 5000 -s
1000 -T 20 -p groupA -p groupB -f phased --popsFile $popfile -g
$input.geno -o $output.csv'
```

### 3.4 Recombination Analysis

#### 3.4.1 Linkage Disequilibrium Decay

1. Extract the variants for each chromosome using VCFtools:

```
'for i in {1..8};do vcftools --vcf $final.vcf --chr
${i} --out Chr${i} --recode;done'
```

2. Calculate LD decay for group A by chromosome using PopLDdecay:

```
'for i in {1..8};do PopLDdecay -InVCF Chr${i}.recode.
vcf -Subpop $A.ids -OutStat A${i}.stat.gz -MaxDist 150;
done'
```

3. Draw the LD decay figure for this group A with multiple chromosomes using the perl script *Plot_OnePop.pl* in the *bin* folder of PopLDdecay. *Code:*

```
'ls A${i}.stat.gz > A_PlotInput.list && perl bin/Plot_OnePop.
pl -inList A_PlotInput.list -output FigA -maxX 150 -bin1
20 -bin2 100 -break 50 -keepR '
```

4. For multiple populations, first draw the LD decay for each population as above. Then write the file path of each population result ( **.bin*) and the population flag into *Multipop.bin.list*, such as:

```
FigA.bin A
FigB.bin B
```

   Draw the figure using the perl script *Plot_MultiPop.pl* in the *bin* folder of PopLDdecay. *Code:*

```
'perl bin/Plot_MultiPop.pl -inList Multipop.bin.list -output
Compare'
```

#### 3.4.2 Phylogenetic Network

Create alignments by concatenating SNPs in the VCF file using the *vcf2phylip.py* script as described in Subheading 3.2.1. Open the SplitsTree app and import the alignments. By default, the SplitsTree app computes *P-distances* and then runs *Neighbor Net* to obtain a split network. The existence of connected networks that differ from phylogenetic trees would indicate recombination events between subclasses.

#### 3.4.3 ABBA-BABA Statistics

1. Prepare a text file with one individual per row and a tab separating the name of the individual from the name of the population to which it belongs. Calculate the *D* and *f4-ratio* statistics for all possible trios of populations using the Dtrios of Dsuite. *Code:*

```
'Dsuite Dtrios $final.vcf SETS.txt'
```

2. For *D* statistics and $f_d$ admixture estimates across the genome, use the Python script *ABBABABAwindows.py* from genomics_general. The input file should also be in the same file

format as described in Subheading 3.3. At least three populations (A, B, and C) and one outgroup (D) should be included in the *$popfile*. *Code:*

```
'python ABBABABAwindows.py -g $input.geno -f phased
-o $output.csv -w 25000 -m 100 -s 10000 -P1 A -P2 B -P3 C
-O D -T 10 --minData 0.5 --popsFile $popfile --writeFailed-
Windows --polarize'
```

## 4 Notes

1. There are many alternative software packages and scripts that can be used for this analysis. For example, read mapping can be done using bowtie2, PCR duplicates can be removed using the MarkDuplicates tool in Picard, and model-based clustering can be estimated using AdmixTools. Due to the length of this section, it is not possible to list all available methods.
2. We have also posted the relevant code on GitHub for better presentation (https://github.com/TpengWang/Evolutionary-genomic-analyses-of-Cryptosporidium).
3. Both samtools/bcftools and GATK pipeline are commonly used for variant calling. When filtering variants with poor quality, the filtering thresholds need to be flexible. In particular, the MAF and the missing rate should be decided based on the number of variants and samples.
4. Information of the sites used for PCA is written to the *$pca.usedSNPs* file with two columns. Only the second column is needed and corresponds to the order of the SNPs in the *$final.vcf* file.
5. The Python script *chooseK.py*, which is built into fastStructure, returns two suggestions of K values for model complexity. One is the maximized marginal likelihood of model complexity, and the other is the model components used to explain the structure. It is suggested to compare the population structures with different K values and also to refer to the results of the phylogenetic tree and PCA.

## Acknowledgments

This study was funded by the National Natural Science Foundation of China (No. 32402919).

**Competing Interest Statement** The authors have no conflict of interests to declare that are relevant to the content of this chapter.

## References


1. Agyabeng-Dadzie F, Xiao R, Kissinger JC (2024) *Cryptosporidium* genomics – current understanding, advances, and applications. Curr Trop Med Rep 11(2):92–103. https://doi.org/10.1007/s40475-024-00318-y
2. Huang W, Guo Y, Lysen C, Wang Y, Tang K, Seabolt MH, Yang F, Cebelinski E, Gonzalez-Moreno O, Hou T, Chen C, Chen M, Wan M, Li N, Hlavsa MC, Roellig DM, Feng Y, Xiao L (2023) Multiple introductions and recombination events underlie the emergence of a hyper-transmissible *Cryptosporidium hominis* subtype in the USA. Cell Host Microbe 31(1): 112–123. e114. https://doi.org/10.1016/j.chom.2022.11.013
3. Corsi GI, Tichkule S, Sannella AR, Vatta P, Asnicar F, Segata N, Jex AR, van Oosterhout C, Caccio SM (2023) Recent genetic exchanges and admixture shape the genome and population structure of the zoonotic pathogen *Cryptosporidium parvum*. Mol Ecol 32(10):2633–2645. https://doi.org/10.1111/mec.16556
4. Tichkule S, Caccio SM, Robinson G, Chalmers RM, Mueller I, Emery-Corbin SJ, Eibach D, Tyler KM, van Oosterhout C, Jex AR (2022) Global population genomics of two subspecies of *Cryptosporidium hominis* during 500 years of evolution. Mol Biol Evol 39(4). https://doi.org/10.1093/molbev/msac056
5. Wang T, Guo Y, Roellig DM, Li N, Santin M, Lombard J, Kvac M, Naguib D, Zhang Z, Feng Y, Xiao L (2022) Sympatric recombination in zoonotic *Cryptosporidium* leads to emergence of populations with modified host preference. Mol Biol Evol 39(7). https://doi.org/10.1093/molbev/msac150
6. Tichkule S, Jex AR, van Oosterhout C, Sannella AR, Krumkamp R, Aldrich C, Maiga-Ascofare O, Dekker D, Lamshoft M, Mbwana J, Rakotozandrindrainy N, Borrmann S, Thye T, Schuldt K, Winter D, Kremsner PG, Oppong K, Manouana P, Mbong M, Gesase S, Minja DTR, Mueller I, Bahlo M, Nader J, May J, Rakotozandrindrain R, Adegnika AA, Lusingu JPA, Amuasi J, Eibach D, Caccio SM (2021) Comparative genomics revealed adaptive admixture in *Cryptosporidium hominis* in Africa. Microb Genom 7(1). https://doi.org/10.1099/mgen.0.000493
7. Nader JL, Mathers TC, Ward BJ, Pachebat JA, Swain MT, Robinson G, Chalmers RM, Hunter PR, van Oosterhout C, Tyler KM (2019) Evolutionary genomics of anthroponosis in *Cryptosporidium*. Nat Microbiol 4(5):826–836. https://doi.org/10.1038/s41564-019-0377-x

# Chapter 11

# A Novel Method to Silence Genes in *Cryptosporidium*

## Alejandro Castellanos-Gonzalez

## Abstract

A major bottleneck for developing new drugs and vaccines against cryptosporidiosis has been the lack of methods to study gene function in this parasite. Silencing of genes using small interference RNA (siRNA) is a powerful methodology to investigate gene function in multiple organisms and has been widely used to identify drug targets for multiple pathogens. *Cryptosporidium* does not possess the enzymes of the classical siRNA pathway. Therefore, its genes cannot be silenced by standard siRNA technology. To circumvent that problem, we developed a strategy to knock down *Cryptosporidium* mRNA by reconstituting the effector arm of the siRNA pathway using recombinant enzymes and synthetic single-strand RNA (ssRNA). We have induced silencing of several genes in *Cryptosporidium* by transfecting parasites with hybrid complexes formed between recombinant human Argonaute (hAgo2) and *Cryptosporidium* ssRNA. We have used this methodology to study the role of selected genes in host–parasite interactions. It also can be used to identify potential targets for chemotherapy. Standardized protocols to silence genes in *Cryptosporidium* parasites are described in this chapter. In addition, we described the use of ssRNA/Ago complexes as a novel strategy to reduce *Cryptosporidium* infection.

**Key words** *Cryptosporidium*, Gene silencing, Argonaute, Ago 2, RNAi, siRNA

## 1 Introduction

***Gene Silencing by RNAi Pathway*** Silencing of genes using small interference RNA (siRNA) via the RNAi pathway is a powerful method to study gene function in both prokaryotic and eukaryotic organisms [1]. Gene silencing mediated by the RNA interference (RNAi) pathway was described in 1998 [2]. Since then, siRNA has been widely used to understand the biology of multiple parasites and pathogens [3]. Some of these studies have led to identification of targets for drug and vaccine development [4, 5]. In addition, siRNA also has been used in therapeutics to treat multiple infections [6, 7].

Jan R. Mead and Michael J. Arrowood (eds.), *Cryptosporidium: Methods and Protocols*, Methods in Molecular Biology, vol. 2978, https://doi.org/10.1007/978-1-0716-4824-7_11,

RNAi technology is based on the use of small double-stranded RNA (siRNA) complementary to an mRNA target; once the siRNA is transfected to the cells, it is processed by the RNA-induced silencing complex (RISC). During this process, the sense strand of the siRNA is removed, and then, the antisense strand is used as guide to bind specifically to the mRNA target. After binding, gene expression is suppressed by inducing degradation of the mRNA [1, 2]; target degradation is mediated by the "slicer protein" Argonaute (Ago) [8]. Using dose-responsive gene silencing, it is possible to study the function of essential genes by titrating mRNA transcript levels and studying resulting phenotypes. Dose-responsive gene silencing can also be used to study nonessential genes, for example, to study the effects of different extents of silencing of a gene of interest and to determine whether other genes can compensate when a gene of interest is silenced. However, some organisms do not have the RISC enzymes and, thus, siRNA is not feasible [9–11]. The genomes of *Cryptosporidium* species do not have RISC machinery [10, 11]. To circumvent this problem, we developed an alternative method for silencing genes in this parasite by reconstituting the effector arm of the siRNA pathway using recombinant enzymes and synthetic single-strand RNA (ssRNA) [12].

The major gap addressed by this methodology was the lack of methods to identify novel drug targets, which are desperately needed for the treatment of cryptosporidiosis. The novel aspects of our method include the use of preassembled hybrid complexes of human Ago2 and *Cryptosporidium* ssRNA to suppress gene expression and the transfection of *Cryptosporidium*. The major advantage of our method over existing "Clustered Regularly Interspaced Short Palindromic Repeats" (CRISPR) technology is the simplicity and speed of analysis of the silencing. The use of preloaded Ago–ssRNA complexes "ready-to-silence" completely bypasses the transcription–translation process associated with gene expression, reducing the time until the cells can be assayed from days to less than 2 h. This may allow scale-up of the method for high-throughput analysis [12]. In this chapter, we describe protocols for gene silencing in *Cryptosporidium* that include (i) in vitro assembling of ssRNA/Ago complexes, (ii) parasite transfection, and (iii) silencing experiments. We also describe the experiments necessary to evaluate anti-cryptosporidial activity of ssRNA/Ago complexes.

***Gene Silencing in Cryptosporidium*** Previous studies demonstrated that human Argonaute (hAgo2) associated with human ssRNA-induced specific degradation of mRNA targets in vitro [8, 13]. We hypothesized that recombinant hAgo2 could be pre-loaded with *Cryptosporidium* ssRNA to form a minimal RISC complex and introduced into live parasites to specifically target

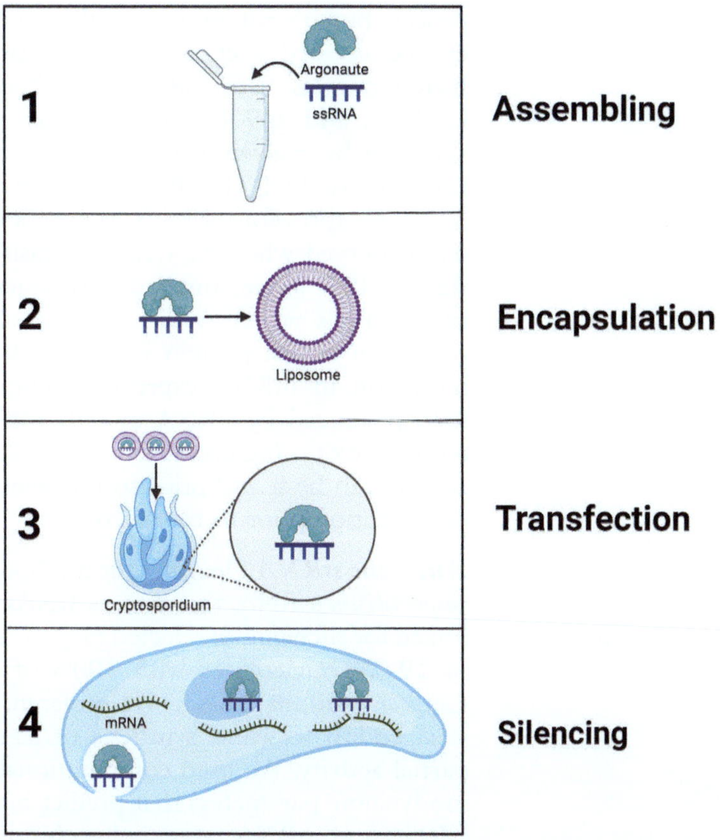


**Fig. 1** Silencing genes in *Cryptosporidium.* Assembling (**a**): hAgo2 protein is loaded with *Cryptosporidium* ssRNA complementary to mRNA target. Encapsulation (**b**): hAgo2–ssRNA complex is encapsulated within liposomes (protein transfection reagent). Transfection (**c**): *Cryptosporidium* oocysts are transfected with complexes. Silencing (**d**): hAgo2–ssRNA binds to mRNA target, translation is blocked, and mRNA target is sliced by hAgo2; then, expression of target is reduced. (Figure created in BioRender. Ortega-Méndez, J. (2024) BioRender.com/h10y771)

mRNA. The result would be degradation of specific mRNAs due to the slicer activity of enzyme Ago2 and downregulation of the targeted protein (*see* Fig. 1). Since bioactive proteins can be transported efficiently into cells using protein transfection reagents (PTR) [14, 15], we demonstrated the feasibility PTR to introduce labeled proteins to parasites within oocysts under excystation [12]. These initial experiments also showed that transfection of ssRNA/Ago complexes into parasites did not affect parasite viability or sporozoites excystation. Then, we investigated the feasibility of silencing specific genes in *Cryptosporidium* [12]. In initial experiments, we selected sporozoite genes as targets: *C. parvum* surface protein of 15 kDa (Cp15), calcium-dependent kinase 1 (CDPK1), and GP900 [12]. These proteins play key roles during the invasion process but are not involved with excystation. Thus, gene silencing can be assessed after excystation in the invasion stage. We transfected oocysts with ssRNA–hAgo2 complexes targeting Cp15, CDPK1, and gp900, and 4 h later, we evaluated the effect of silencing by analyzing the expression of mRNA target by reverse transcriptase PCR (RT-PCR). All three ssRNAs–hAgo2 complexes reduced expression of the target genes (~70–90%). No reduction was observed when we treated parasites with scrambled ssRNA/Ago or PRT alone. In these experiments, we conducted Western blot analysis which confirmed that silencing correlates with the reduction of the protein [12]. We demonstrated up to 70–90% reduction in mRNA expression, comparable with the silencing level produced by siRNA in cells with intact RISC proteins. An advantage of using preloaded complexes is that assembly and slicer activity can be tested prior to transfection, which allows for initial optimization prior to transfection.

***Antisense ssRNA Design*** Our method implicates the use of *Cryptosporidium* ssRNA; therefore, ssRNAs must be designed and optimized for silencing of selected targets. We design ssRNA sequences of 19–29 nucleotides with 100% of complementarity to target *Cryptosporidium* mRNA. We preferentially target regions close to start codon to avoid synthesis of large peptides that may show partial activity. We used computational software to calculate thermodynamic parameters that predict binding probabilities between ssRNA and mRNA target (oligo binding energy). The secondary structure of mRNA target is analyzed to confirm the accessibility of ssRNAs. To calculate thermodynamical parameters, we use Sfold a software for statistical folding of nucleic acid [16]. To analyze secondary structure of *Cryptosporidium* mRNA, we use RNAfold software (Institute for Theoretical Chemistry, University of Vienna) [17]. All designs are based on computational predictions and silico analysis. To optimize activity, we developed 3–5 ssRNAs per target. All ssRNAs were compared in vitro to determine silencing levels; optimal silencing levels range from 80% to 99% [18].

***Parasite Transfection*** Lipofection uses cationic lipids that comprise hydrophilic and hydrophobic domains, enabling interaction between the DNA and the cell membrane that results in uptake by endocytosis [19]. We have demonstrated that oocysts and intracellular parasites are susceptible to protein transfection using a commercial cationic lipid-based carrier system (protein transfection reagent [PTR] Pierce™, Thermo Fisher Scientific) [12, 18, 20, 21]. This kind of liposome attaches to negatively charged surfaces and fuses directly to membranes. Since *Cryptosporidium* organisms do not have caveolins, most likely the ssRNA/Ago complexes are introduced by clathrin-mediated endocytosis where the liposomes are fused with parasitophorous membrane and surface membranes of extracellular forms (e.g., merozoites). Confocal studies confirmed that ssRNA/Ago complexes cross-membranes of intracellular and extracellular parasites [21]. Therefore, *Cryptosporidium* transfection can be conducted in permeabilized oocyst before excystation to study molecules involved in attachment.

***Biological Assays to Evaluate Silenced Genes*** To evaluate the role of targeted genes in a biological process, we have developed assays to study excystation, invasion, proliferation, and egress in vitro during the infection of a human epithelial cell line (HCT-8) cultured in the laboratory [18]. The hypothesis was that silencing of essential genes would block excystation, entrance, proliferation, or egress of the parasite to the cell which would lead into a reduction in the number of parasites in HCT-8. Previous reports showed that inhibition (with antibodies and drugs) of *Cryptosporidium* surface protein Cp15 and calcium-dependent kinase 1 (CDPK1) lead to a reduction of parasite invasion in host cells [22, 23]. We tested ssRNA/Ago complexes targeting these two genes. As predicted, silencing of Cp15 and CDPK1 led to a 70% and 60% reduction in the number of parasites found inside host cells compared with infection with non-transfected parasites [12]. By contrast, silencing of other targets (e.g., gp900) did not reduce parasite numbers. Overall, these data demonstrate that our methodology could be used to evaluate the phenotype of targeted genes during infection. In subsequent studies, we knocked down expression of the four selected genes: actin (Act), Apicomplexan DNA-binding protein (AP2), rhomboid protein 1 (ROM 1), and nucleoside diphosphate kinase (NDK) [18]. After gene silencing, we evaluated the role of each target on parasite development using in vitro models of excystation, invasion, proliferation, and egress. For these studies, we evaluated the number of parasites by RT-PCR and flow cytometry in differing biological samples, for example, excysted sporozoites, supernatants (2, 24 h), and lysates of infected cells. We demonstrated that silencing of Act, AP2, NDK, and Rom1 reduced invasion, proliferation, and egress of *Cryptosporidium*. However, silencing of NDK markedly inhibited *Cryptosporidium* proliferation (~70%) [18]. We also demonstrated that silencing SUB1 and PKG allowed parasite invasion and proliferation, but blocked parasites egress from host cells [20, 24].

***Gene Silencing Therapeutics*** Currently, nitazoxanide is the only drug FDA approved for the treatment of cryptosporidiosis and it does not work well in the groups at highest risk of adverse outcomes. A number of groups have worked on developing effective compounds for the treatment of cryptosporidiosis [25–27]. Most have failed to progress to clinical trials due to unanticipated adverse effects [28]. We considered the use of RNA interference (RNAi) as a drug. Compared to other small-molecule drugs or antibody-based treatments, ssRNA/Ago complexes can potentially have improved specificity determined by complementary base pairing. Transcriptomic data from *Cryptosporidium hominis* and *C. parvum* is available and can be compared to human data. Then, ssRNA can be designed for potential targets and be tested in vitro for anti-cryptosporidial activity. Recent studies support the feasibility of using RNAi against infectious agents in vivo [29].

Previous studies demonstrated that siRNA encapsulated in lipid nanoparticles is efficiently delivered orally to mouse intestines [30]. We hypothesized that ssRNA/Ago could be delivered in vivo using cationic lipids previously tested in our in vitro experiments. We used labeled ssRNA/Ago complexes to track nucleic acid delivery on mice intestine; after treatment with encapsulated and free ssRNA, we conducted experiments to evaluate ssRNA uptake by intestinal cells. These experiments confirmed that cationic lipids enhance significantly delivery of complexes into intestinal epithelial cells, the host cells for *Cryptosporidium* [21]. To evaluate anti-*Cryptosporidium* activity in vivo, we used an immunocompromised mouse model to test whether treatment with nucleoside diphosphate kinase (NDK)-ssRNA/Ago could reduce acute infection. Our results demonstrated that oral treatment with complexes led to a significant reduction of parasites by day 3 in all mice, with four of five mice showing no infection [21]. However, further optimization of dose, schedule, and duration of therapy may be needed prior to initiating studies in large animal models.

Multiple experimental studies have confirmed that siRNAs can be used therapeutically to block the synthesis of disease-causing proteins and to reduce proliferation of infectious agents [31]. Chemically synthesized siRNAs are relatively simple and quick to generate and scale up can reduce the costs for its manufacturing [32]. Thus, ssRNA/Ago complexes targeting essential proteins could be an alternative to treat cryptosporidiosis. However, there are obstacles to overcome. Delivery of RNA–protein complexes by the oral route needs to be optimized in order to address potential hurdles, including stability of the RNA and cellular uptake. Delivery of orally administered antibodies using nanoparticles have been demonstrated [33]. In a recent study, we used enteric capsules to deliver drug–antibody complexes in mice intestines. Thus, enteric capsules or next-generation liposomes could be explored as an alternative to deliver RNA/Ago complexes to intestinal tissues.

## 2 Materials

1. Recombinant protein human Argonaute 2 hAgo2 from human (Sino Biologicals, North Wales, PA).
2. RNase-free water.

3. Assembling buffer: 30 mM 4-(2-hydroxyethyl)-1-piperazineethanesulfonic acid (HEPES), pH 7.4, 150 mM KOAc, 2 mM $Mg(OAc)_2$, 5 mM dithiothreitol (DTT).
4. Protein transfection reagent (PTR): Pierce™ Protein Transfection Reagent (Thermo Fisher Scientific Inc., Waltham, MA).
5. HEPES buffered saline (HBS) for PTR: 10 mM HEPES, 150 mM NaCl, pH 7.0.
6. Acidic water: $H_2O$ at pH 2.5–3.
7. Excystation media: RPMI-1640 medium, 1× antibiotic/antimycotic mix, and 0.8% taurocholic acid sodium salt hydrate. The media is filtered through a 22-μm filter.
8. qRT-PCR Kit: qScriptTM One-Step SYBR® Green qRT-PCR Kit, Low ROXTM (Quanta BioSciences/VWR, Radnor, PA).
9. PCR amplification plate: MicroAmp® Fast 96-well (Applied Biosystems, Foster City, CA).
10. PCR machine: qPCRs in a 7500 Fast Real-Time PCR System (Applied Biosystems).
11. PCR optical adhesive film for 96-well PCR plates.
12. RNA purification kit: RNeasy Plus Mini Kit (QIAGEN, Germantown, MD).

## 3 Methods

### 3.1 Preparation and Design of Single-Stranded RNA (ssRNA)

1. We design the ssRNA using computational software and synthetize the ssRNA from a commercial vendor (*see* **Note 1**).
2. Resuspend ssRNA at 100 mM with RNase-free water (*see* **Note 2**).
3. Dilute the sample 1:10 with RNase-free water and store at −20 °C until use.
4. Single-strand (ssRNA) sequences used as controls for silencing experiments are indicated in Table 1.

### 3.2 Complex Assembling (ssRNA-hAgo2)

1. Linearize the ssRNA: incubate the tube (with the diluted ssRNA) for 1 min at 95 °C.
2. Assembling ssRNA–hAgo2 complexes (Table 2): mix in a micro centrifuge tube 250 ng (2.5 μL) of recombinant hAgo2 (*see* **Note 3**) and 2.5 μL of the respective ssRNA previously diluted (*see* Subheading 3.1); add 15 μL of the assembling buffer and mix well with pipette (triturate).
3. Incubate the sample at room temperature for 1 h.

**Table 1**
**List of synthetic antisense RNAs**

| Single-strand antisense RNA | Sequence |
|---|---|
| ssCDPK1[a,b] | 5′ rCrArCrUrUrCrCrUrCrUrCrUrCrUrCrCrUrCrCrC TT 3′ |
| ssActin[b,c] | 5′ rCrArUrArUrGrArArCrUrArCrCrArGrArUrGrGrC 3′ |
| ssAp2[b,d] | 5′ rGrCrUrArGrUrGrGrArUrArUrArGrGrGrArUrGrC 3′ |
| ssNDK[b,e] | 5′ rArArArCrUrArGrCrCrCrArCrArArUrCrCrArCrG 3′ |
| ssRom1[f] | 5′rUrArArCrGrArCrUrArCrUrUrGrArArUrGrGrG TT 3′ |
| Scrambled (negative control)[g] | 5′ rGrGrGrUrCrArCrGrUrGrCrGrArArCrArUrGrA 3′ |

[a]ssCDPK1: ssRNA for calcium-dependent kinase 1
[b]ssRNA sequences that induce silencing >70% by RT-PCR
[c]ssActin: ssRNA for actin
[d]ssAp2: ssRNA for Apicomplexan transcription factor 2
[e]ssNDK: ssRNA for nucleoside diphosphate kinase
[f]ssROM1: ssRNA for rhomboid protein 1
[g]ssRNA scr: scrambled RNA used as negative control

**Table 2**
**Assembling reaction**

| Components | Positive control (μL) | Negative control (μL) | Sample (μL) |
|---|---|---|---|
| ssRNA (1:10) | 2.5 (*see* Table 1) | 2.5 (*see* Table 1) | 2.5 |
| Human Argonaute protein 2 | 2.5 (250 ng) | 2.5 (250 ng) | 2.5 (250 ng) |
| Assembling buffer | 15 | 15 | 15 |
| Total volume | 20 | 20 | 20 |

**Table 3**
**Encapsulation reaction**

| Components | Positive control (μL) | Negative control (μL) | Sample (μL) |
|---|---|---|---|
| Assembling reaction components | 20 | 20 | 20 |
| Protein transfection reagent | 15 | 15 | 15 |
| Adjust to 35 μL with RNase-free water | 0 | 15 | 15 |
| Total volume | 35 | 35 | 35 |

### 3.3 SsRNA–hAgo2 Complex Encapsulation with Protein Transfection Reagent (PTR)

1. Add 15 μL of PTR into the 1.5-mL tube that contains the assembling reaction (Table 3) prepared as above (*see* Subheading 3.2) and mix the sample by triturating.
2. Incubate the sample for 30 min at room temperature.

**Table 4**
**Transfection reaction**

| Components | Positive control (μL) | Negative control (μL) | Sample (μL) |
|---|---|---|---|
| Parasites | $5 \times 10^5$ (5 μL) | $5 \times 10^5$ (5 μL) | $5 \times 10^5$ (5 μL) |
| Acidic water, pH 2.5 | 25 | 25 | 25 |
| Excystation media | 250 | 250 | 250 |
| Total volume | 280 | 280 | 280 |

### 3.4 SsRNA/Ago Transfection in Pre-Sensitized Oocysts

1. Transfer $5 \times 10^5$ *Cryptosporidium* oocysts (*see* **Note 4**) to a 1.5-mL tube, add equal volumes of water (free RNase), and triturate.
2. Centrifuge the sample at 500× *g* for 10 min. Discard the supernatant in a plastic container containing ~250 mL of water containing 6% bleach.
3. Resuspend the pellet in the required amount of RNase-free water. For each tested sample, use the ration of $5 \times 10^5$ oocysts resuspended in 5 μL and then multiply this volume for the total number of samples. For example, for 20 samples, a total volume of water added to the pellet would be 5 μL × 20 = 100 μL.
4. For transfection reaction (Table 4): Transfer $5 \times 10^5$ oocysts (5 μL) to a reaction tube (1.5-mL tube) that contains the assembled complex and the protein transfection reagent described above. To allow permeabilization and transfection, incubate the sample for 2 h at room temperature.
5. To activate the silencing, incubate the sample for 2 h at 37 °C.
6. In preparation for RNA extraction or excystation assays, centrifuge parasites for 1 min at maximum speed and discard the supernatant.
7. Silencing analysis is performed by extracting the total RNA. Lyse the parasites for RNA extraction by adding 350 μL of RLT buffer contained in the RNeasy Plus Mini Kit as shown in Subheading 3.6 (*see* **Note 5**).

### 3.5 Parasite Excystation

1. Resuspend the pellet of treated parasites in 25 μL of acidic water (pH 2.5) and incubate on ice for 10 min.
2. Add 250 μL of excystation media to the tube with the parasites.
3. Activate the excystation by incubating the samples for 1 h at 37 °C.

### 3.6 RNA Extraction

1. Resuspend transfected oocysts in 350 μL of RLT buffer included in the RNA isolation kit (*see* **Note 6**).
2. Before RNA extraction, heat the sample for 2 min at 95 °C.

3. Removed genomic DNA by transferring the heated 350 μL to the genomic DNA (gDNA) separator columns provided in the RNA extraction kit.
4. Centrifuge the samples for 30 s at >8000× *g* and recover the elution in a collection tube. Discard the genomic DNA column and save the flow-through in collection tube for further processing.
5. Add 350 μL ethanol (70%) to the flow-through in the collection tube and mix well (pipette up/down).
6. After mixing, transfer a maximum of 700 μL to the RNeasy column and centrifuge the sample at >8000× *g* for 30 s. Discard the elution from collection tube (provided in the kit).
7. Place column in elution tube.
8. Wash the column with 700 μL of washing membrane-bound RNA (RWI) buffer and centrifuge as before. Discard the elution.
9. Wash the column with 500 μL of RNA Pre-Elution (RPE) buffer and centrifuge as before. Discard the elution.
10. Wash the column a final time with 500 μL RPE buffer, but then centrifuge at >8000× *g* for 2 min. Discard the eluate.
11. Transfer the empty RNeasy column to a new collection tube and centrifuge for 1 min at maximum speed to remove ethanol traces.
12. Transfer the RNeasy column to a new 1.5-mL microcentrifuge tube and add 100 μL of RNase-free water to detach and elute the RNA from the silica membrane.
13. Measure the RNA concentration of the sample by spectrophotometry or using a NanoDrop.
14. Dilute RNA samples at 5 ng/μL (*see* **Note 7**).

### 3.7 Primer Design for mRNA Target Amplification by RT-PCR

1. For designing primers, we use standard computational software (*see* **Note 8**). Targets and primers sequences of ssRNAs used as controls for silencing are indicated in Table 5.

### 3.8 Evaluation of Silencing in Cryptosporidium by RT-qPCR

1. We place all RT-PCR reagents (Table 6) and RNA sample on ice.
2. Add a total of 2 μL of purified RNA template to a 96-well reaction plate (0.1 mL).
3. Prepare RT-PCR mix as follow: 5 μL of the One-Step SYBR Green Master Mix, Low Rox (2×), 0.25 μL of each primer at a 10 μM concentration, 0.25 μL of the reverse transcriptase, and 4.25 μL of nuclease-free water for a total of 10 μL of mix per sample.

**Table 5**
**Target genes and primer sequences used for RT-PCR analysis**

| Target | Accession number | Primer sequence |
|---|---|---|
| GAPDH[a] | XM_627718 | Forward: 5′-TGA CCC ATC AAA ATT CAA CG-3′<br>Reverse: 5′-ACG GAA ACA TCT GGA GTT GG-3′ |
| CP23[a] | XM_625938 | Forward: 5′-CAA TCA GCA ACC AAG CTC AA-3′<br>Reverse: 5′-TTG TTG AGC AGC AGG TTC AG-3′ |
| CDPK1[b] | XM_001388059 | Forward: 5′-GCA GTA GGG AAT ACA GGA ACA A-3′<br>Reverse: 5′-CCG CTA CCA CTC TGA AGA AA-3′ |
| Actin[b] | FX115681 | Forward: 5′-CAT TGC TTT AGA TTA CGA GGA AGA-3′<br>Reverse: 5′-TTG GAA TAA AGC CTC AGG ACA-3′ |
| Ap2[b] | XM_627219 | Forward: 5′-TAA ACA AGC CTG AGC CAA CC-3′<br>Reverse: 5′-TCG GTT CCA ATT TCA TGT CC-3′ |
| NDK[b] | FX115659 | Forward: 5′-GCA ACT CGA CTG AAT CGT CA-3′<br>Reverse: 5′-CTT TCC TGC ATG ATC GGA AT-3′ |
| GP900[c] | AF068065 | Forward: 5′-GCA CCATTT GAG TTG GAT GTT AG-3′<br>Reverse: 5′-AGT TGG GAG ACC AGT GAT AGA-3′ |

[a]Primers for housekeeping genes (non-silenced) used for normalization
[b]Primers for positive ssRNA controls used for silencing
[c]Primers for nontargeted gene during silencing (i.e., silencing with ssNDK does not affect GP900 expression in treated parasites versus wild type parasites.)

**Table 6**
**Evaluation of silencing by RT-PCR**

| Components | Positive control (μL) | No RNA control (μL) | Sample (μL) |
|---|---|---|---|
| RNA sample | 2 (*see* Table 1) | 2 (*see* Table 1) | 2 |
| One-Step SYBR Green Mix, Low ROX (2X) | 5 | 5 | 5 |
| qScript One-step reverse transcriptase | 0.25 | 0.25 | 0.25 |
| Forward primer | 0.25 | 0.25 | 0.25 |
| Reverse primer | 0.25 | 0.25 | 0.25 |
| Nuclease-free water | 4.25 | 4.25 | 4.25 |
| Total volume | 12 | 12 | 12 |

4. After RNA is loaded, add 10 μL of the reaction mix to each well containing the 2 μL of RNA sample, for a total of 12 μL per well.
5. Seal the plate with PCR optical adhesive film.
6. Run the reactions on a real-time PCR system with the following cycling conditions:

   (a) 50 °C for 15 min
   (b) 95 °C for 5 min
   (c) (95 °C for 15 s, 63 °C for 1 min) for 50 cycles
   (d) Melting point analysis (95 °C for 15 s, 60 °C for 1 min, 95 °C for 15 s, and 60 °C for 15 s).

7. Normalize the results for each target with a housekeeping gene Glyceraldehyde-3-phosphate dehydrogenase (GAPDH, *see* Table 1) to probe for parasite counts.
8. For analysis, use the "Delta-Delta threshold cycle" (ddCt) calculation method to compute the relative gene expression in all samples. Results are shown by the average for each target with the respective standard deviation. The calculation steps are as follows:
   (a) Determine the averages and standard deviation from the PCR replicates for all samples.
   (b) Normalize all threshold cycles (Ct) values of the reference gene (e.g., GAPDH) in treated samples (ssRNA/Ago) and control samples (untreated parasites or treated with scramble ssRNA or unrelated ssRNA):

$$dCt_{GAPDH} = Ct_{GAPDH}\ Control - Ct_{GAPDH}\ treated$$

   (c) Calculate the ddCt by subtracting the $dCt_{Target}$ Control from the $dCt_{Target}$ Gene.

$$ddCt_{Target}\ Gene = dCt_{Target}\ Gene - dCt_{Target}\ Control$$

   (d) Calculate the fold change. FC = 2^- $ddCt_{Target}$.

### 3.9 Labeled Complexes Delivered to Mouse Intestines

1. Assemble selected *Cryptosporidium* ssRNA/Ago complexes as described in Subheading 3.2 but using Fluorescein amidites (FAM)-labeled ssRNA.
2. Encapsulate complexes in PTR (as in Subheading 3.3) containing FAM-labeled ssRNA/Ago (ssRNA [100 nM]/Ago [125 ng] diluted in 100 μL of phosphate-buffered saline [PBS]).
3. Dilute encapsulated ssRNA–FAM complexes in 100 μL of PBS.
4. Load the diluted complexes to a 1-mL syringe and place the oral gavage needle.
5. Place the gavage needle in the mouse's mouth and then gently advance it along the upper palate until it reaches the esophagus.
6. Once the tube is in place, attach a syringe and inject the ssRNA complexes (100 μL).
7. Gently remove the needle.
8. To evaluate ssRNA delivery, the mice are sacrificed 4 h after the oral dose and the terminal ileum is collected.

9. 2–4-mm sections of ileum were transferred to PBS-ethylenediaminetetraacetic acid (EDTA) solution.
10. Intestinal crypts are isolated by incubating ileum samples in 30 mL of crypto isolation media (2 mM EDTA in PBS). After incubation, the sample is vortexed vigorously for 5 min.
11. Crypts are washed twice with PBS and fixed with 4% paraformaldehyde in PBS for 30 min.
12. Quantify cellular uptake of FAM-labeled ssRNA/Ago by fluorometric assays.
13. For fluorometry: 500 crypts are placed in 5 mL of PBS.
14. Cells were washed twice with PBS and lysed in 400 μL lysis buffer (1% Triton X-100, 2% SDS in PBS) on ice for 30 min.
15. Lysates are centrifuged (15 min, 14,000× *g*, 4 °C) to remove cell debris.
16. To evaluate fluorescence in intestinal crypts, 200 μL of the supernatant is transferred to a black 96-well plate to measure the fluorescence using a BMG FLUOstar Omega Microplate Reader (BMG Labtech Inc., Cary, NC).
17. For determination of the mean fluorescence intensity, fluorescent signals are corrected for the amount of protein in the samples.

### 3.10 ssRNA/Ago Treatment in Mice Model and Parasite Quantification by qPCR

1. Four days after infection, mice are treated with ssRNA–Ago complexes (prepared as in **step 3.9**) by oral gavage daily for 3 days.
2. After infection, stool samples (three to five pellets) are collected every day during acute phase of infection (15 days).
3. For parasite quantification, approximately 20 mg of stool (two pellets) resuspended in 1 mL of water and sample is vortexed vigorously for 5 min.
4. The samples are frozen at −20 °C until DNA extraction.
5. DNA from frozen samples is extracted using the QIAamp Fast DNA Stool Mini Kit (QIAGEN).
6. After extraction, the concentration of DNA in samples is determined by spectrophotometry.
7. The parasite burden is determined by qPCR using a Real-Time PCR System.
8. For qPCR, we use iTaq Universal SYBR Green Supermix Kit (Bio-Rad, Hercules, California) with 18 s primers for *C. parvum* as described.
9. For quantification, a standard curve is generated from serial dilutions of DNA extracted from a known number of parasites ($1 \times 10^6$) spiked in mouse stool. Total numbers of parasites are calculated extrapolating experimental CT values to standard curve.

## 4 Notes

1. We used computational software to design ssRNA of 17–23 nt complementary to *Cryptosporidium* mRNA targets (Table 1). We synthetize the ssRNA with the following modifications: 5′ prime is phosphorylated, and 3′ prime is tailed with deoxythymidine (dTdT). Negative controls may include (17–22 nt) unrelated ssRNA sequences, sense RNA, or scrambled RNA. Positive controls are listed in Table 1.
2. To prepare ssRNA stock at 100 mM: centrifuge the tube with the lyophilized ssRNA for 5 min at maximum speed. To calculate the amount of water to resuspend ssRNA at 100 mM, first, we identify the number of nmol indicated in the tube, and then, we use the following formula: number of mM = μL of RNase-free water to be added. To prepare ssRNA: centrifuge the tube with the lyophilized ssRNA for 5 min at maximum speed and resuspend the ssRNA at 100 mM. To calculate the amount of water, we use the number of nmol indicated in the tube and multiplied it for 10; then, we add the same amount (μL) of RNase-free water.
3. For assembling protocol, resuspend the lyophilized hAgo2 at 125 mg/μL with 160 μL ultra-pure RNase-free water. Aliquot the protein and store at −80 °C until use.
4. For silencing experiments, we recommend using fresh *C. parvum* oocysts (up to 1 month old). On the day of the experiments, the viability of the oocyst is confirmed by excystation assay and observing motility of the sporozoites by microscopy. We only use oocysts with >80% excystation rate. For the transfection experiments, the parasites can be prepared during the encapsulation time (30 min) as described in the encapsulation section (*see* Subheading 3.3). The stock of *Cryptosporidium* oocysts ($1 \times 10^8$) is diluted in 1 mL of PBS/Tween 20 solution. We calculate the total number of parasites for our experiment as follows: For each tested sample, we use $5 \times 10^5$ oocysts; then, we multiply this number by the total number of treated samples (including duplicates, untreated samples, and negative controls). For example, the total number of oocysts needed for 20 samples would be calculated as follows: ($5 \times 10^5$ oocysts) $\times$ 20 = $100 \times 10^5$ = $1 \times 10^7$ total oocysts.
5. For silencing experiments, the effect can be evaluated before the excystation and/or after excystation. If the RNA is isolated after the excystation, then skip to the excystation section (*see* Subheading 3.5).

6. RNA extraction is conducted with the QIAGEN's RNeasy Plus Mini Kit following the instructions listed in the quick protocol included in the kit.
7. After extraction, the sample can be stored at −20 °C until qPCR analysis.
8. For designing, we analyze full *Cryptosporidium* mRNA sequences obtained from GenBank NCBI. For primer positions, we select regions that flank a 17–22 nucleotide slicing region. Primers for qPCR of positive controls are indicated in Table 5.

## Acknowledgments

This work was supported by grants from the National Institutes of Health (R21AI12627502, R21AI151901) and the Bill & Melinda Gates Foundation (OPP1161026).

**Competing Interest Statement** The authors have no conflict of interests to declare that are relevant to the content of this chapter.

### References

1. Agrawal N, Dasaradhi PV, Mohmmed A, Malhotra P, Bhatnagar RK, Mukherjee SK (2003) RNA interference: biology, mechanism, and applications. Microbiol Mol Biol Rev 67(4):657–685. https://doi.org/10.1128/MMBR.67.4.657-685.2003
2. Fire A, Xu S, Montgomery MK, Kostas SA, Driver SE, Mello CC (1998) Potent and specific genetic interference by double-stranded RNA in *Caenorhabditis elegans*. Nature 391(6669):806–811. https://doi.org/10.1038/35888
3. Batista TM, Marques JT (2011) RNAi pathways in parasitic protists and worms. J Proteome 74(9):1504–1514. https://doi.org/10.1016/j.jprot.2011.02.032
4. Perwitasari O, Bakre A, Tompkins SM, Tripp RA (2013) siRNA genome screening approaches to therapeutic drug repositioning. Pharmaceuticals (Basel) 6(2):124–160. https://doi.org/10.3390/ph6020124
5. Manzano-Roman R, Oleaga A, Perez-Sanchez-R, Siles-Lucas M (2012) Gene silencing in parasites: current status and future prospects. Adv Parasitol 78:1–55. https://doi.org/10.1016/B978-0-12-394303-3.00002-5
6. Fopase R, Panda C, Rajendran AP, Uludag H, Pandey LM (2023) Potential of siRNA in COVID-19 therapy: emphasis on in silico design and nanoparticles based delivery. Front Bioeng Biotechnol 11:1112755. https://doi.org/10.3389/fbioe.2023.1112755
7. Chang YC, Yang CF, Chen YF, Yang CC, Chou YL, Chou HW, Chang TY, Chao TL, Hsu SC, Ieong SM, Tsai YM, Liu PC, Chin YF, Fang JT, Kao HC, Lu HY, Chang JY, Weng RS, Tu QW, Chang FY, Huang KY, Lee TY, Chang SY, Yang PC (2022) A siRNA targets and inhibits a broad range of SARS-CoV-2 infections including Delta variant. EMBO Mol Med 14(4):e15298. https://doi.org/10.15252/emmm.202115298
8. Rivas FV, Tolia NH, Song JJ, Aragon JP, Liu J, Hannon GJ, Joshua-Tor L (2005) Purified Argonaute2 and an siRNA form recombinant human RISC. Nat Struct Mol Biol 12(4):340–349. https://doi.org/10.1038/nsmb918
9. Lye LF, Owens K, Shi H, Murta SM, Vieira AC, Turco SJ, Tschudi C, Ullu E, Beverley SM (2010) Retention and loss of RNA interference pathways in trypanosomatid protozoans. PLoS Pathog 6(10):e1001161. https://doi.org/10.1371/journal.ppat.1001161
10. Xu P, Widmer G, Wang Y, Ozaki LS, Alves JM, Serrano MG, Puiu D, Manque P, Akiyoshi D,

Mackey AJ, Pearson WR, Dear PH, Bankier AT, Peterson DL, Abrahamsen MS, Kapur V, Tzipori S, Buck GA (2004) The genome of *Cryptosporidium hominis*. Nature 431(7012): 1107–1112. https://doi.org/10.1038/nature02977

11. Abrahamsen MS, Templeton TJ, Enomoto S, Abrahante JE, Zhu G, Lancto CA, Deng M, Liu C, Widmer G, Tzipori S, Buck GA, Xu P, Bankier AT, Dear PH, Konfortov BA, Spriggs HF, Iyer L, Anantharaman V, Aravind L, Kapur V (2004) Complete genome sequence of the apicomplexan, *Cryptosporidium parvum*. Science 304(5669):441–445. https://doi.org/10.1126/science.1094786
12. Castellanos-Gonzalez A, Perry N, Nava S, White AC Jr (2016) Preassembled single-stranded RNA-Argonaute complexes: a novel method to silence genes in *Cryptosporidium*. J Infect Dis 213(8):1307–1314. https://doi.org/10.1093/infdis/jiv588
13. Lima WF, Prakash TP, Murray HM, Kinberger GA, Li W, Chappell AE, Li CS, Murray SF, Gaus H, Seth PP, Swayze EE, Crooke ST (2012) Single-stranded siRNAs activate RNAi in animals. Cell 150(5):883–894. https://doi.org/10.1016/j.cell.2012.08.014
14. Chiang WC, Geel TM, Altintas MM, Sever S, Ruiters MH, Reiser J (2010) Establishment of protein delivery systems targeting podocytes. PLoS One 5(7):e11837. https://doi.org/10.1371/journal.pone.0011837
15. Weill CO, Biri S, Adib A, Erbacher P (2008) A practical approach for intracellular protein delivery. Cytotechnology 56(1):41–48. https://doi.org/10.1007/s10616-007-9102-3
16. Ding Y, Chan CY, Lawrence CE (2004) Sfold web server for statistical folding and rational design of nucleic acids. Nucleic Acids Res 32 (Web Server issue):W135–W141. https://doi.org/10.1093/nar/gkh449
17. Schramm G, Ramey R (2005) siRNA design including secondary structure target site prediction. Nat Methods 2(8):632–632. https://doi.org/10.1038/nmeth780
18. Castellanos-Gonzalez A, Martinez-Traverso G, Fishbeck K, Nava S, White AC Jr (2019) Systematic gene silencing identified *Cryptosporidium* nucleoside diphosphate kinase and other molecules as targets for suppression of parasite proliferation in human intestinal cells. Sci Rep 9(1):12153. https://doi.org/10.1038/s41598-019-48544-z
19. Nayerossadat N, Maedeh T, Ali PA (2012) Viral and nonviral delivery systems for gene delivery. Adv Biomed Res 1:27. https://doi.org/10.4103/2277-9175.98152
20. Nava S, White AC Jr, Castellanos-Gonzalez A (2019) *Cryptosporidium parvum* subtilisin-like serine protease (SUB1) is crucial for parasite egress from host cells. Infect Immun 87(5). https://doi.org/10.1128/IAI.00784-18
21. Castellanos-Gonzalez A, Sadiqova A, Ortega-Mendez J, White AC Jr (2022) RNA-based therapy for *Cryptosporidium parvum* infection: proof-of-concept studies. Infect Immun: e0019622. https://doi.org/10.1128/iai.00196-22
22. Castellanos-Gonzalez A, Sparks H, Nava S, Huang W, Zhang Z, Rivas K, Hulverson MA, Barrett LK, Ojo KK, Fan E, Van Voorhis WC, White AC Jr (2016) A novel calcium-dependent kinase inhibitor, bumped kinase inhibitor 1517, cures cryptosporidiosis in immunosuppressed mice. J Infect Dis 214(12):1850–1855. https://doi.org/10.1093/infdis/jiw481
23. Tilley M, Upton SJ (1994) Both cp15 and cp25 are left as trails behind gliding sporozoites of cryptosporidium parvum (apicomplexa). FEMS Microbiol Lett 120(3):275–278
24. Nava S, Sadiqova A, Castellanos-Gonzalez A, White AC Jr (2020) *Cryptosporidium parvum* cyclic GMP-dependent protein kinase (PKG): an essential mediator of merozoite egress. Mol Biochem Parasitol 237:111277. https://doi.org/10.1016/j.molbiopara.2020.111277
25. Janes J, Young ME, Chen E, Rogers NH, Burgstaller-Muehlbacher S, Hughes LD, Love MS, Hull MV, Kuhen KL, Woods AK, Joseph SB, Petrassi HM, McNamara CW, Tremblay MS, Su AI, Schultz PG, Chatterjee AK (2018) The ReFRAME library as a comprehensive drug repurposing library and its application to the treatment of cryptosporidiosis. Proc Natl Acad Sci USA 115(42):10750–10755. https://doi.org/10.1073/pnas.1810137115
26. Chavez MA, White AC Jr (2018) Novel treatment strategies and drugs in development for cryptosporidiosis. Expert Rev Anti-Infect Ther 16(8):655–661. https://doi.org/10.1080/14787210.2018.1500457
27. Wang B, Castellanos-Gonzalez A, White AC Jr (2020) Novel drug targets for treatment of cryptosporidiosis. Expert Opin Ther Targets 24(9):915–922. https://doi.org/10.1080/14728222.2020.1785432
28. Iroh Tam P, Arnold SLM, Barrett LK, Chen CR, Conrad TM, Douglas E, Gordon MA, Hebert D, Henrion M, Hermann D, Hollingsworth B, Houpt E, Jere KC, Lindblad R, Love MS, Makhaza L, McNamara CW, Nedi W, Nyirenda J, Operario DJ, Phulusa J, Quinnan GV, Sawyer LA, Thole H, Toto N, Winter A, Van Voorhis WC (2021)

Clofazimine for treatment of cryptosporidiosis in human immunodeficiency virus infected adults: an experimental medicine, randomized, double-blind, placebo-controlled phase 2a trial. Clin Infect Dis 73(2):183–191. https://doi.org/10.1093/cid/ciaa421

29. Sajid MI, Moazzam M, Cho Y, Kato S, Xu A, Way JJ, Lohan S, Tiwari RK (2021) siRNA therapeutics for the therapy of COVID-19 and other coronaviruses. Mol Pharm 18(6): 2105–2121. https://doi.org/10.1021/acs.molpharmaceut.0c01239
30. Ball RL, Bajaj P, Whitehead KA (2018) Oral delivery of siRNA lipid nanoparticles: fate in the gi tract. Sci Rep 8(1):2178. https://doi.org/10.1038/s41598-018-20632-6
31. Wittrup A, Lieberman J (2015) Knocking down disease: a progress report on siRNA therapeutics. Nat Rev Genet 16(9):543–552. https://doi.org/10.1038/nrg3978
32. Levanova A, Poranen MM (2018) RNA interference as a prospective tool for the control of human viral infections. Front Microbiol 9: 2151. https://doi.org/10.3389/fmicb.2018.02151
33. Tashima T (2021) Delivery of orally administered digestible antibodies using nanoparticles. Int J Mol Sci 22(7). https://doi.org/10.3390/ijms22073349

# Chapter 12

# Use of ncRNAs to Study Host Cell–Parasite Interactions

**Marion L. Graham and Xian-Ming Chen**

## Abstract

Noncoding RNAs (ncRNAs) are RNA molecules that do not undergo the process of translation into protein. High-throughput sequencing technology has shown that a large portion of the human genome is transcribed into ncRNAs. ncRNAs can be divided into different categories, including two main categories based on their size: microRNAs (miRNAs), which represent a subclass of endogenous small ncRNAs with an average 22 nucleotide length, and lncRNAs, which represent a subclass of ncRNAs over 200 nucleotides in length. Both of which have been identified in both mammalian and nonmammalian cells. miRNAs and lncRNAs are an essential part of the complex regulatory networks that control numerous biological processes and may play an important role in host defense and/or microbial offense during host–parasite interactions. Here, several methodologies to explore the role for miRNAs and lncRNAs in host–parasite interactions are briefly summarized, including the detection, quantification, and intracellular localization of miRNAs, identification and validation of miRNA targets, functional manipulation of specific miRNAs, mapping of lncRNA occupancy genome wide, and detection of the association between lncRNAs and proteins.

**Key words** ncRNAs, lncRNA, miRNA, RNA interference, Host defense, Parasite, PCR, In situ hybridization, Luciferase reporter assay, siRNA, Northern blot, Western blot, ChIRP, RIP

## 1 Introduction

Noncoding RNAs (ncRNAs) are functional RNA molecules that are not translated into proteins. They represent a significant proportion of the human genome and have been shown to play a key role in many cellular processes including cell proliferation and development, gene transcription and translation, epigenetic modifications, and chromatin remodeling [1]. In general, ncRNAs can be classified based on transcript length, with short noncoding transcripts containing less than 200 nucleotides (nt) and long noncoding RNAs (lncRNA) containing greater than 200 nucleotides [1].

Jan R. Mead and Michael J. Arrowood (eds.), *Cryptosporidium: Methods and Protocols*, Methods in Molecular Biology, vol. 2978, https://doi.org/10.1007/978-1-0716-4824-7_12,

MicroRNAs (miRNAs) refer to a subclass of endogenous small noncoding RNAs (~24-nt) [2, 3]. Most miRNAs are initially transcribed as primary transcripts (termed pri-miRNAs) by Poly II and processed by the RNase III Drosha (in the nucleus to produce pre-miRNAs) and a second RNase III Dicer (in the cytoplasm) to generate mature miRNA molecules. In the cytoplasm, miRNAs function via base-pairing with complementary sequences within messenger RNA molecules (mRNAs), mostly at the 3′-untranslated regions (UTR) of mRNAs, resulting in mRNA cleavage or translational suppression [2, 3]. Therefore, miRNAs resemble the small interfering RNAs (siRNAs) of the RNA interference (RNAi) pathway through the Argonaute-containing RNA-induced silencing complex (RISC). It has been proposed that a single miRNA can repress hundreds of target transcripts, and multiple miRNAs may target the same RNA transcript [2, 3].

LncRNAs represent the largest class of ncRNAs over 200 nt in length. Like miRNAs, lncRNAs are predominantly transcribed by Poly II, but unlike miRNAs, most lncRNAs are 5′-capped, poly A-tailed, and post-transcriptionally modified, mostly by splicing [4]. LncRNAs mechanistic function is highly diverse, and many can employ more than one mechanism of action. lncRNAs have been shown to be involved in virtually all levels of genome organization including mRNA transcription, splicing, translation, export, import, and stability [5]. They have been shown to function as recruiters, scaffolds, and molecular decoys for transcriptional factors as well as act as transcriptional coactivators [6, 7]. LncRNAs have also been shown to be involved in chromatin looping and nuclear body formation and function [6]. In addition to transcriptional regulation, lncRNAs can also regulate mRNA splicing, inhibit translation, act as miRNA sponges, and compete for mRNA binding [6]. Furthermore, lncRNAs can regulate protein or transcript trafficking and shuttling [8]. This diverse functionality can be attributed to lncRNAs ability to form complex secondary structures with DNA, RNA, and RNA-binding proteins. Additionally, a single lncRNA may contain multiple binding loops capable of binding to nucleic acids via base-pairing or to proteins via RNA binding motifs further diversifying their functions [4].

Both miRNAs and lncRNAs have been identified in mammalian and nonmammalian cells, including plants, animals, and microorganisms [9, 10]. While lncRNAs play a crucial role in regulating gene expression in parasites [11], not all parasites encode miRNA genes. The lack of coding genes for Dicer and Argonaute proteins (key factors of RISC) in the genome of a number of protozoan parasites, such as *Cryptosporidium parvum*, implies the absence of miRNA-induced regulation of gene expression [12]. Nevertheless, alignment analysis of Argonaute showed that the expression modulation by small RNAs is extensively present across parasitic cestodes and nematodes [13]. These ncRNAs of different lengths, including small- and intermediate-size ncRNAs and lncRNAs, have recently been identified in *Cryptosporidium* spp. [14, 15].

Accumulating data indicate that ncRNAs are an essential part of the complex regulatory networks that control numerous biological processes, including host antimicrobial immune responses and microbial offense. The human genome may encode thousands of ncRNAs which appear to target about 60% of the mRNAs in humans. Host ncRNAs in mammalian cells have been shown to regulate cellular responses to infection by diverse pathogens, including viruses, parasites, and bacteria [16–18]. On the other hand, pathogens, in particular viruses, may deliver their ncRNAs into infected host cells to modulate host cell functions to benefit pathogen replication [17]. Unraveling the regulatory functions of ncRNAs in host–parasite interactions will not only yield new insights into our understanding of the pathogenesis of parasite diseases but also provide a basis for the development of new therapeutic interventions.

Here, we briefly summarize several methodologies which have recently been used to explore miRNAs and lncRNAs in host–parasite interactions, including the detection, quantification, and intracellular localization of miRNAs (via PCR, in situ hybridization, Northern blot), identification and validation of targeted mRNAs by specific miRNAs (luciferase assay and Western blot validation), functional manipulation of each miRNA (loss- and gain-of function methods), genome-wide mapping of lncRNA occupancy (chromatin isolation by RNA purification [ChIRP]), and detection of the association of lncRNAs with individual proteins (RNA immunoprecipitation [RIP]).

## 2 Materials

### 2.1 miRNA Extraction and Quantification Kits

1. miRNA extraction: mirVana miRNA Isolation Kit (Thermo Fisher Scientific Inc., Waltham, MA), ACS grade 100% ethanol.
2. Preparation of solutions from mirVana miRNA Isolation Kit: (A) add 21 mL ethanol to miRNA Wash Solution 1; (B) add 40 mL ethanol to Wash Solution 2/3.
3. RNA reverse transcription: TaqMan MicroRNA Reverse Transcription Kit (Applied Biosystems, Foster City, CA).
4. Quantitative real-time PCR (qRT-PCR): TaqMan Universal PCR Master Mix (Applied Biosystems).

### 2.2 Plasmids and Reagents for Luciferase Reporter Assay

1. Luciferase construct establishment and transfection: pGL3 Basic Vector (Promega), T4 DNA Ligase (Thermo Fisher Scientific), and Lipofectamine™ 2000 (Invitrogen, Catalog number: 11668019, Thermo Fisher Scientific).

2. Luciferase assay: 5X Luciferase Lysis Buffer (PR-E3971, Thermo Fisher Scientific; or Promega E3971, Thermo Fisher Scientific), diluted into 1X luciferase lysis buffer with $ddH_2O$.
3. Assessment of luciferase activity: Luciferase Assay Reagent (Promega PR-E1483, Thermo Fisher Scientific).
4. β-gal assay: 2X β-gal assay buffer: 200 mM sodium phosphate buffer pH 7.3, 2 mM $MgCl_2$, 100 mM β-mercaptoethanol, and 4 mg/mL o-Nitrophenyl-β-D-galactopyranoside (ONPG).

### 2.3 ChIRP Buffers

1. Cross-linking cells: 1% glutaraldehyde and 2.5 M glycine.
2. ChIRP lysis buffer: 50 mM Tris–HCl pH 7.0, 10 mM Ethylenediaminetetraacetic acid (EDTA), Sodium dodecylsulphate (SDS) 1%, and Diethylpyrocarbonate (DEPC) $H_2O$.
3. Hybridization buffer: 750 mM NaCl, Tris–HCl 50 mM pH 7.0, 1 mM EDTA, SDS 1%, formamide 15%, and DEPC $H_2O$.
4. Wash buffer: Saline-Sodium Citrate (SSC) 2X (20X SSC: 3 M NaCl, 0.3 M trisodium citrate; adjust the pH to 7.0 with NaOH or HCl; dilute to 2X SSC with $ddH_2O$), SDS 0.5%, and DEPC $H_2O$.
5. DNA elution buffer: 50 mM $NaHCO_3$, SDS 1%, and DEPC $H_2O$.
6. Dynabeads M-280 Streptavidin (Invitrogen 11205, Thermo Fisher Scientific).
7. Protease inhibitors (PIs): leupeptin (1 mg/mL), pepstatin (1 mg/mL), sodium fluoride (1 mg/mL), Phenylmethylsulfonyl fluoride (PMSF) (1 mg/mL), sodium orthovanadate (1 mg/mL), aprotinin (1 mg/mL), dithiothreitol (DTT) (1 mg/mL), and proteinase K (1 mg/mL).
8. RNase inhibitor (100 U/mL), RNase A (10 mg/mL), and RNase H (5 U/mL).
9. Diluted probes.
10. GlycoBlue™ coprecipitant (Thermo Fisher Scientific).

### 2.4 RIP Buffers

1. Cross-linking cells: 37% formaldehyde and 2.5 M glycine.
2. Whole cell extraction buffer: 20 mM 4-(2-Hydroxyethyl)piperazine-1-ethanesulfonic acid (HEPES), pH 7.4, 0.2 M NaCl, Triton X-100 0.5%, glycerol 10%, 1 mM ethylene glycol tetraacetic acid (EGTA), 1 mM EGTA, 1 mM DTT, and DEPC $H_2O$.
3. Protease inhibitors: 1 mM PMSF, 1 mM NaF, $Na_3VO_2$, 1 mM pepstatin A (1 μg/mL), leupeptin (1–2 μg/mL), and aprotinin (1–2 μg/mL).
4. Magnetic beads (Millipore, 16-663).

5. IgG antibody (depending on the species needs) and the desired antibody.
6. ChIRP proteinase K buffer: 100 mM NaCl, 10 mM Tris–HCl, pH 7.0, 1 mM EDTA, and 0.5% SDS.
7. DNase Enzyme Kit (90083, Thermo Fisher Scientific).

## 3 Methods

### 3.1 miRNA Extraction and Quantification by Real-Time PCR (qRT-PCR) (See Note 1)

While some pri-miRNAs or pre-miRNAs can be up to thousands of nucleotides long, the mature miRNAs are quite small (~24 nt). Therefore, it is critical to retain the small RNA fraction in the RNA extraction, particularly for mature miRNA detection from either the host or parasites. Total RNA extraction is usually adequate as it can retain both long and small RNA transcripts. Accordingly, qRT-PCR can quantify the amount of miRNA transcripts, pri-miRNA, pre-miRNA, or mature miRNAs, depending on the specificity of the designed PCR primers [19]. For example, the TaqMan™ microRNA Assays are qRT-PCR assays and are designed to detect and accurately quantify mature miRNAs by annealing a short RT priming sequence to the 3′ portion of the miRNA [20]. Many of such PCR primers are commercially available. Material and methods have been previously reported [19].

### 3.2 In Situ Hybridization for Detection of miRNAs

In situ hybridization has been used to characterize the intracellular distribution of miRNAs, either the pre-miRNAs or mature form, depending on the specificity of designed probes. Of note, probes for the mature forms should also be able to detect the corresponding pri-miRNAs and pre-miRNAs. In addition, hybridization signals may be weak for these miRNAs if they are expressed at a very low level. Material and methods have been previously reported [21].

### 3.3 Northern Blot

Northern blot can detect the amounts of pri-miRNAs, pre-miRNAs, or mature miRNAs, depending on the specificity of designed probes. In addition, it can detect the ratio of pre-miRNA versus its mature form, reflecting the mature process of a given miRNA [19]. Material and methods have been previously reported [21].

### 3.4 Knockdown miRNAs and miRNA Mimic

Several technologies have been developed to silence miRNAs, including anti-miRNA antisense inhibitor oligoribonucleotide (AMO) technology, multiple-target AMO technology (MT-AMO technology), miRNA sponge technology, miRNA-masking antisense oligonucleotide technology, sponge miR-mask technology, and miRNA knockout technology [22]. Moreover, synthetic siRNA targeting pre-miRNA loop has been reported to knockdown pre-miRNA transcript [22]. The following technologies have been used for achieving gain of function of miRNAs by forced expression or overexpression: synthetic canonical miRNA technology, miRNA mimic technology, multi-miRNA hairpins technology, multi-miRNA mimics technology, and miRNA transgene technology [22]. Most miRNA mimics are small, double-stranded RNA molecules, designed to mimic endogenous mature miRNA molecules when transfected into cells. Antagomirs are powerful tools to silence specific miRNAs *in vivo* [22]. Here, we list the approach to use AMOs to knockdown miRNAs and miRNA mimics to overexpress miRNAs in cultured cells.

1. Query the National Center for Biotechnology Information (NCBI) Gene database (Entrez Gene) to obtain miRNA sequence (http://www.rnaiweb.com/RNAi/siRNA_Design/).
2. Use BLOCK-iT™ RNAi Designer to design specific siRNA (https://rnaidesigner.invitrogen.com/rnaiexpress).
3. Use NCBI basic local alignment search tool (BLAST) for homology search (*see* **Note 2**).

### 3.5 Clustered Regulatory Interspersed Short Palindromic Repeat (CRISPR)/Cas9 lncRNA Knockout and Overexpression Stable Cell Line

A powerful tool for genetic manipulation of lncRNAs is the clustered regulatory interspersed short palindromic repeat (CRISPR) system. The CRISPR system has emerged as a revolutionary tool in molecular biology for genomic editing due to its ease and flexibility. This technique relies primarily on guide RNAs to direct the catalytic activity of the Cas enzymes [23]. The adaptability of the various Cas enzymes allows targeting of the gene locus as well as the RNA transcripts. This versatility presents the ability to target nuclear or cytoplasmic lncRNAs, disrupt the genome to selectively interact with adjacent genetic material, and upregulate or downregulate transcription all without degrading the lncRNA through knockout, knockdown, or overexpression approaches [24]. The methods of using CRISPR to investigate lncRNAs has been previously described, and data using CRISPR lncRNA knockout cells are shown in reference [25].

### 3.6 Identification and Validation of miRNA Targets

Considerable effort has gone into developing different computational algorithms that identify miRNA targets [26]. Currently, there are more than a dozen algorithms which claim to predict miRNA targets, including miRanda, PicTar, MicroTar, RNA22, TargetScanS, and RNAhybrid [27]. These algorithms are based on several parameters of which homology between the 5′-seed sequence of the miRNA and a complementary region in the 3′-UTR of the target mRNA is an important component. However, a major drawback of these prediction models is a substantial false positive rate and an inevitable bias due to reliance on the few known miRNA/target mRNA interactions. Perfectly matched miRNA seeds are neither necessary nor sufficient for all functional miRNA–target interactions [26]. There is a clear need for methodologies that identify and validate the functional targets of specific miRNAs. Experimental methods for target validations are difficult, expensive, and time-consuming. Here, we briefly describe the luciferase reporter assay to identify miRNA targets based on computational algorithm prediction, followed by validation using real-time PCR (at RNA level) and Western blot (at the protein level) for the targets. There are many other technologies for miRNA target identification and validation, such as a functional assay based on a 3'UTR-enriched library and a positive/negative selection strategy [28].

#### 3.6.1 Luciferase Reporter Constructs

Predicted region of the target mRNA, for example, based on the above computational algorithm prediction, will be ligated into the pMIR-REPORT Luciferase vector (Ambion). The basic pMIR-REPORT Luciferase vector (empty vector) and the pMIR-REPORT Luciferase construct containing a mutant at the putative seed region of the target mRNA are often used as control. β-gal is commonly used as a control for transfection efficiency. Constructs will be confirmed by sequencing. Material and methods have been previously reported [21].

### 3.7 ChIRP

Previous studies mapping RNA occupancy at chromatin have revealed important insights into the functions of lncRNAs, however, only a single gene locus at a time [29]. The occupancy sites of most lncRNAs are not known, and the function has been mostly reasoned from the indirect effects of lncRNA perturbation. ChIRP is a strategy used to map lncRNA occupancy genome-wide at a high resolution. Using an affinity capture of as target complex of lncRNAs and chromatin by tiling antisense oligos, ChIRP allows scientists to generate a map of genomic binding sites at a resolution of several hundred bases with high sensitivity and low background. ChIRP is also applicable to many lncRNAs because the design of the affinity probes only requires the RNA sequence and no knowledge of the RNA's structure or functional domains [30].

#### 3.7.1 Probe Design

1. Design antisense oligo probes using the online probe design at https://www.biosearchtech.com/chirp-designer
2. Parameters: RNA target sequence length: <200 bp, oligonucleotide length: 20, masking level: 2, target GC% = 45, spacing length: 80 (if target RNA is too long, space length can be up to 200).
3. Order an antisense DNA probes with BiotinTEG at 3-prime end.
4. Dilute probes to 200 μM with Tris-EDTA (TE) buffer and store at −80 °C.

*3.7.2 Cell Collection and Fixation*

1. Grow cells to confluency in culture plate or flasks. Rinse with phosphate buffered solution (PBS) once and trypsinize. Quench trypsin with >2X volume of media and dislodge cells. Collect resuspension into a 15-mL falcon tube. Centrifuge and decant cell pellet. Resuspend in 1 mL media; count cells as usual.
2. After counting, centrifuge cells at 1000 rpm for 3 min and decant medium.
3. Prepare 10 mL 1% glutaraldehyde in room temperature PBS per $1 \times 10^7$ cells and resuspend cell pellet, shaking for 10 min at 37 °C.
4. Quench cross-linking reaction with 1/10th volume of 2.5 M glycine. Incubate room temperate for 5 min. Centrifuge 3000 rpm for 4 min and decant pellet.
5. Wash cell pellet with 10 mL cold 1X PBS twice.
6. After the second wash, resuspend the cell pellet in $1 \times 10^7$ cell/mL concentration in a 1.5-mL Eppendorf tube and centrifuge 3000 rpm for 4 min. Remove as much PBS as possible without disturbing the pellet.

*3.7.3 Sheer DNA by Sonification*

1. Add 500 μL of ChIRP lysis buffer (with added PIs (10 μL/mL if 1 mg/mL) and RI (100 U/mL) to resuspend cell pellet.
2. Sonicate (Branson sonifier 150, power 5, output power WATTS >10; Emerson Electric Company, St. Louis, MO): 15 s on, 45 s off: 20 cycles
3. Centrifuge at 13,200 rpm 10 min at 4 °C.
4. Transfer supernatant to 2-mL tube and measure protein concentration (keep volume at 500 μL, if not add ChIRP lysis buffer with PIs and RI to reach 500 μL).
5. Take 10 μL of supernatant and save for DNA input.

*3.7.4 Hybridization*

1. Add 1 mL of hybridization buffer (with added PIs and RI) and appropriate volume of diluted probes (100 pmol probe per 1 mL sample), mix well.
2. Incubate 37 °C for 4 h with shaking.

*3.7.5 Preparation of Dynabeads M-280 Streptavidin*

Start this preparation 15–20 min before hybridization is complete.

1. Take 50 μL of beads out for each sample.
2. Wash beads with 1 mL of ChIRP lysis buffer three times using the magnet to separate the beads from the buffer.
3. After the third wash, resuspend beads in original volume of ChIRP lysis buffer (with added PIs and RI): For example, if you originally took 50 μL of beads out, resuspend in 50 μL of ChIRP lysis buffer.

#### 3.7.6 Post-Hybridization

1. Add 50 μL of washed beads to each post hybridization reaction. Incubate 37 °C for 30 min with shaking.
2. Using the magnet, keep the beads and discard the supernatant.
3. Wash beads with 1 mL 37 °C wash buffer (with added PMSF), five times at 37 °C for 5 min with shaking.

#### 3.7.7 DNA Isolation

1. Resuspend beads in 150 μL freshly prepared DNA elution buffer (with added RNase A (10 μL) and RNase H (20 μL)); incubate 37 °C for 30 min with shaking.
2. Use the magnet to transfer buffer to new 1.5-mL tube.
3. Add another 150 μL of DNA elution buffer to beads, incubate 37 °C for 30 min with shaking.
4. Use the magnet to transfer buffer to same 1.5-mL tube from **step 3**.
5. Resuspend DNA input in 290 μL DNA elution buffer; incubate 37 °C for 1 h with shaking (this can be done at the same time as the samples).
6. Add 15 μL proteinase K to each sample; incubate 50 °C for 45 min with shaking.
7. Pre-spin down yellow phase-lock gel tubes (5PRIME 2302810) at 13200 rpm for 1 min.
8. Transfer DNA samples to yellow phase-lock gel tubes (~300 μL), add 300 μL phenol/chloroform, vortex, and shake vigorously for 10 min.
9. Centrifuge 13,200 rpm at 4 °C for 5 min.
10. Take aqueous phase from the top (~300 μL) and add 3 μL GlycoBlue™ coprecipitant, 30 μL NaOAc, and 900 μL 100% ethanol. Mix well and store −20 °C overnight.
11. Centrifuge 13,200 rpm at 4 °C for 30 min to pellet DNA.
12. Wash sample pellets twice with 70% ethanol.
13. Resuspend pellet in 30 μL TE buffer, 1:3–4 dilution for qPCR (Fig. 1b).

### 3.8 RIP

RNA–protein interactions play a critical role in modulating mRNA and ncRNA function. These interactions play an important role in several processes including control of gene expression via ncRNA binding to histone modifying complexes, transcription factors, and other regulatory proteins [31, 32]. Several methods have been developed to map RNA–protein interactions including RIP, which examines the physical associations between individual proteins and RNA molecules in vivo. This approach uses a specific antibody raised against the protein of interest to pull down the

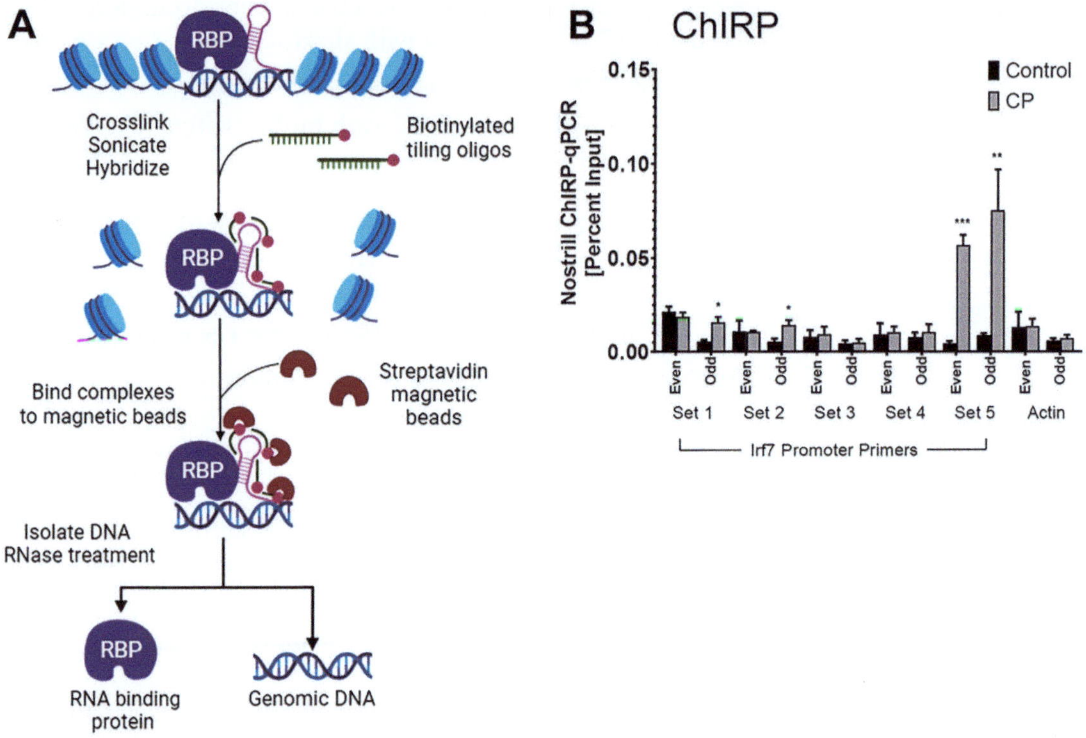


**Fig. 1** Recruitment of Nostrill to the *Irf7* promoter following *C. parvum* infection. (**a**) Schematic of ChIRP protocol created with BioRender.com. (**b**) Intestinal epithelial cells were infected with *C. parvum* for 24 h, followed by ChIRP analysis using two pools of tiling oligonucleotide probes specific to Nostrill and the PCR primer sets as designed. Probes 1, 3, 5, and 7 were mixed as the "Odd" pool, and probes 2, 4, 6, and 8 were mixed as the "Even" probes. *$p < 0.05$, **$p < 0.01$, and ***$p < 0.001$ versus control (Reproduced with permission [34]).

RNA-binding protein and target–RNA complexes. Any RNA associated with the protein complex will be isolated and can be further analyzed by polymerase chain reaction, hybridization, or sequencing [33].

*3.8.1 Cell Collection and Fixation*

1. Grow cells to confluency in culture plate or flasks. Rinse with PBS once and trypsinize. Quench trypsin with >2X volume of media and dislodge cells. Collect resuspension into 15-mL falcon tube. Centrifuge and decant cell pellet. Resuspend in 1 mL media, count cells as usual, and resuspend to $2 \times 10^6$ cells/mL in media.
2. Fix cells with 37% formaldehyde per mL of media; gently rock samples at 37 °C for 10 min.
3. Quench cross-linking reaction with 1/10th volume of 2.5 M glycine. Incubate room temperate for 5 min. Centrifuge 3000 rpm for 4 min and decant pellet.

4. Wash cell pellet with 10 mL cold 1xPBS twice.
5. After the second wash, resuspend the cell pellet in $1 \times 10^7$ cell/mL concentration in 1.5-mL Eppendorf tube and centrifuge 3000 rpm for 4 min. Remove as much PBS as possible without disturbing the pellet.

*3.8.2 Whole Cell Protein Isolation*

1. Resuspend pellet in 500 μL of whole cell extraction buffer (with added PIs and RI).
2. Sonicate (Branson sonifier 150, power 5, output power WATTS >10): 20 s on, 2 min off: 10 cycles.
3. Centrifuge at 13,200 rpm 10 min at 4 °C.
4. Transfer supernatant to new 1.5-mL tube and measure protein concentration.
5. Transfer 1100 μg of protein lysate (supernatant) to a new tube.

*3.8.3 Preclean*

1. Take out 20 μL of Millipore magnetic beads for each sample.
2. Wash beads with 500 μL of 1X PBS, twice, using the magnet to separate the beads from the PBS.
3. Add whole cell extraction buffer (with added PIs and RI) to the protein lysate to reach 1100 μL (protein concentration is now 1 μg/μL).
4. Remove 50 μg of protein lysate for input. Store −80 °C.
5. Transfer remaining lysate to washed beads.
6. Add 1 μg/mL of IgG antibody; rotate 30 min at 4 °C.
7. Briefly centrifuge, using the magnet, collect the supernatant, and discard the beads.

*3.8.4 RIP Binding*

1. Split the supernatant form each tube into two different tubes (500 μL for IgG and 500 μL for desired antibody).
2. Add 1 μg of IgG antibody to one tube and 1 μg of desired antibody to the other and rotate overnight at 4 °C.
3. Take 20 μL of Millipore magnetic beads per sample and wash beads with 500 μL of 1X PBS, twice, using the magnet to separate the beads from the PBS.
4. Transfer supernatants into the washed beads and rotate for 1 h at 4 °C.
5. Remove the supernatant using the magnet and keep the beads (now containing the RNA–protein complex bindings).
6. Wash beads four times with 500 μL of whole cell extraction buffer (with added PIs and RI) for 5 min at 4 °C.

*3.8.5 Protein Digestion*

1. Add 100 μL of ChIRP proteinase K buffer (with freshly added proteinase K and RI) to washed beads.
2. Fill protein lysate input samples from previous step to 100 μL with ChIRP proteinase K buffer (with freshly added proteinase K and RI).
3. Shake tubes at 50 °C, 400 rpm for 45 min.
4. Reverse cross-link by shaking 400 rpm, 65 °C for 2–4 h.

*3.8.6 RNA Extraction*

1. Add 500 μL TRIzol; incubate 1–5 min at RT.
2. Add 350 μL chloroform, vortex, and incubate 5 min at RT.
3. Centrifuge 13,200 rpm at 4 °C for 5 min.
4. Transfer supernatant (~350 μL) to new tube.
5. Add 350 μL isopropanol, 35 μL NaAc, and 1.5 μL GlycoBlue. Incubate 30 min at RT.
6. Precipitate at −20 °C, overnight.
7. Centrifuge 13,200 rpm at 4 °C for 20 min.
8. Wash pellets with 500 μL of 70% ethanol, twice. After the second wash, let pellet completely air-dry.

*3.8.7 Prepare RNA for Real-Time qPCR*

1. Add 21 μL DEPC $H_2O$ to resuspend RNA
2. Add 2.4 μL DNase treatment buffer (10X) and 0.5 μL DNase enzyme; incubate 37 °C for 30 min.
3. Add 2.4 μL DNase inactivating buffer (10X); incubate 2 min at RT.
4. Centrifuge 1000 rpm for 1.5 min.
5. Transfer 11.5 μL of RNA to use for reverse transcriptase.
6. Following reverse transcriptase, dilute cDNA to 1:5 to carry out qPCR (Fig. 2b).

## 4 Notes

1. To globally assess miRNA expression in host cells following parasite infection, a microarray analysis or whole genomic RNA-Seq (e.g., The Exiqon miRCURY LNA microRNA arrays) of mature miRNA expression in cells could be applied.
2. The siRNA targets on the miRNA sequence should be specific. Homology with other genes or sequences will result in off-target effects.

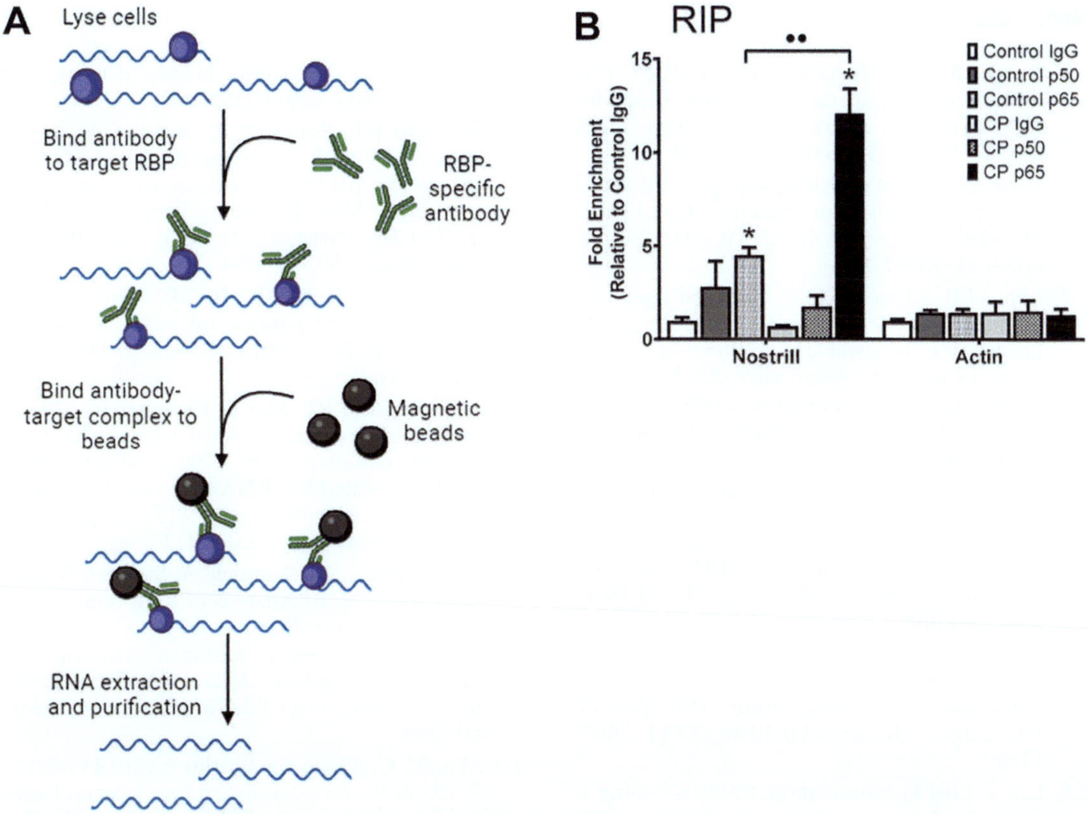


**Fig. 2** The interaction between Nostril, a host lncRNA, and NF-κB p65 in intestinal epithelial cells measured via RIP assay. (**a**) RIP protocol schematic created with BioRender.com. (**b**) Intestinal epithelial cells were infected with *C. parvum* for 24 h followed by RIP analysis using anti-p65 or p-50 antibodies. The presence of Nostril was measured via qRT-PCR, and the fold-enrichment method with normal IgG used as a control. Data represent means ± SEM from three independent experiments. *$p < 0.05$ versus control IgG (reproduced with permission [34]).

## Acknowledgments

This work was supported by funding from the National Institutes of Health (AI071321, AI095532, AI116323, AI136877, AI141325, AI156370, and AI177188) to XMC.

**Competing Interest Statement** The authors have no conflict of interests to declare that are relevant to the content of this chapter.

## References


1. Charles Richard JL, Eichhorn PJA (2018) Platforms for investigating LncRNA functions. SLAS Technol 23(6):493–506. https://doi.org/10.1177/2472630318780639
2. Bartel DP (2004) MicroRNAs: genomics, biogenesis, mechanism, and function. Cell 116(2): 281–297. https://doi.org/10.1016/s0092-8674(04)00045-5
3. Lim LP, Lau NC, Garrett-Engele P, Grimson A, Schelter JM, Castle J, Bartel DP, Linsley PS, Johnson JM (2005) Microarray analysis shows that some microRNAs downregulate large numbers of target mRNAs. Nature 433(7027):769–773. https://doi.org/10.1038/nature03315
4. Hwang J, Kang X, Wolf C, Touma M (2023) Mapping chromatin occupancy of Ppp1r1b-lncRNA genome-wide using chromatin isolation by RNA purification (ChIRP)-seq. Cells 12(24). https://doi.org/10.3390/cells12242805
5. Kornienko AE, Guenzl PM, Barlow DP, Pauler FM (2013) Gene regulation by the act of long non-coding RNA transcription. BMC Biol 11: 59. https://doi.org/10.1186/1741-7007-11-59
6. Lau E (2014) Non-coding RNA: zooming in on lncRNA functions. Nat Rev Genet 15(9): 574–575. https://doi.org/10.1038/nrg3795
7. Wang KC, Chang HY (2011) Molecular mechanisms of long noncoding RNAs. Mol Cell 43(6):904–914. https://doi.org/10.1016/j.molcel.2011.08.018
8. Richard JLC, Eichhorn PJA (2018) Deciphering the roles of lncRNAs in breast development and disease. Oncotarget 9(28):20179–20212. https://doi.org/10.18632/oncotarget.24591
9. Scaria V, Hariharan M, Maiti S, Pillai B, Brahmachari SK (2006) Host-virus interaction: a new role for microRNAs. Retrovirology 3:68. https://doi.org/10.1186/1742-4690-3-68
10. Winter J, Jung S, Keller S, Gregory RI, Diederichs S (2009) Many roads to maturity: microRNA biogenesis pathways and their regulation. Nat Cell Biol 11(3):228–234. https://doi.org/10.1038/ncb0309-228
11. Batugedara G, Lu XM, Hristov B, Abel S, Chahine Z, Hollin T, Williams D, Wang T, Cort A, Lenz T, Thompson TA, Prudhomme J, Tripathi AK, Xu G, Cudini J, Dogga S, Lawniczak M, Noble WS, Sinnis P, Le Roch KG (2023) Novel insights into the role of long non-coding RNA in the human malaria parasite, *Plasmodium falciparum*. Nat Commun 14(1):5086. https://doi.org/10.1038/s41467-023-40883-w
12. Militello KT, Refour P, Comeaux CA, Duraisingh MT (2008) Antisense RNA and RNAi in protozoan parasites: working hard or hardly working? Mol Biochem Parasitol 157(2): 117–126. https://doi.org/10.1016/j.molbiopara.2007.10.004
13. Zheng Y, Cai X, Bradley JE (2013) microRNAs in parasites and parasite infection. RNA Biol 10(3):371–379. https://doi.org/10.4161/rna.23716
14. Li Y, Baptista RP, Mei X, Kissinger JC (2022) Small and intermediate size structural RNAs in the unicellular parasite *Cryptosporidium parvum* as revealed by sRNA-seq and comparative genomics. Microb Genom 8(5). https://doi.org/10.1099/mgen.0.000821
15. Li Y, Baptista RP, Sateriale A, Striepen B, Kissinger JC (2020) Analysis of long non-coding RNA in *Cryptosporidium parvum* reveals significant stage-specific antisense transcription. Front Cell Infect Microbiol 10:608298. https://doi.org/10.3389/fcimb.2020.608298
16. Maudet C, Mano M, Eulalio A (2014) MicroRNAs in the interaction between host and bacterial pathogens. FEBS Lett 588(22): 4140–4147. https://doi.org/10.1016/j.febslet.2014.08.002
17. Piedade D, Azevedo-Pereira JM (2016) The role of microRNAs in the pathogenesis of Herpesvirus infection. Viruses 8(6). https://doi.org/10.3390/v8060156
18. Staedel C, Darfeuille F (2013) MicroRNAs and bacterial infection. Cell Microbiol 15(9): 1496–1507. https://doi.org/10.1111/cmi.12159
19. Zhou R, Hu G, Liu J, Gong AY, Drescher KM, Chen XM (2009) NF-kappaB p65-dependent transactivation of miRNA genes following *Cryptosporidium parvum* infection stimulates epithelial cell immune responses. PLoS Pathog 5(12):e1000681. https://doi.org/10.1371/journal.ppat.1000681
20. Schmittgen TD, Lee EJ, Jiang J, Sarkar A, Yang L, Elton TS, Chen C (2008) Real-time PCR quantification of precursor and mature microRNA. Methods 44(1):31–38. https://doi.org/10.1016/j.ymeth.2007.09.006
21. Xu JH, Mathy NW, Chen XM (2020) Use of miRNAs to study host cell-parasite interactions. Methods Mol Biol 2052:205–218. https://doi.org/10.1007/978-1-4939-9748-0_12

22. Wang Z (2009) MicroRNA interference technologies. Springer-Verlag, Heidelberg. https://doi.org/10.1007/978-3-642-00489-6
23. Lino CA, Harper JC, Carney JP, Timlin JA (2018) Delivering CRISPR: a review of the challenges and approaches. Drug Deliv 25(1): 1234–1257. https://doi.org/10.1080/10717544.2018.1474964
24. Zibitt MS, Hartford CCR, Lal A (2021) Interrogating lncRNA functions via CRISPR/Cas systems. RNA Biol 18(12):2097–2106. https://doi.org/10.1080/15476286.2021.1899500
25. Li J, Jin K, Li M, Mathy NW, Gong AY, Deng S, Martins GA, Sun M, Strauss-Soukup JK, Chen XM (2021) A host cell long noncoding RNA NR_033736 regulates type I interferon-mediated gene transcription and modulates intestinal epithelial anti-*Cryptosporidium* defense. PLoS Pathog 17(1): e1009241. https://doi.org/10.1371/journal.ppat.1009241
26. Bentwich I (2005) Prediction and validation of microRNAs and their targets. FEBS Lett 579(26):5904–5910. https://doi.org/10.1016/j.febslet.2005.09.040
27. Hamzeiy H, Allmer J, Yousef M (2014) Computational methods for microRNA target prediction. Methods Mol Biol 1107:207–221. https://doi.org/10.1007/978-1-62703-748-8_12
28. Gaken J, Mohamedali AM, Jiang J, Malik F, Stangl D, Smith AE, Chronis C, Kulasekararaj AG, Thomas NS, Farzaneh F, Tavassoli M, Mufti GJ (2012) A functional assay for microRNA target identification and validation. Nucleic Acids Res 40(10):e75. https://doi.org/10.1093/nar/gks145
29. Carter D, Chakalova L, Osborne CS, Dai YF, Fraser P (2002) Long-range chromatin regulatory interactions in vivo. Nat Genet 32(4): 623–626. https://doi.org/10.1038/ng1051
30. Chu C, Qu K, Zhong FL, Artandi SE, Chang HY (2011) Genomic maps of long noncoding RNA occupancy reveal principles of RNA-chromatin interactions. Mol Cell 44(4): 667–678. https://doi.org/10.1016/j.molcel.2011.08.027
31. Chen Y, Fu X, Li Z, Peng L, Zhuo L (2021) Prediction of lncRNA-protein interactions via the multiple information integration. Front Bioeng Biotechnol 9:647113. https://doi.org/10.3389/fbioe.2021.647113
32. Hu H, Zhu C, Ai H, Zhang L, Zhao J, Zhao Q, Liu H (2017) LPI-ETSLP: lncRNA-protein interaction prediction using eigenvalue transformation-based semi-supervised link prediction. Mol BioSyst 13(9):1781–1787. https://doi.org/10.1039/c7mb00290d
33. Gagliardi M, Matarazzo MR (2016) RIP: RNA immunoprecipitation. Methods Mol Biol 1480:73–86. https://doi.org/10.1007/978-1-4939-6380-5_7
34. Mathy NW, Deng S, Gong AY, Li M, Wang Y, Burleigh O, Kochvar A, Whiteford ER, Shibata A, Chen XM (2022) The long non-coding RNA Nostrill regulates transcription of Irf7 through interaction with NF-kappaB p65 to enhance intestinal epithelial defense against *Cryptosporidium parvum*. Front Immunol 13:863957. https://doi.org/10.3389/fimmu.2022.863957

# Chapter 13

## Genetic Manipulation of *Cryptosporidium parvum*

**Sebastian Shaw, Adam Sateriale, Mattie C. Pawlowic, Sumiti Vinayak, Carrie F. Brooks, Jessica H. Byerly, and Boris Striepen**

### Abstract

*Cryptosporidium parvum* can be reliably genetically manipulated using CRISPR/Cas9 driven homologous repair coupled to in vivo propagation in immunodeficient mice. This chapter will guide through foundational procedures for excystation, transfection, infection, collection, and purification of transgenic *Cryptosporidium parvum*. The genetic tools for *Cryptosporidium* research were expanded significantly over the last 5 years. While we cannot cover all advances in detail, we will highlight novel selection markers, conditional mutagenesis strategies, and genetic crosses.

**Key words** Transfection, Transgenesis, CRISPR, Cas9, Genetics

## 1 Introduction

Cas9 is an RNA-guided endonuclease and part of the bacterial immunity pathway clustered regularly interspaced short palindromic repeats (CRISPR) [1]. When provided with an RNA guide sequence, the Cas9 protein will scan DNA for a precise match prior to cutting, facilitating directed double-strand breaks in the target genome. In many eukaryotic cells, double-strand breaks in genomic DNA are quickly repaired by nonhomologous end joining; DNA is trimmed back from the break site, then re-ligated [2]. *Cryptosporidium* lacks the machinery for nonhomologous end joining; therefore, any break in the DNA must be repaired by homologous recombination [3]. This can be exploited for genome engineering by providing homologous DNA stretches flanking a selection cassette. The combination of a targeted cut by Cas9 with matching homologous flanks will direct recombination of the desired locus with high fidelity, and transgenic parasites can be drug selected for during an in vivo infection [4]. To enrich and propagate these transgenic *C. parvum* strains, we use IFN-γ-deficient mice, as they yield the highest level of infection.

Jan R. Mead and Michael J. Arrowood (eds.), *Cryptosporidium: Methods and Protocols*, Methods in Molecular Biology, vol. 2978, https://doi.org/10.1007/978-1-0716-4824-7_13,

The success of this protocol depends heavily on the transfection efficiency achieved. For this reason, we recommend optimization of transfection in vitro prior to attempting to make stable transgenics. Transfection of *Cryptosporidium* was first pioneered using a BTX electroporator (Harvard Apparatus, MA, US), and then optimized using an AMAXA 4D-Nucleofector (Lonza, Basel, Switzerland). We have included the steps for transfection using the AMAXA 4D-Nucleofector, as this delivers superior efficiency (*see* Pawlowic et al. [5] for detail on other instruments). To pilot transfection, we recommend using a simple *Cryptosporidium* Nanoluciferase expression vector [4]. The sensitivity of the Nanoluciferase assay allows the detection of even low levels of transfection. As transfection efficiency declines as the parasites age, we recommend using oocysts that were recently isolated from infected animals and kept at 4 °C, ideally within the past 4 months. In stable transfection experiments, the age of the mice used is also important. Younger mice are more susceptible to infection, enhancing recovery. For the widely used Iowa II strain, we prefer 4- to 6-week-old mice pretreated with drinking water containing ampicillin, streptomycin, and vancomycin for 5 days. Note that when using more virulent strains, age and antibiotic treatment may result in overly vigorous infections and mortality, and adjustments may be needed.

## 2 Materials

### 2.1 Optimizing Transfection with Transient Expression of Nanoluciferase

#### 2.1.1 Excystation

1. $1 \times 10^7$ viable *Cryptosporidium parvum* oocysts.
2. Phosphate-buffered saline (PBS), pH 7.4.
3. Household bleach, diluted 1:4 in deionized water.
4. Sodium taurodeoxycholate (Sigma-Aldrich Co., St. Louis, MO, US): 0.2 mM dissolved in PBS.
5. Water bath or incubator set to 37 °C.
6. Microcentrifuge.
7. 1.5-mL microcentrifuge tubes.
8. Hemocytometer.

   *Optional (for HCl protocol* [6, 7]):
9. 0.01 M hydrochloric acid (HCl) in deionized water.

   *Optional (for Watson protocol* [8]):
10. 1% sodium hypochlorite in deionized water.
11. 0.75% sodium taurocholate (Merck) in RPMI-1640 (1% fetal bovine serum).

#### 2.1.2 Transfection

1. AMAXA 4D-Nucleofector (Lonza, Basel, Switzerland).
2. SF Cell Line 4D-Nucleofector X Kit S (Lonza, Basel, Switzerland).

3. 10–50 μg *Cryptosporidium* Nanoluciferase expression vector [4].
4. 10–50 μg Cas9/guide expression vector.

*2.1.3 Cell Monolayer Infection*

1. 24-well tissue culture plate with HCT-8 cells (ATCC) grown to 70% confluency.
2. Infection media: RPMI or DMEM with 1% fetal bovine serum.

*2.1.4 Measuring Nanoluciferase*

1. NanoLuc lysis buffer (50 mM Tris-HCL pH 7.6, 2 mM DTT, 2 mM EDTA, 10% glycerol, 1% Triton-X100).
2. Nano-Glo Luciferase Assay System (Promega, WI, US, cat. N1110).
3. White 96-well plates for reading luminescence (Corning, NY, US, cat. 3789).
4. Plate reader capable of measuring luminescence at the wavelength suitable for Nanoluciferase, which is different from firefly luciferase [9].

## 2.2 Isolation of Stable Transgenic Cryptosporidium parvum

*2.2.1 Excystation*

1. *See* Subheading 2.1.1.

*2.2.2 Transfection*

1. AMAXA 4D-Nucleofector (Lonza, Basel, Switzerland).
2. SF Cell Line 4D-Nucleofector X Kit L (Lonza, Basel, Switzerland).
3. 50 μg Cas9/gRNA expression vector [4] (*see* **Notes 1** and **2**).
4. 50 μg PCR segment for homologous repair (*see* **Notes 1** and **3**).

*2.2.3 Mouse Infection*

1. 1-mL syringes with needle tips for gavage.
2. 0.8% w/v sodium bicarbonate in water.
3. One cage of four, 4- to 6-week-old IFNγ KO mice (Jackson Labs, ME, US, cat. 002287, B6.129S7-*Ifng*$^{tm1Ts}$/J).
4. Ampicillin (Sigma-Aldrich, cat. A9518), Streptomycin (Thermo Fisher Scientific, Waltham, MA, US, cat. 11860-038), and Vancomycin (Gold Bio, MO, US, cat. V-200-25).
5. 16 mg/mL paromomycin sulfate salt (Gemini Bio-Products, CA, US, cat. 400-155P) dissolved in sterile water (for use with Neo).

   *Optional*: BRD7929 [10] in 70% PEG400 + 30% glucose in water (5%) for use with PheRS (there is currently no commercial source, but a detailed guide to its synthesis can be found in the supplementary materials of [6]), DDD01510706 in 0.5% hydroxypropylmethylcellulose, 0.4% Tween-80, 0.5% benzyl alcohol for use with KRS (there is currently no commercial source, but compound can be obtained from the University of Dundee MRC Unit's Reagents and Services, https://mrcppureagents.dundee.ac.uk).

#### 2.2.4 Collection

1. DietGel Boost, calorie supplement for mice (Clear $H_2O$, ME, US, cat. 72-04-5022).
2. Dedicated forceps for fecal collection.
3. Optional: Wire mesh flooring for cage collection (Ancare, NY, US, cat. R20SSRWF).

#### 2.2.5 Purification

1. Tap water.
2. LabGen 125 homogenizer (Cole-Parmer, Vernon Hills, IL, US).
3. Saturated sucrose flotation solution: 1.33 specific gravity (756 g sucrose dissolved in 483 mL deionized water produces 1 L).
4. Saline solution: 0.85% NaCl in deionized water.
5. CsCl solution: 1.25 M (21.75 g dissolved in 103.25 mL deionized water).
6. 50-mL conical tubes.
7. 500-mL centrifuge tubes.
8. 2-mL low binding microcentrifuge tubes.
9. 1.5-mL microcentrifuge tubes.
10. Cell scrapers.
11. 250-μm wire mesh (Bel-Art, South Wayne, NJ, US, cat. 37845-1000 Micro Sieve Set).
12. Table-top centrifuge.
13. Microcentrifuge.
14. Hemocytometer.

## 3 Methods

### 3.1 Optimizing Transfection with Transient Expression of Nanoluciferase

#### 3.1.1 Excystation (Bleach [4, 11])

To transfect *Cryptosporidium*, it is necessary to rupture the protective oocyst wall and release the motile sporozoites. We typically use the well-established combination of sodium taurodeoxycholate and elevation of temperature to 37 °C to induce excystation [12]. However, there are strain and species differences that significantly impact excystation rates and viability, and we thus suggest to explore different methods. We provide modifications below that you may want to test [6, 8]. To test your electroporator, transfect sporozoites, using the same amount for control parasites transfected without DNA. We recommend starting with $1 \times 10^7$ oocysts.

1. Resuspend $1 \times 10^7$ oocysts in 500 μL of cold diluted household bleach (1:4). Incubate on ice for 10 min. Oocysts are naturally resistant to chlorine treatment and this incubation is intended to kill any bacterial contaminants. Bleach treatment also enhances wall permeability and can enhance excystation.
2. Centrifuge at 16,000× *g* for 3 min, then resuspend in 1 mL cold PBS.
3. Repeat the step twice to remove residual bleach.
4. Centrifuge at 16,000× *g* for 3 min then resuspend oocysts in 400 μL of 0.2 mM sodium taurodeoxycholate.
5. Incubate on ice for 10 min.
6. Add 1 mL PBS and incubate a 37 °C for 1 h.
7. Use a disposable cell counting chamber/hemocytometer to monitor excystation. Excystation rates vary depending on the viability of the oocysts, and we normally see excystation rates between 50 and 90%. Incubation at 37 °C can be prolonged up to 2 h to achieve a higher excystation rates, if needed. If you are experiencing low excystation rates, consider making a fresh stock of sodium taurodeoxycholate as well as the age of your oocysts.
8. Count sporozoites and aliquot two new 1.5-mL tubes with $1 \times 10^7$ each.

#### 3.1.2 Excystation (HCl [7, 6])

1. Centrifuge $1 \times 10^7$ oocysts at 16,000× *g* for 3 min at 4 °C. Remove supernatant.
2. Resuspend the pellet in 400 μL of 0.01 M HCl.
3. Incubate for 45 min at 37 °C.
4. Add 600 μL PBS and centrifuge at 16,000× *g* for 3 min at 4 °C.
5. Remove the supernatant, leaving approximately 50 μL of PBS above the pellet.
6. Add 1 mL of PBS and centrifuge at 16,000× *g* for 3 min at 4 °C.
7. Repeat **steps 5** and **6**.
8. Resuspend oocysts in 1 mL 0.2 mM sodium taurodeoxycholate and 20 mM sodium bicarbonate in PBS.
9. Incubate for 1 h at 37 °C.
10. Remove 10–20 μL to check for excystation.

11. Centrifuge at 16,000× *g* for 3 min at 4 °C. Resuspend in 1 mL of PBS.
12. Repeat wash **step 11** two times.
13. Removed supernatant on the last wash.
14. Proceed to Subheading 3.1.4 for transfection.

#### 3.1.3 Excystation (Taurocholate in RPMI [8])

1. Resuspend 1 × $10^7$ oocysts in 1% sodium hypochlorite on ice for 5 min.
2. Centrifuge at 16,000× *g* for 3 min, then resuspend in 1 mL cold PBS.
3. Repeat the step twice to remove residual bleach.
4. Centrifuge at 16,000× *g* for 3 min then resuspend oocysts in 400 μL of 0.75% sodium taurocholate in RPMI 1640 medium with 1% fetal bovine serum.
5. Incubate at 37 °C for 10 min.
6. Proceed to Subheading 3.1.4 for transfection.

#### 3.1.4 Transfection

1. As discussed in the introduction, we recommend optimizing transfection with a simple *Cryptosporidium* Nanoluciferase expression vector. The protocol below is specific to the AMAXA 4D-Nucleofector. Resuspend 4 × $10^7$ excysted sporozoites in 15 μL complete SF buffer (12.3 μL Cell Line Solution +2.7 μL Supplement 1).
2. Add 10–30 μg of *Cryptosporidium* Nanoluciferase expression vector resuspended in 5 μL of nuclease free water.
3. Transfer sporozoites and DNA into one well of the 16-well cuvette strip and electroporate using the program "EH100."
4. Move transfected sporozoites into a new 1.5-mL tube containing 80 μL PBS and keep on ice until infection.

#### 3.1.5 Cell Monolayer Infection

1. Replace growth medium of a 70% confluent 24-well HCT-8 culture with infection medium.
2. Distribute sporozoites equally between three wells of the 24-well plate.
3. Incubate for 24 h at 37 °C.

#### 3.1.6 Measuring Nanoluciferase

1. Remove medium from wells and replace with 110 μL NanoLuc lysis buffer.
2. Incubate at 37 °C for 10 min.
3. Scrape the bottom of the wells with your pipette tip. Mix carefully by pipetting up and down while avoiding the formation of bubbles.
4. Transfer 100 μL into a 96-well luciferase plate.

5. Combine Nano-Glo Luciferase Assay Substrate and Nano-Glo Luciferase Assay Buffer at a 1:50 ratio and add 100 μL to the samples.
6. Measure luminescence.

### 3.2 Establish Stable Transgenic Cryptosporidium parvum Parasites

#### 3.2.1 Excystation

To create a stable transgenic strain of *Cryptosporidium*, we typically infect a cage of four mice. Each mouse should be infected with $5 \times 10^7$ transfected sporozoites. Please follow one of the three excystation protocols outlined in Subheadings 3.1.1, 3.1.2, and 3.1.3 based on the results of your preliminary experiments. If you have not piloted this, begin by using the most widely used bleach protocol outline in Subheading 3.1.1.

#### 3.2.2 Transfection

This protocol is specific to the AMAXA 4D-Nucleofector.

1. Resuspend sporozoites in 75 μL complete SF buffer (65.6 μL Cell Line Solution +14.4 μL Supplement 1).
2. Add 25 μL of concentrated DNA (50 μg Cas9/gRNA expression plasmid +50 μg PCR segment for homologous repair).
3. Transfer to cuvette and electroporate using the program "EH100" (note that Watson et al. [8] use the program "FL115" instead).
4. Move transfected sporozoites into a new 1.5-mL tube containing 500 μL PBS and keep on ice until infection (total volume 600 μL, 100 μL per mouse). Keep transfected sporozoites on ice until infection.

#### 3.2.3 Mouse Infection

Once egressed from their protective oocyst, sporozoites are less able to survive the harsh environment and acidity of the host stomach. The most effective way to bypass this problem is to neutralize the stomach with sodium bicarbonate prior to infection.

To enhance infection, pretreat mice with antibiotics (ampicillin (1 g/L), streptomycin (1 g/L), and Vancomycin (0.5 g/L) in drinking water ad libitum for 1 week). Mice are provided normal drinking water upon infection.

1. Gavage each mouse with 100 μL of 0.8% sodium bicarbonate and wait 5 min.
2. Gavage each mouse with 100 μL of transfected sporozoites.
3. Allow parasites to establish infection overnight prior to initiating drug selection.

#### 3.2.4 Drug Selection for Stable Transgenic Parasites

The amino 3′-glycosylphosphotransferase neo confers resistance to paromomycin to *Cryptosporidium* and is now widely used by many investigators. The selection is typically robust, and the reasonable cost of paromomycin (when bought in bulk) permits its delivery in drinking water, which is convenient. Note however, that in our experience not all strains are equally susceptible, and elevated doses and prolonged treatments may be required to achieve homogeneous populations. Building on identification of compounds targeting different *Cryptosporidium* tRNA synthetases [13, 10], selection schemes using mutated drug-resistant versions of lysine (KRS [14]) and phenylalanine (PheRS [6]) tRNA synthase have been developed. As compound supply is more limiting, we deliver these drugs by gavage. While both markers can be used in trans, in our hands, targeting the resistance locus in cis is most reliable (*see* detailed discussion in [6]). For genetic crosses, infect mice with two strains that carry different drug resistance markers, then treat with both drugs starting on Day 1 postinfection as detailed [6, 14].

*Treatment regimens according to selection marker used:*

1. Neo [4]: 16 g/L paromomycin in water throughout the entire infection.
2. PheRS [6]: 10 mg/kg/day BRD7929 for 7 days, vehicle: 70% PEG400 + 30% glucose in water (5%).
3. KRS [14]: 30 mg/kg/day DDD01510706 for 4 days, vehicle: 0.5% hydroxypropylmethylcellulose, 0.4% Tween-80, 0.5% benzyl alcohol.

#### 3.2.5 Collection of Oocysts

Mice will typically begin to shed transgenic parasites between the 5th and 14th day postinfection. Parasite shedding can be monitored from collected fecal samples by qPCR using a *Cryptosporidium* specific set of probe and primers [15]. More conveniently, the Nanoluciferase reporter component of our standard selection cassette can be monitored in collected fecal samples without need for DNA isolation, thus greatly reducing the workload associated with establishing parasite shedding from multiple cages. Mice should be monitored daily for signs of distress including hunching, lethargy, and significant weight loss. To mitigate weight loss during infection, we supplement mouse diet with DietGel 76A at the onset of parasite shedding.

Once mice begin to shed transgenic parasites, start daily fecal collections. Take extra precautions to avoid cross-contamination of transgenic strains. Each strain should have its own collection cage and forceps, and mouse handlers should change gloves before and after handling each infected mouse cage. The biosafety cabinet should also be cleaned between each infected cage. Surgical drapes placed as a barrier between the cage and the cabinet's work surface help minimize cabinet contamination. Moving mice to cages with wire mesh floors during the day is a simple way to maximize collections. Wet paper towels can be spread at the bottom of the collection cage to prevent fecal material from drying out. As mice in these cages lack bedding we provide a plastic "mouse house" to help them maintain body temperature. Store all collected fecal samples at 4 °C until purification and avoid freezing.

*3.2.6 Measure Nanoluciferase Activity in the Feces*

1. Suspend 20 mg fecal material in 1 mL lysis buffer (50 mM Tris-HCl, 10% glycerol, 1% Triton X-100, 2 mM dithiothreitol, 2 mM ethylenediaminetetraacetic acid [EDTA]), add four to six glass beads and vortex rigorously.
2. Short-spin to clear the supernatant.
3. Transfer 100 μL into a 96-well luciferase plate.
4. Combine Nano-Glo Luciferase Assay Substrate and Nano-Glo Luciferase Assay Buffer at a 1:50 ratio and add 100 μL to each sample.
5. Measure luminescence.

*3.2.7 Oocyst Purification*

To purify oocysts, we use sucrose flotation followed by CsCl gradient centrifugation. There are many other purification methods for *Cryptosporidium* oocysts, including sucrose gradients, Nycodenz gradients, ether treatment, and sodium chloride flotation, and you may want to optimize this for your specific requirements [16]. Keep in mind the need to balance oocyst yield, purity, and viability. To maximize viability, all purifications should be done on ice with cold reagents.

1. Gather collected fecal samples (approximately 5–10 days of collection per purification) into a 50-mL conical tube and add 35 mL cold tap water.
2. Thoroughly blend samples. We use a mounted homogenizer for ease, but this can be done manually.
3. Filter samples through a 250-μm wire mesh and into a 500-mL centrifuge tube. Pour 450 mL cold water over the sample and stir to suspend.
4. Centrifuge at 1000× *g* for 10 min at 4 °C.
5. Resuspend pellet in 50 mL cold tap water and split into two 50-mL conical tubes.
6. Add 25 mL (1:1) cold saturated sucrose solution to conical tubes and invert several times to mix.
7. Centrifuge at 1000× *g* for 5 min at 4 °C.
8. Carefully decant supernatant into a new 500-mL centrifuge tube and add 450 mL cold tap water.
9. Centrifuge at 1500× *g* for 15 min at 4 °C.
10. Resuspend pellet in 5 mL cold saline solution.
11. Slowly overlay 1 mL of sample onto 1 mL of cold CsCl solution. For this, we use low-binding 2-mL centrifuge tubes.
12. Centrifuge at 16,000× *g* for 3 min at 4 °C.

13. Oocysts should be visible white layer at the saline–CsCl interface. Slowly pipette the oocyst from the interface and transfer 0.75 mL into a new 1.5-mL tube. Add 0.75 mL saline solution.
14. Centrifuge at 16,000× *g* for 3 min at 4 °C.
15. Discard supernatant and combine pellets using 1 mL cold saline solution.
16. Count oocysts using a hemocytometer.

## 4 Notes

1. *Concentration of DNA for transfection.* The protocols for transient and stable transfection call for highly concentrated DNA. We recommend using ethanol precipitation to concentrate plasmid DNA and/or PCR products prior to transfection.
2. *Design of Cas9 guide RNA.* There are a number of resources to help design an appropriate guide RNA. Eukaryotic Pathogen CRISPR guide RNA/DNA Design Tool (EuPaGDT) has the genome of *Cryptosporidium parvum* preloaded and allows for adjustable on-target and off-target search parameters [17]. For a walk—through of cloning guide RNA into a *Cryptosporidium* Cas9/U6 vector, *see* [5]. Keep in mind that recombination efficiency increases when the Cas9 cut site is located in close proximity to either of the homologous flanks.
3. *Design of PCR segments for homologous repair for different applications.* The amount of homologous DNA required for recombination in *Cryptosporidium* is surprisingly low, around 50 bp. With this low requirement, you can create homologous regions simply by adding them to the 3′ end of primers used to amplify the selection cassette. When designing your selection cassette, remember that you must include neomycin phosphotransferase, which provides resistance to paromomycin (or PheRS or KRS instead). Your parasites will be constantly exposed to the selection drug during mouse infection, so drive the expression of this resistance marker with a constitutive *Cryptosporidium* promoter. We have successfully used the promoters for actin, tubulin, enolase, and aldolase. Adding a luminescent reporter such as Nanoluciferase to your selection cassette will make it easier and faster to measure shedding of transgenic parasites. We commonly use a translational fusion of these two genes for our selection cassettes.
4. *Cryptosporidium* has a haploid genome throughout most of its lifecycle. To study the impact of loss of essential gene function, multiple conditional systems have been evaluated [18–20]. Some of these were found best suited to study the impact of gain of function [19]. These mutants can be constructed in a

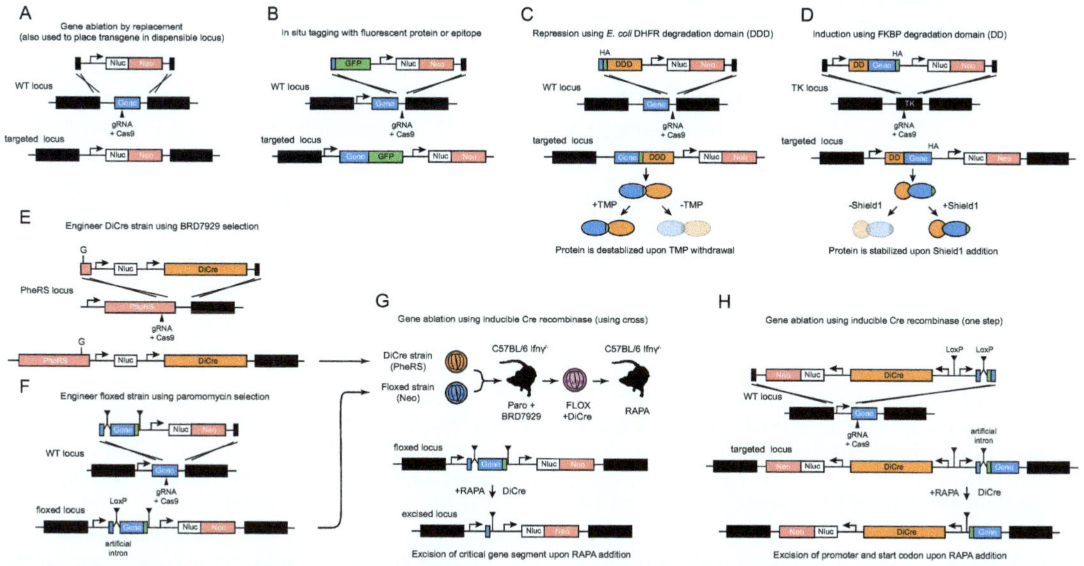


**Fig. 1** Schematic representation of different strategies to genetically modify *Cryptosporidium* (not drawn to scale). All strategies are based on placing a CRISPR/Cas9 mediated double strand break into the target locus (highlighted here as "gene" in blue). (**a**) A marker cassette is inserted into a locus to disrupt a gene and/or to express a transgene [4]. (**b**) A protein is tagged by appending the coding region with a sequence encoding a C-terminal fluorescent protein or an epitope like HA in translational fusion [21]. There are multiple options for conditional gene expression. (**c**) Translational fusion of the target gene with an *E. coli* dihydrofolate reductase derived destabilization domain (DDD) [20]. The fusion protein is stable in the presence of trimethoprim but ablated upon drug withdrawal. (**d**) A similar strategy using the FKB12 derived DD domain best suited for the induced overexpression of a transgene [19]. The transgene is placed into a dispensable locus, and the resulting fusion protein is stabilized by addition of Shield1 (note that this is restricted to experimentation in culture). (**e–h**) Essential genes can be conditionally ablated using rapamycin inducible split Cre recombinase (DiCre) (*see* **Notes 4**). Two strategies are shown. DiCre is introduced as a transgene using one selection marker (**e**), and the target is "floxed" by inserting the LoxP recombination motives recognize by Cre using a different marker in a second strain (**f**). Artificial introns can be used to place LoxP sites into the coding sequence without disruption of the protein. (**g**) A strain containing both modifications is then established by genetic cross [6] (using dual drug selection), and the floxed portion of the gene is excised upon addition of rapamycin (RAPA, importantly, this can be done in culture and in infected animals) [19]. (**h**) This strategy can also be used employing a single construct and selection marker, for example, by floxing the promoter of a gene [18]. Note that the large size of DiCre may limit the size of the parasite genome that can be floxed. Several groups have applied DiCre-based strategies recently, and changes in the promoter strength or expression of the DiCre fragments from distinct promoters instead of the use of a 2A skip peptide have been reported to lower the level of DiCre activity in the absence of rapamycin [8, 22]

single step with a single plasmid or for some (e.g., the inducible Cre/Lox system) elements can be engineered individually into separate parasite strains, which are then subjected to a genetic cross to combine both traits. We lack room to describe all of these approaches in full, but Fig. 1 shows a schematic overview of several important strategies and provides reference to the primary publication for technical detail.

## Acknowledgments

Work on *Cryptosporidium* in the Striepen lab is currently supported by multiple grants from the US National Institutes of Health to BS.

**Competing Interest Statement** The authors have no conflict of interests to declare that are relevant to the content of this chapter.

### References

1. Jinek M, Chylinski K, Fonfara I, Hauer M, Doudna JA, Charpentier E (2012) A programmable dual-RNA-guided DNA endonuclease in adaptive bacterial immunity. Science 337(6096):816–821. https://doi.org/10.1126/science.1225829
2. van Gent DC, Hoeijmakers JH, Kanaar R (2001) Chromosomal stability and the DNA double-stranded break connection. Nat Rev Genet 2(3):196–206. https://doi.org/10.1038/35056049
3. Abrahamsen MS, Templeton TJ, Enomoto S, Abrahante JE, Zhu G, Lancto CA, Deng M, Liu C, Widmer G, Tzipori S, Buck GA, Xu P, Bankier AT, Dear PH, Konfortov BA, Spriggs HF, Iyer L, Anantharaman V, Aravind L, Kapur V (2004) Complete genome sequence of the apicomplexan, *Cryptosporidium parvum*. Science 304(5669):441–445. https://doi.org/10.1126/science.1094786
4. Vinayak S, Pawlowic MC, Sateriale A, Brooks CF, Studstill CJ, Bar-Peled Y, Cipriano MJ, Striepen B (2015) Genetic modification of the diarrhoeal pathogen *Cryptosporidium parvum*. Nature 523(7561):477–480. https://doi.org/10.1038/nature14651
5. Pawlowic MC, Vinayak S, Sateriale A, Brooks CF, Striepen B (2017) Generating and maintaining transgenic *Cryptosporidium parvum* parasites. Curr Protoc Microbiol 46:20B.22.21–20B.22.32. https://doi.org/10.1002/cpmc.33
6. Shaw S, Cohn IS, Baptista RP, Xia G, Melillo B, Agyabeng-Dadzie F, Kissinger JC, Striepen B (2024) Genetic crosses within and between species of *Cryptosporidium*. Proc Natl Acad Sci USA 121(1):e2313210120. https://doi.org/10.1073/pnas.2313210120
7. Kato S, Jenkins MB, Ghiorse WC, Bowman DD (2001) Chemical and physical factors affecting the excystation of *Cryptosporidium parvum* oocysts. J Parasitol 87(3):575–581. https://doi.org/10.1645/0022-3395(2001)087[0575:CAPFAT]2.0.CO;2
8. Watson LC, Sala KA, Bernitz N (2025) Targeted CRISPR screens reveal genes essential for Cryptosporidium survival in the host intestine. Nat Commun 16, 7749. https://doi.org/10.1038/s41467-025-63012-1
9. Hall MP, Unch J, Binkowski BF, Valley MP, Butler BL, Wood MG, Otto P, Zimmerman K, Vidugiris G, Machleidt T, Robers MB, Benink HA, Eggers CT, Slater MR, Meisenheimer PL, Klaubert DH, Fan F, Encell LP, Wood KV (2012) Engineered luciferase reporter from a deep sea shrimp utilizing a novel imidazopyrazinone substrate. ACS Chem Biol 7(11):1848–1857. https://doi.org/10.1021/cb3002478
10. Vinayak S, Jumani RS, Miller P, Hasan MM, McLeod BI, Tandel J, Stebbins EE, Teixeira JE, Borrel J, Gonse A, Zhang M, Yu X, Wernimont A, Walpole C, Eckley S, Love MS, McNamara CW, Sharma M, Sharma A, Scherer CA, Kato N, Schreiber SL, Melillo B, Striepen B, Huston CD, Comer E (2020) Bicyclic azetidines kill the diarrheal pathogen *Cryptosporidium* in mice by inhibiting parasite phenylalanyl-tRNA synthetase. Sci Transl Med 12(563). https://doi.org/10.1126/scitranslmed.aba8412
11. Gut J, Nelson RG (1999) *Cryptosporidium parvum*: synchronized excystation in vitro and evaluation of sporozoite infectivity with a new lectin-based assay. J Eukaryot Microbiol 46(5):56S–57S
12. Robertson LJ, Campbell AT, Smith HV (1993) In vitro excystation of *Cryptosporidium parvum*. Parasitology 106:13–19
13. Baragana B, Forte B, Choi R, Nakazawa Hewitt S, Bueren-Calabuig JA, Pisco JP, Peet C, Dranow DM, Robinson DA, Jansen C, Norcross NR, Vinayak S, Anderson M, Brooks CF, Cooper CA, Damerow S, Delves M, Dowers K, Duffy J, Edwards TE, Hallyburton I, Horst BG, Hulverson MA, Ferguson L, Jimenez-Diaz MB, Jumani RS, Lorimer DD, Love MS, Maher S,

Matthews H, McNamara CW, Miller P, O'Neill S, Ojo KK, Osuna-Cabello M, Pinto E, Post J, Riley J, Rottmann M, Sanz LM, Scullion P, Sharma A, Shepherd SM, Shishikura Y, Simeons FRC, Stebbins EE, Stojanovski L, Straschil U, Tamaki FK, Tamjar J, Torrie LS, Vantaux A, Witkowski B, Wittlin S, Yogavel M, Zuccotto F, Angulo-Barturen I, Sinden R, Baum J, Gamo FJ, Maser P, Kyle DE, Winzeler EA, Myler PJ, Wyatt PG, Floyd D, Matthews D, Sharma A, Striepen B, Huston CD, Gray DW, Fairlamb AH, Pisliakov AV, Walpole C, Read KD, Van Voorhis WC, Gilbert IH (2019) Lysyl-tRNA synthetase as a drug target in malaria and cryptosporidiosis. Proc Natl Acad Sci USA 116(14):7015–7020. https://doi.org/10.1073/pnas.1814685116

14. Hanna JC, Corpas-Lopez V, Seizova S, Colon BL, Bacchetti R, Hall GMJ, Sands EM, Robinson L, Baragana B, Wyllie S, Pawlowic MC (2023) Mode of action studies confirm on-target engagement of lysyl-tRNA synthetase inhibitor and lead to new selection marker for *Cryptosporidium*. Front Cell Infect Microbiol 13:1236814. https://doi.org/10.3389/fcimb.2023.1236814
15. Jothikumar N, da Silva AJ, Moura I, Qvarnstrom Y, Hill VR (2008) Detection and differentiation of *Cryptosporidium hominis* and *Cryptosporidium parvum* by dual TaqMan assays. J Med Microbiol 57(Pt 9):1099–1105. https://doi.org/10.1099/jmm.0.2008/001461-0
16. Arrowood MJ, Donaldson K (1996) Improved purification methods for calf-derived *Cryptosporidium parvum* oocysts using discontinuous sucrose and cesium chloride gradients. J Eukaryot Microbiol 43(5):S 89
17. Peng D, Tarleton R (2015) EuPaGDT: a web tool tailored to design CRISPR guide RNAs for eukaryotic pathogens. Microb Genom 1(4):e000033. https://doi.org/10.1099/mgen.0.000033
18. Tandel J, Walzer KA, Byerly JH, Pinkston B, Beiting DP, Striepen B (2023) Genetic ablation of a female-specific Apetala 2 transcription factor blocks oocyst shedding in *Cryptosporidium parvum*. MBio 14(2):e0326122. https://doi.org/10.1128/mbio.03261-22
19. Walzer KA, Tandel J, Byerly JH, Daniels AM, Gullicksrud JA, Whelan EC, Carro SD, Krespan E, Beiting DP, Striepen B (2024) Transcriptional control of the *Cryptosporidium* life cycle. Nature. https://doi.org/10.1038/s41586-024-07466-1
20. Choudhary HH, Nava MG, Gartlan BE, Rose S, Vinayak S (2020) A conditional protein degradation system to study essential gene function in *Cryptosporidium parvum*. mBio 11(4). https://doi.org/10.1128/mBio.01231-20
21. Tandel J, English ED, Sateriale A, Gullicksrud JA, Beiting DP, Sullivan MC, Pinkston B, Striepen B (2019) Life cycle progression and sexual development of the apicomplexan parasite *Cryptosporidium parvum*. Nat Microbiol 4(12):2226–2236. https://doi.org/10.1038/s41564-019-0539-x
22. Huang Y, Li J, Pei S, You H, Liu H, Guo Y, Xu R, Li N, Feng Y, Xiao L (2024) Optimization of a DiCre recombinase system with reduced leakage for conditional genome editing of *Cryptosporidium*. Parasit Vectors 17(1):352. https://doi.org/10.1186/s13071-024-06431-1

# Chapter 14

# Mouse Models for Use in *Cryptosporidium* Infection Studies, Quantification of Parasite Burden Using Flow Cytometry, qPCR and Histopathology, and Confocal Imaging of Oocysts

**Karine Sonzogni-Desautels, Jackson F. Chen, Jan R. Mead, and Momar Ndao**

## Abstract

Cryptosporidiosis threatens the lives of young children in developing countries and newborn calves around the world. No vaccine or therapy can prevent or cure this diarrhea-inducing enteric disease caused by *Cryptosporidium* spp. protozoan parasites. There is an essential need to discover new therapeutic drugs efficient in reducing parasite burden in infected individuals. Research therefore relies on reliable small animal models of cryptosporidiosis. Here, we present excellent mouse models which can efficiently mimic pathogenesis of human and bovine cryptosporidiosis. We also describe methods to purify *Cryptosporidium parvum* oocysts from stool and intestine of infected mice to facilitate oocyst quantification. Moreover, we present protocols using flow cytometry, quantitative polymerase chain reaction, and histopathology to accurately quantify parasite burden in stool or intestine samples. Also, we are presenting a high-resolution confocal microscopy protocol that constitutes a tool to image *C. parvum* sporozoites within oocysts. Polyclonal and monoclonal antibodies against *Cryptosporidium* p23 protein can be used to detect sporozoites within permeabilized or partially excysted oocysts. Additionally, our novel qPCR method, targeting the HSP70 genes, offers enhanced sensitivity for *Cryptosporidium* detection and quantification.

**Key words** *Cryptosporidium parvum*, Oocysts, Interferon gamma knock-out mice, Interferon gamma receptor knock-out mice, IL-12 knock-out mice, SCID mice, Flow cytometry, Confocal microscopy, Quantitative polymerase chain reaction, Histopathology

## 1 Introduction

Cryptosporidiosis causes transient diarrhea in immunocompetent individuals and life-threatening diarrhea and dehydration in immunocompromised individuals [1]. In the United States alone, there are an estimated 748,000 cases annually, with incidence rates by region/jurisdiction ranging from 1.3 to 6.3 cases per 100,000 people [2–5]. This enteric disease affects hundreds of thousands of infants and toddlers in sub-Saharan Africa and South Asia [6–8]

Jan R. Mead and Michael J. Arrowood (eds.), *Cryptosporidium: Methods and Protocols*, Methods in Molecular Biology, vol. 2978, https://doi.org/10.1007/978-1-0716-4824-7_14,

and has become a leading cause of moderate-to-severe diarrhea in children under 5 years old [2, 9]. From these *Cryptosporidium*-positive cases of moderate-to-severe diarrhea, 78% were associated to *Cryptosporidium hominis* and 10% to *Cryptosporidium parvum*, the zoonotic species [10].

No mouse model can sustain persistent *C. hominis* infection [11]. However, several mouse models of *C. parvum* infection mimic clinical signs and pathogeny of human and bovine cryptosporidiosis [1, 12–16]. While there is no ideal animal model to study cryptosporidiosis, small rodent models provide a mechanism for studying infection dynamics, host–parasite interactions and for evaluating drugs and other therapeutics [17]. Neonatal mice, less than 3 weeks old, are very susceptible to infection and have been used to study immune responses and for some drug testing and may provide some insight to diseases in child [18, 19]. Severe infections develop in 7-day-old BALB/c mice infected with *C. parvum* oocysts, with peak oocyst shedding occurring 7–10-day postinfection [20]. However, these mice are more difficult to dose due to their size and rapidly changing weight, and samples could be more limiting [21–23]. Adult wild-type mice are not usually suitable to evaluate drugs, because infections with *C. parvum* are transient and generate low oocyst output [24].

For drug testing, mice that are genetically immunocompromised provide a consistent model to study host–parasite interactions and for drug evaluation. Acute models include interferon gamma receptor knockout (IFNγR-KO) mice [13, 14, 16], interferon gamma knockout (IFNγ-KO) mice [12, 15, 25, 26], and IL-12 knockout (IL-12 KO) mice [12, 15, 27, 28]. IFNγR-KO and IFNγ-KO mice on a C57 black/6 background (C57BL/6) are highly susceptible to infection [13, 14, 16, 26] mimicking the high susceptibility to infection of infants and newborn calves [18]. Effectively, IFNγ is the most important cytokine to fight cryptosporidiosis via an innate immune response involving natural killer cells and an adaptive $T_H1$ cell-mediated immune response [29]. In our hands, these mice died or had to be euthanized between 8- and 11-day postinfection [14, 26]. However, other studies have observed that infections in these mice are not lethal [30], which may reflect differences in *C. parvum* strains or intestinal flora. In fact, virulence of *C. parvum* strains can differ and pathogenicity needs to be determined for each strain. For example, the Iowa strain is typically highly virulent, and 4000 oocysts is a lethal dose threatening life of IFNγR-KO mice in 9- to 11-day postinfection [14]. However, other *C. parvum* strains such as oocysts purified from naturally infected calves [31, 32] may induce a less severe infection in mice.

Likewise, infections in IL-12 KO mice are acute, increasing several logs over a couple of days, eventually recovering by 10- to 14-day postinfection (*see* Fig. 1) [27].

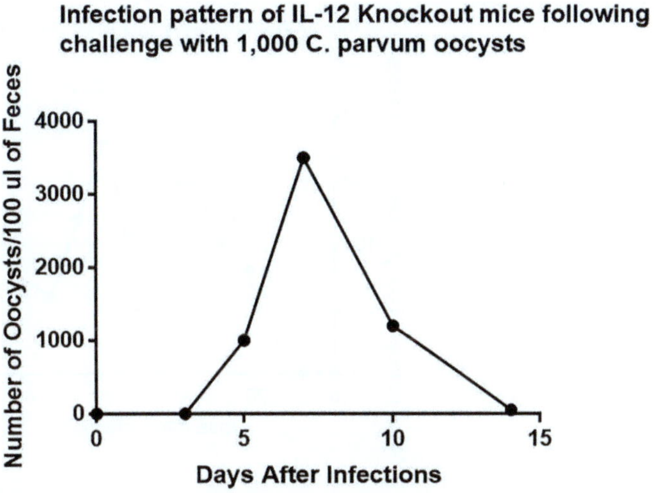


**Fig. 1** Infection pattern of IL-12 KO mice infected with 1000 oocysts. The x-axis represents days after infection ranging from 0 to 15, and the y-axis shows number of oocysts per 100 ul of feces ranging from 0 to 4000. Note that infections are not detectable before 4-day postinfection. Oocyst numbers increase rapidly, peaking around 7-day postinfection. Infection levels decrease, and oocysts are usually not detectable beyond 14-day postinfection

In contrast, *C. parvum* infections are chronic in severely immunodeficient mouse models. Several chronic infection models exist which include nude mice [24, 33], SCID mice [24, 34] and NOD SCID mice [35]. As in wild-type mice, neonatal immunodeficient mouse models develop rapidly intensifying infections but differ in the frequently lethal outcome [24]. Even with high inoculum doses, infections in the adult immunodeficient models are initially lower (*see* Fig. 2) than in the acute models due to intact innate immune system responses. Over time (8–12 weeks), infections can become disseminated, and weight loss is observed in the late stages; mice eventually succumb to infection [24, 36]. Depending on the desired infection level and whether extra-intestinal sites are involved, timing of treatment or experimental procedure is important [23, 24, 37].

To sum up, genetically immunocompromised mice (such as IFNγR-KO, IFNγ-KO, and IL-12KO mice) can provide consistency in a model of cryptosporidiosis. While chronic models may be useful for certain objectives (e.g., disseminated disease and eradication confirmation), the acute models are a challenging and rapid model for drug testing [14, 38, 39]. Candidate compounds proven efficient to inhibit *C. parvum* infection in mammalian cell culture [14, 37, 40] can be tested in vivo in these mouse models [14].

Accurate methods of detection allowing quantification of oocysts in stool and intestine samples are then essential to compare

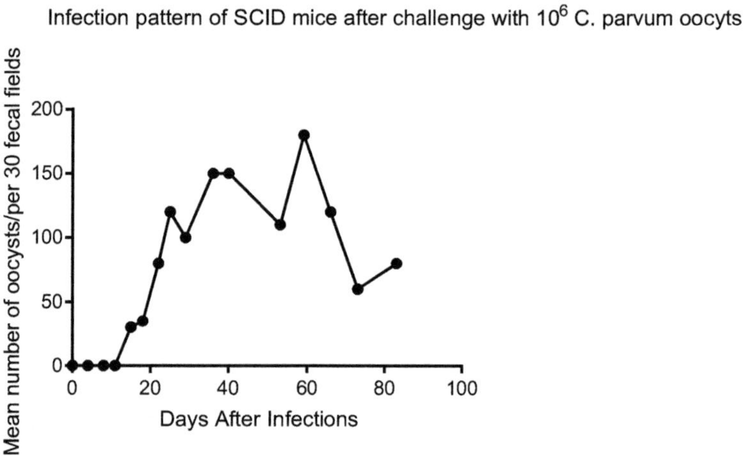


**Fig. 2** Infection pattern of SCID mice infected with $10^6$ oocysts. The x-axis represents days after infection, ranging from 0 to 100. The y-axis shows the mean number of oocysts per 30 fecal fields, ranging from 0 to 200. The graph depicts fluctuations in oocyst count. Note that infections are not detectable before 10- to 12-day postinfection. Oocyst numbers increase slowly over 8–10 weeks, waxing and waning. Oocyst count is peaking around day 60, followed by a decline. Decline in oocyst shedding levels is usually the result of morbidity or mortality late in infection

parasite burden of treated versus control mice. For the detection of oocysts, we describe flow cytometric methods, using staining antibodies or a two-step gating strategy relying on the inherent morphological and optical properties of oocysts, for the quantification of *Cryptosporidium parvum* oocysts in stool and intestine specimens. These flow cytometric methods were originally developed to replace conventional microscopic immunofluorescent assays [41] and constitute a relatively rapid and simple alternative to microscopy.

Confocal microscopy has been described as a sophisticated method to identify and localize intracellular molecules, and other manuscripts thoroughly describe the methods of staining as well as the settings of the microscopes [42]. To validate this method, we utilized polyclonal sera targeting the p23 protein, a well characterized sporozoite antigen that has been shown to be immunogenic and protective and is involved in the attachment and invasion of sporozoites into host cells [43] The combination of antibodies targeting *C. parvum* proteins (such as the anti-p23 polyclonal antibodies) and confocal microscopy can identify *C. parvum* parasites and localize specific proteins on/in *C. parvum* sporozoites/oocysts [44–47]. This high-resolution imaging technique also offers several advantages, including the ability to create three-dimensional reconstructions [48, 49].

For in-depth analysis, we also present quantitative PCR and histopathology as protocols to obtain quantitative or semiquantitative, respectively, evaluation of parasite burden in mouse stool during the study and intestine at the time of necropsy. Our qPCR protocol specifically focuses on the heat shock protein 70 (HSP70) gene [50].

But, first, we describe methods of oocyst purification from stool and intestine samples using either discontinuous sucrose gradients or a saturated NaCl/distilled water layer. Using these protocols to purify *C. parvum* oocysts, techniques of flow cytometry, qPCR, and histopathology allow investigating the potential effect of candidate compounds on oocyst shedding in stool and parasite burden in intestine in acute mouse models of infection [13, 14, 51].

## 2 Materials

*C. parvum* is a zoonotic parasite and precautions (lab coat, gloves, washing hands, etc.) should be taken to avoid infection of laboratory personnel (*see* **Note 1**). All experiments using *C. parvum* oocysts or infected samples should be carried out inside Biological Safety Cabinets by trained personnel.

Proper waste management is critical. Disposable material that was in contact with stool or intestine of infected mice or purified oocysts (e.g., specimen containers, microcentrifuge tubes, pipet tips, etc.) should be autoclaved before disposal. Nondisposable material and equipment potentially contaminated (e.g., hood, biological safety cabinet, homogenizer, hemocytometer, magnetic stir bars, surgical tools, etc.) should be disinfected using hydrogen peroxide [52] or treated with heat to inactivate oocysts [53, 54]. Oocyst-containing liquids (e.g., supernatant from stool and intestine oocyst purification protocols) should be treated with heat to inactivate oocysts (This is the preferred alternative to chemical disinfection, which can be challenging in the presence of significant organic loads.) Infected mice should be kept in filter-top cages and housed in Biological Level 2 facilities. Mice should be handled in Biological Safety Cabinets, and bedding should be autoclaved before disposal to ensure complete inactivation of oocysts [55]. All personnel handling *C. parvum* must be trained in biosafety protocols and emergency procedures.

### 2.1 Kinyoun Stain

1. Carbol fuchsin.
2. 50% ethanol.
3. 2% sulfuric acid.
4. Methylene blue.

### 2.2 Infection with C. parvum Oocysts

1. Phosphate-buffered saline (PBS) 1X: 0.01 M, pH 7.2 (or equivalent).
2. Hemocytometer.
3. Luer lock 1-mL syringes.
4. 20G or 22G, disposable or autoclavable, bulb gavage needles.

### 2.3 Collection of Stool Pellet and Intestine Samples

1. 1.7-mL low-binding microcentrifuge tubes.
2. 60-mL specimen container.
3. 2.5% (w/v) potassium dichromate ($K_2Cr_2O_7$) in sterile, cold distilled water.
4. 0.02% (v/v) Tween 20 in PBS 1X.

### 2.4 Purification of Oocysts from Stool

1. 1.7- and 2.2-mL low-binding microcentrifuge tubes.
2. 1:2 and 1:4 sucrose solutions (*see* **Note 2**).
3. Solution of saturated NaCl.

### 2.5 Purification of Oocysts from Intestine

1. Stir bars and magnetic stirrers.
2. 50-mL conical polypropylene centrifuge tubes and 2.2-mL microcentrifuge tubes.
3. Tissue cell scrapers.
4. 150 W handheld tissue homogenizer.
5. 0.02% (v/v) Tween 20 in distilled water.
6. Sputasol (dry mixture of 10% DTT, 76% NaCl, 2% KCl, 10% $Na_2HPO_4$, and 2% $KHPO_4$).
7. Diethyl ether (or substitute ethyl acetate, where required).
8. Solution of saturated NaCl.

### 2.6 Flow Cytometry

1. Microtiter tubes and flow tubes.
2. 96-well plate.
3. 8% paraformaldehyde.
4. Counting beads.
5. Oocyst-specific monoclonal antibody labeled with fluorescein isothiocyanate (OW50-FITC) (also available in the MERIFLUOR *Cryptosporidium/Giardia* kit (Meridian Bioscience, Inc., Cincinnati, OH).
6. Anti-*Cryptosporidium* Alexa Fluor® 488 monoclonal antibody.
7. Flow cytometers such as BD LSRFortessa or BD LSRFortessa X-20 flow cytometer (BD Biosciences, San Jose, CA).
8. Flow cytometry analysis software such as FlowJo v.10 (FlowJo LLC, Ashland, OR).

### 2.7 Confocal Microscopy

1. *C. parvum* oocysts in 2.5% w/v $K_2Cr_2O_7$ at 4 °C.
2. PBS 1X.
3. Hemacytometer.
4. 4% paraformaldehyde in PBS.
5. Imaging eight-chamber slide.
6. 5% goat serum in PBS.
7. Anti-p23 polyclonal antibodies.
8. 1% BSA in PBS.
9. Alexa Fluor 647 conjugate goat anti-rabbit secondary antibody.
10. Alexa Fluor 488 conjugate monoclonal IgG mouse anti-*Cryptosporidium* clone BEL 0126.
11. DAPI stain.
12. Confocal microscope such as Zeiss LSM780-NLO laser scanning confocal with IR-OPO lasers microscope (Zeiss, Oberkochen, Baden-Württemberg, Germany).

### 2.8 Genomic DNA Extraction and Purification

1. Fresh *C. parvum* oocysts stored in 2.5% w/v $K_2Cr_2O_7$.
2. PBS 1X.
3. 1.7-mL microcentrifuge tubes.
4. Liquid nitrogen.
5. ≥56 °C water bath.
6. Room temperature water.
7. Thermometer.
8. QIAamp DNA Extraction Mini Kit (Qiagen, Germantown, MD).
9. Tris-EDTA Buffer, pH 7.1.
10. NanoDrop Spectrophotometer (for DNA quantification) such as the Thermo Fisher NanoDrop Microvolume Spectrophotometer (Thermo Fisher Scientific, Eugene, OR).

### 2.9 qPCR Analysis

1. Qiagen 200 μL PCR grade tubes (Qiagen, Venlo, the Netherlands).
2. PCR-grade water.
3. 10 μM forward primer (5′-AACTTTAGCTCCAGTTGAGAAAGTACTC-3′).
4. 10 μM reverse primer (5′-CATGGCTCTTTACCGTTAAAGAATTCC-3′).
5. 2 μM 5′ 6-carboxyfluorescein reporter dye (FAM) probe with a 3′ tetramethylrhodamine (TMR) quencher dye (5′-AATACGTGTAGAACCACCAACCAATACAACATC-3′).

6. Qiagen QuantiNova Multiplex PCR Kit (Qiagen).
7. *C. parvum* genomic DNA.
8. Qiagen Rotor-Gene Q 5plex (Qiagen).

### 2.10 Histopathology

1. Histology cassettes.
2. 10% formalin.
3. 100% ethanol, 80% ethanol, 70% ethanol.
4. Xylene.
5. Hematoxylin.
6. Eosin-phloxine (eosin Y, phloxine B, 95% ethanol, acetic acid).
7. Mounting media.
8. Microscope slides and coverslips.
9. Microtome.
10. Tissue processor such as Leica ASP300S tissue processor (Leica Biosystems Inc., Buffalo Grove, IL).
11. Paraffin embedding station such as Leica EG1150 station (Leica Biosystems).

## 3 Methods

### 3.1 Kinyoun Staining to Identify Sporozoite-Containing C. parvum Oocysts

The Kinyoun stain is a method commonly used for detecting *Cryptosporidium* oocysts in fecal samples [56]. This technique is particularly useful for identifying the proportion of sporozoite-containing oocysts versus "empty shells," which affects the virulence of *C. parvum* parasites (*see* Fig. 3). Sporozoites being the infectious form of the parasite, oocyst "empty shells," which correspond to oocysts that released their four sporozoites, are noninfectious. >80% sporozoite-containing oocysts, as determined by a Kinyoun stain, characterize a stock of good quality (*see* **Note 3**).

1. If oocysts were stored in 2.5% $K_2Cr_2O_7$, wash oocysts with 1 mL of PBS 1X.
2. Centrifuge microcentrifuge tube at 10,000× *g* for 5 min at 4 °C.
3. Discard supernatant and repeat **steps 1** and **2** for a total of three washes.
4. Put one drop of oocyst solution on a microscope slide.
5. Fix with methanol for 1 min and air dry.
6. To stain, apply carbol fuchsin for 5 min.
7. To decolorize, apply 50% ethanol for 1 min, and then 2% sulfuric acid for 30 s.

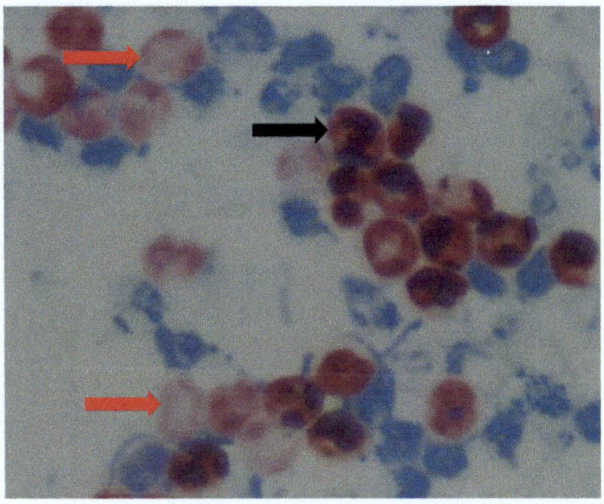

**Fig. 3** Kinyoun stain of *C. parvum* oocysts showing sporozoite-containing oocysts (black arrow) and "empty shells" (red arrows) corresponding to oocysts that released their four sporozoites. This staining method enables visual distinction between intact and empty oocysts, providing an indirect assessment of oocyst viability

8. Rinse with tap water for 1 min.
9. To counterstain, apply methylene blue for 1 min.
10. Rinse with tap water, drain, and air-dry.

### *3.2 Infection of Mice with* C. parvum *Oocysts*

*Cryptosporidium parvum* oocysts can be kept in 2.5% $K_2Cr_2O_7$ or water for up to 6 months with reasonable viability [57, 58], but viability decreases with age. However, in our laboratory, we experienced that *C. parvum* oocysts kept in 2.5% $K_2Cr_2O_7$ maintained viability longer (*see* **Note 4**). Also, oocysts less than 3 months old should have >80% viability.

1. In a sterile microcentrifuge tube, transfer from the oocyst stock a volume containing oocyst numbers that exceed the minimum required to infect all mice.

   *If oocysts were kept in 2.5%* $K_2Cr_2O_7$*, perform steps 2–4. If oocysts were kept in water, directly proceed with step 5.*

2. Wash oocysts with 1 mL of PBS 1X.
3. Centrifuge microcentrifuge tube at 10,000× *g* for 5 min at 4 °C.
4. Repeat **steps 2** and **3** for a total of three washes. If oocysts are still orange-colored due to trace of $K_2Cr_2O_7$, perform one supplemental wash.
5. Add PBS 1X for a total volume of 1 mL.
6. In a separate tube, mix 10 μL of oocysts with 10 μL of Trypan blue and load a chamber of a hemocytometer.

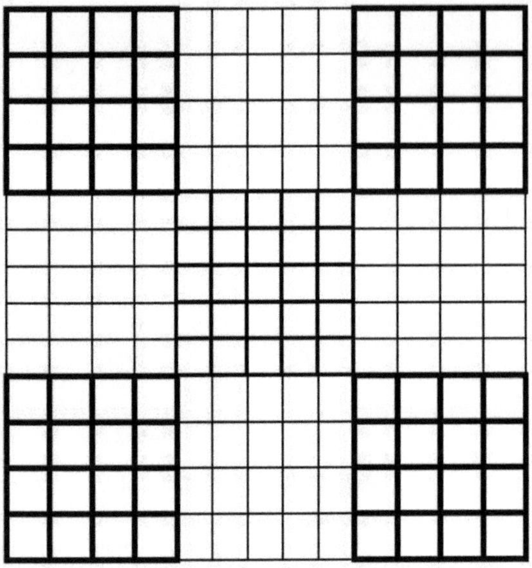

**Fig. 4** Hemocytometer grid presenting a large square divided into smaller squares. The grid is composed of a central section surrounded by four quadrants; each of the four quadrants contains a 4 × 4 grid of smaller squares outlined with bold lines. The central area consists of a 5 × 5 grid without bold outlines. The four quadrants (containing 4 × 4 smaller squares) at the four corners of the grid are the designated areas for oocyst counting

7. Counts intact oocysts in each of the four big squares (containing 4 × 4 small squares) at the four corners of the grid of the hemocytometer (*see* Fig. 4) using a light microscope.
8. Calculate the mean of the counts of the four big squares and multiply by the factor of dilution with Trypan blue (×2) and by the factor of the hemocytometer according to manufacturer protocol (×10,000) to obtain the concentration of oocysts/mL.
9. Dilute oocysts with PBS 1X to a final concentration of 30,000 oocysts/mL in a final volume exceeding the volume needed to infect all mice (3000 oocysts in 100 μL per mouse).
10. Keep oocysts on ice for the transport to the animal facility.
11. Fill 1-mL syringe with oocyst solution using 20G bulb gavage needle for mice weighing between 20 and 25 g or 22G bulb gavage needle for mice weighing between 15 and 20 g.
12. Eliminate any bubble and set plunger to a 100-μL increment.
13. Pick up the mouse to be gavaged by the scruff of the neck with index finger and thumb tightly grasping the skin around the neck (*see* **Note 5**).
14. Tuck the tail under the small finger to prevent movement of the lower body.

15. Gently insert the bulb gavage needle into the mouse esophagus by angling the needle along the roof of the mouth and back of the throat, keeping slightly to the left of center.
16. Wait for the mouse to swallow the bulb of the needle and allow the needle to descend slightly further in the esophagus (*see* **Note 6**).
17. Gently depress the syringe plunger to administer 100 μL of oocysts (*see* **Note 7**).
18. Carefully remove the gavage needle and return the mouse to the cage.
19. Check that the mouse is breathing and moving normally after gavage to ensure that the procedure was done correctly, and that no solution was aspirated.

### 3.3 Treatment of Mice

Usually, compounds are resuspended/diluted in distilled water, PBS, saline, or DMSO according to the chemical properties of each compound. Then, compounds are administered orally using a feeding needle, as mentioned above or via the parenteral route (often intraperitoneally). The dose, route of administration and treatment regimen must be optimized for each compound. We routinely prepare compounds fresh daily and administer them by oral gavage for 10 days starting at 3-day postinfection [14].

### 3.4 Monitoring of Infected Mice

Apart from signs of compound toxicity which are specific to each compound, mice should also be monitored for any sign of *C. parvum* infection adverse effects. Each institution will have their own endpoint criteria. Typical end points which may lead to euthanasia include weight loss (>20% weight lost from initial body weight), reluctance to move, hunched posture and/or severe sign of dehydration by pinching the skin, or drop in temperature. Clinical signs of cryptosporidiosis such as reduced activity and soft stool may be observed in more susceptible models after 7-day postinfection.

### 3.5 Evaluation of Treatment Efficacy

First, clinical signs of infection should be monitored closely to compare treated infected mice with untreated infected mice (positive controls). Weight curves and survival curves can also provide significant insights on the capacity of the compound to prevent frailty and lethality, respectively. Moreover, both oocyst shedding during the course of infection and intestinal parasite burden at the time of necropsy illustrate the effect of compounds on parasite survival and replication. Mice can be sacrificed either at peak of oocyst shedding, on Day 10 or Day 11 postinfection [16], or when they reach an interventional point requiring euthanasia as mentioned above. Oocysts from stool or intestine samples can be quantified by visual counting using a hemocytometer, by flow cytometry or by qPCR. In addition, histopathology presents a semiquantitative representation of oocyst burden in ileum of infected mice.

### 3.6 Collection of Mouse Fecal Pellet

1. Collect fresh stool in preweighed 1.7-mL microcentrifuge tube (*see* **Note 8**).
2. Weigh microcentrifuge tube containing fecal pellet to calculate weight of stool sample for normalization of count of oocysts.
3. Add 1 mL 2.5% $K_2Cr_2O_7$ or sterile cold distilled water into the tube.
4. Store stool samples at 4 °C until processed.

### 3.7 Collection of Mouse Intestine Sample

1. Collect intestine (except ileum which is kept for histology) at time of necropsy.
2. Cut intestine in 0.5 cm pieces in a preweighed 60-mL specimen container containing 10 mL 0.02% Tween 20 in PBS.
3. Weigh specimen container containing intestine for normalization of count of oocysts.
4. Store specimen container until oocyst purification at 4 °C (maximum 48 h).

### 3.8 Purification of Oocysts from Stool Sample Using Discontinuous Sucrose Gradients

1. Add 400 μL 2.5% $K_2Cr_2O_7$ to each pellet of approximately 20 mg.
2. Incubate pellet overnight at 4 °C.
3. Homogenize the fecal mixture using a wooden application stick and vortex.
4. In a separate 2.2-mL microcentrifuge tube, add 750 μL of 1:4 sucrose solution using a glass pipe, then add 750 μL of 1: 2 sucrose solution beneath the first solution (*see* **Note 9**).
5. Vortex the fecal sample again and let large debris settle for approximatively 30 s.
6. Add 200 μL of the supernatant at the top layer to the microscale sucrose gradients tube and store the remaining original fecal solution in 4 °C.
7. Centrifuge the microscale sucrose gradients tube for 20 min at 1000× *g*.
8. Aspirate the supernatant down to the 1.25-mL mark and transfer into a new tube.
9. Add PBS 1X to the 2-mL mark and centrifuge tube 16,000× *g* for 3 min at 4 °C.
10. Aspirate down to the 0.5-mL mark and add PBS 1X to the 2-mL mark.
11. Centrifuge tube 16,000× *g* for 3 min at 4 °C.
12. Repeat **steps 10** and **11** for four washes or until the sample is completely clear.
13. Aspirate down to the 0.25-mL mark and adjust volume to 1000 μL with PBS 1X or water.

**3.9 Purification of Oocysts from Stool Sample Using NaCl/Distilled Water Layer**

1. Add 1 mL of sterile cold distilled water to fecal pellet and store at 4 °C until process.
2. Homogenize stool sample by vortexing and doing up and down using pipet tips.
3. Centrifuge microcentrifuge tube 10,000× *g* for 5 min at 4 °C and discard supernatant.
4. Wash stool pellet with sterile cold distilled water and discard supernatant.
5. Resuspend stool pellet in 1 mL of saturated NaCl.
6. Vortex thoroughly stool sample and overlay with 250 μL of sterile cold distilled water.
7. Centrifuge microcentrifuge tube 2000× *g* for 10 min at 4 °C.
8. Collect supernatant and transfer in a separate 1.7-mL microcentrifuge tube.
9. Repeat **steps 5–8** a second time.
10. Pool two supernatants together in a 2.2-mL microcentrifuge tube.
11. Centrifuge the 2.2-mL microcentrifuge tube 10,000× *g* for 5 min at 4 °C.
12. Discard supernatant and resuspend oocysts in 1000 μL of sterile cold distilled water.
13. Store oocyst-containing sample at 4 °C for short-term storage until analysis.

**3.10 Purification of Oocysts from Intestine Sample Using NaCl/Distilled Water Layer**

1. Homogenize intestine sample using 150 W handheld tissue homogenizer.
2. Incubate intestine sample with 0.05 g sputasol and a stir bar for 90 min at 4 °C on a magnetic stirrer.
3. Transfer intestine sample to 50-mL centrifuge tube using cell scraper.
4. Centrifuge 50-mL centrifuge tube at 2000× *g* for 10 min at 4 °C and discard supernatant.
5. Resuspend the pellet in 8 mL 0.02% (v/v) Tween 20 in distilled water and 4 mL of diethyl ether (or ethyl acetate, where required by local safety guidelines).
6. Centrifuge 50-mL centrifuge tube at 2000× *g* for 10 min at 4 °C.
7. Discard the supernatant and fatty layer and wash the pellet with 20 mL sterile, cold distilled water.
8. Centrifuge 50-mL centrifuge tube at 2000× *g* for 10 min at 4 °C.

9. Discard the supernatant and resuspend pellet in 20 mL saturated NaCl.
10. Vortex thoroughly intestine sample and overlay with 5 mL sterile cold distilled water.
11. Centrifuge 50-mL centrifuge tube at 2000× *g* for 10 min at 4 °C.
12. Collect 1 mL of the interphase in each of four 2.2-mL microcentrifuge tubes prefilled with 1 mL sterile cold distilled water (total 1 mL sample + 1 mL water in each of four tubes/mouse).
13. Invert several times 2.2-mL microcentrifuge tubes to mix sample/water to dilute traces of NaCl.
14. Centrifuge tubes at 16,400× *g* for 30 min at 4 °C (*see* **Note 10**).
15. Discard 1.8 mL of supernatant of each 2.2-mL microcentrifuge tube.
16. Resuspend oocysts in remaining 200 μL in each 2.2-mL microcentrifuge tube.
17. Pool oocyst solutions of the four tubes from the same mouse.
18. Adjust volume of combined oocyst sample to 1000 μL with sterile cold distilled water.
19. Store oocyst-containing sample at 4 °C for short-term storage until analysis.

### 3.11 Quantification of Oocysts by Flow Cytometry Using OW50-FITC Monoclonal Antibody

The oocyst-specific monoclonal antibody labeled with fluorescein isothiocyanate (e.g., OW50-FITC), same as in MERIFLUOR *Cryptosporidium*/*Giardia* test, can be used to count *C. parvum* oocysts by flow cytometry [27].

1. Prepare a 1:10 antibody solution: 1.8 mL PBS + 200 μL OW50-FITC.
2. Cover the antibody solution and samples with foil; work with the hood light off due to OW50-FITC's light sensitivity.
3. Transfer 250 μL of purified stool sample to a 1.7-mL microcentrifuge tube.
4. Add 20 μL of OW50-FITC antibody solution to each sample and vortex.
5. Incubate for at least 20 min.
6. Wash two times with PBS and add 4% paraformaldehyde to fix the sample, bringing the sample volume up to 1 mL (PBS wash steps are not absolutely necessary; paraformaldehyde can be added directly to the oocyst/antibody suspension and then analyzed).

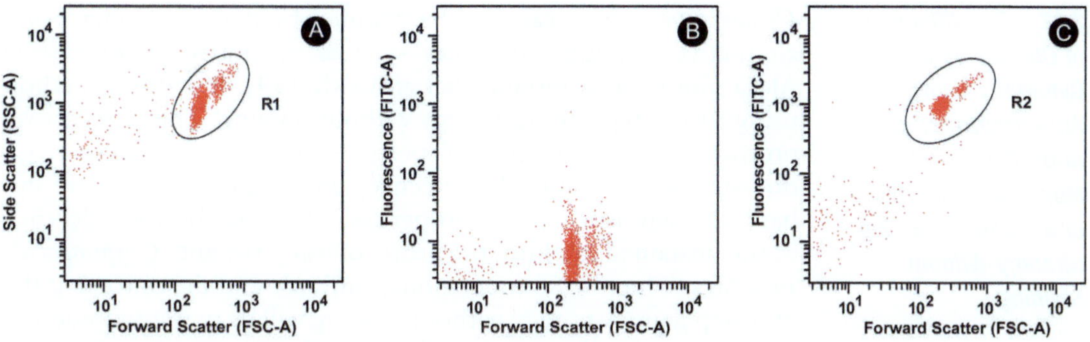


**Fig. 5** Flow cytometric analysis of *C. parvum* oocysts using oocyst-specific monoclonal antibody labeled with fluorescein isothiocyanate OW50-FITC labeling. Three-panel figure showing X-Y charts labeled A, B, and C. Panel A illustrates a purified oocyst sample plotted in two dimensions including forward scatter (size) and side scatter (internal complexity). Axes are presented in log scale, and oocysts are clearly evident in region 1 (R1). Panel B illustrates the same population of particles seen in Panel A using forward scatter versus log fluorescence with the 488-nm laser and the FITC-A filter. Here, the sample has not been labeled with OW50-FITC reagent, and all particles show background fluorescence (nonfluorescent). Panel C illustrates the same sample labeled with OW50-FITC, clearly showing the intense fluorescence of the labeled oocysts and a slight increase in the background fluorescence of the nonoocyst debris. Therefore, panel C is similar to B but includes a cluster labeled R2. Coupling regions 1 and 2 (R1 and R2) into a "logical gate" enables the quantification of antibody labeled oocysts by the flow cytometer instrument and software. Oocyst labeling is effective with specific antibodies and with appropriate fluorescent lectins. Fluorescent labeling is particularly important if excessive debris remains in the sample after the oocyst purification step

7. Vortex and let the sample fix overnight or at least for 1 h.
8. Transfer 250 μL of each sample to a 96-well plate or flow tube.
9. Perform flow cytometry according to the manufacturer's recommendations.
10. Acquire samples in 96-well plate or in flow tubes using a set volume (100 μL) at a particular flow rate or a set number of events to acquire.
11. Run positive controls with labeled and unlabeled oocysts to calibrate region settings necessary to discriminate the labeled oocyst population from the background.
12. Draw "regions" to define the oocyst-containing population ("oocyst region") in the forward scatter (FSC) versus side scatterplot (SSC) (*see* Fig. 5, panel A).
13. Draw a "fluorescent oocyst region" in the fluorescence (FL1) plot to identify the antibody-labeled subset (*see* Fig. 5, panel C).
14. Create a logical "oocyst gate" by combining the "oocyst region" defined in the FSC/SSC plot with the "fluorescent oocyst region" in the FL1 plot.

### 3.12 Quantification of Oocysts by Flow Cytometry Using an Anti-cryptosporidium Alexa Fluor® 488 Monoclonal Antibody or a Two-Step Gating Strategy Without Staining

*C. parvum* oocysts can also be counted by flow cytometry using alternative staining antibodies such as an anti-*Cryptosporidium* Alexa Fluor® 488 monoclonal antibody [59] or a novel method using a two-step gating strategy without staining. This latter, novel protocol uses the inherent properties of *C. parvum* oocysts (e.g., oocysts show no auto-fluorescence, unlike intestinal debris, and have characteristic optical properties). To describe the relevance of the unstained method, the protocol using the anti-*Cryptosporidium* Alexa Fluor® 488 monoclonal antibody and the one using the two-step gating strategy without staining will be presented side-by-side. Using the above protocols to purify oocysts from stool and intestine samples, oocysts purified are a homogeneous population of organisms corresponding to a clear population of events in a FSC-A vs SSC-A plot. One gate surrounding this population defines oocyst-like events which can then be further studied to determine fluorescent events. With the unstained method, the channel usually used for PerCP allows adequate gating between oocysts and debris, because intestinal debris (e.g., vegetal fibers) are auto-fluorescent in the PerCP channel and can easily be differentiated from oocysts which are not auto-fluorescent (*see* **Note 11**). This approach offers a cost-effective alternative when working with highly purified samples [60].

1. Vortex thoroughly oocyst sample kept in sterile cold distilled water at 4 °C.
2. Transfer in microtiter tube 4 μL of oocysts purified from stool with 16 μL of sterile cold distilled water or 20 μL of oocysts purified from intestine.

   *If using the protocol without staining antibody, directly proceed with step 8.*
3. Incubate oocyst sample with 1 μL mouse anti-*Cryptosporidium* Alexa Fluor® 488 monoclonal antibody (final concentration 0.5 μg/mL) for 30 min at room temperature.
4. Wash oocyst sample with 900 μL of 1% w/v bovine serum albumin in PBS 1X.
5. Centrifuge oocyst samples at 10,000× *g* for 5 min at room temperature.
6. Discard supernatant.
7. Resuspend oocysts in 20 μL PBS 1X.
8. Add 60 μL sterile cold distilled water in microtiter tubes.
9. Add 100 μL 8% paraformaldehyde in PBS in microtiter tubes.
10. Incubate samples overnight at room temperature or at 4 °C (*see* **Note 12**).
11. Add 20 μL counting beads (total volume of 200 μL).
12. Vortex samples immediately before acquisition.

13. Acquire samples using a flow cytometer such as BD LSRFortessa or BD LSRFortessa X-20.
14. Use a flow cytometry analysis software such as FlowJo v.10 for data analysis.
15. Define a "bead gate" surrounding counting beads in the FSC-A vs SSC-A plot.
16. Define an "oocyst gate" surrounding the oocyst-like events in the FSC-A vs SSC-A plot.
17. From events in the "oocyst gate" (*see* Fig. 6):
    *Top panel*: method *with* staining antibody: define a gate around Alexa Fluor-488-positive events.

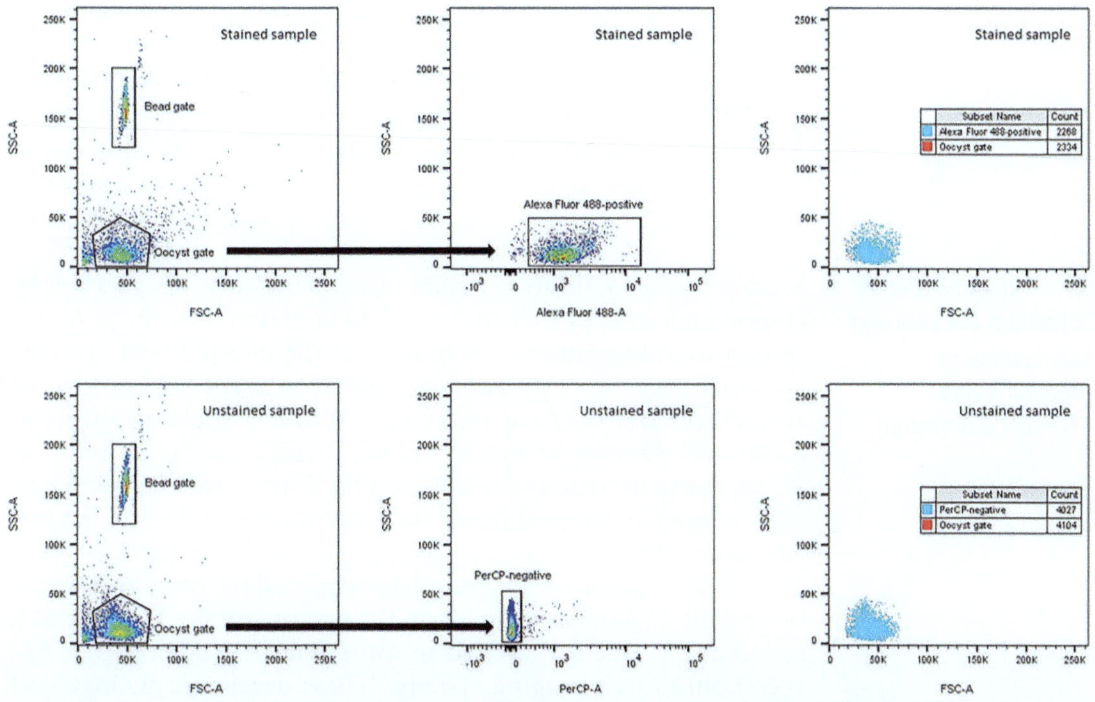


**Fig. 6** Flow cytometry analysis of stained and unstained oocysts from intestine sample of the same infected mouse. Arrows indicate the gating strategy. The top row shows stained samples with three scatter plots. Top panels from left to right: the first plot displays FSC-A vs. SSC-A with a 'Bead gate' and 'Oocyst gate'; the second plot shows Alexa Fluor 488-A vs. SSC-A with an 'Alexa Fluor 488-positive' gate; and the third plot displays FSC-A vs. SSC-A and includes a legend indicating counts for 'Alexa Fluor 488-positive' and 'Oocyst gate'. The third plot displays an overlay of Alexa Fluor-488-positive events over events in oocyst gate (visible extra red events being debris). The bottom row shows unstained samples with similar scatter plots. Bottom panels from left to right: the first plot displays FSC-A vs. SSC-A with a 'Bead gate' and 'Oocyst gate'; the second plot shows PerCP-A vs. SSC-A with a 'PerCP-negative' gate; and the third plot displays FSC-A vs. SSC-A and includes a legend indicating counts for 'PerCP-negative' and 'Oocyst gate'. The third plot displays an overlay of PerCP-negative events over events in oocyst gate (visible extra red events being debris). These two methods provide similar result, but stained method presents lower counts of oocysts, because some were lost during washes

*Bottom panel*: method *without* staining antibody: define a gate around PerCP-negative events.

18. Determine counts of oocysts in acquired sample by applying appropriate formula:
    *Method with staining antibody*: (count of Alexa Fluor-488-positive events in "oocyst gate" X quantity of beads per 20 μL of counting beads ÷ count of events in "bead gate").
    *Method without staining antibody*: (count of PerCP-negative events in "oocyst gate" X quantity of beads per 20 μL of counting beads ÷ count of events in "bead gate").
19. Subtract count of uninfected control (debris) from count of sample as a baseline *(optional)*.
20. Multiply count of oocysts in acquired sample X250 for sample purified from stool (4 μL acquired vs 1000 μL total volume of sample purified from stool) or X50 for sample purified from intestine (20 μL acquired vs 1000 μL total volume of sample purified from intestine).
21. Normalize final count of oocysts using the weight of the stool or the intestine sample (e.g., count of oocysts/g of stool or intestine).

### 3.13 Identification of C. parvum Oocysts and Localization of Proteins Using Confocal Microscopy

Studies on the localization of p23 protein in *C. parvum* parasites by confocal microscopy may contribute to understanding the potential of p23 to induce immune responses in the infected host. Specific antibodies against *C. parvum* proteins (such as anti-p23 polyclonal antibodies) and confocal microscopy afford a triple action: first, identify *C. parvum* parasites; second, localize specific proteins on *C. parvum* parasites; and third, identify if these proteins are accessible in oocysts permeabilized via excystation or other treatments (*see* **Note 13**).

While earlier studies have utilized confocal microscopy to localize specific *C. parvum* proteins on the parasite surface [48, 49, 61], recent advancements have led to more refined protocols [62, 63] and sophisticated imaging systems. These developments may lead to even more detailed and informative results in future research. Herein are protocols to produce anti-p23 polyclonal antibodies needed for assays and confocal microscopy experiments.

#### Preliminary Steps to Obtain Anti-p23 Polyclonal Antibodies Prior to Confocal Microscopy

1. Immunize rabbits with recombinant proteins (5 immunizations at 2- to 3-week intervals, 0.25 mg per immunization, 1 mg/mL of p23 protein in PBS with complete Freund's adjuvant for the first immunization and with incomplete Freund's adjuvant for the following immunizations) [64].

2. Titrate rabbit hyperimmune sera from the fourth production bleed to demonstrate high affinity (high absorbance values at low dilution factors) and high specificity (proportional decrease of absorbance values with increasing dilutions) of polyclonal antibodies for the p23 recombinant protein.
3. Affinity purify the anti-p23 polyclonal antibodies. First, immobilize p23 recombinant protein by covalent attachment on resin. Then, incubate the rabbit hyperimmune sera with resin and elute the anti-p23 polyclonal antibodies.
4. Concentrate the anti-p23 polyclonal antibodies using a filter device and dialyze them against PBS.
5. Dilute anti-p23 polyclonal antibodies at a concentration of 400 μg/mL.

**Confocal Microscopy Protocol to Identify and Localize *C. parvum* Proteins (Adapted from ( [62, 63]))**

1. Wash oocysts stored at 4 °C in 2.5% w/v $K_2Cr_2O_7$ for three times in PBS.
2. Centrifuge (10,000× *g* for 5 min at 4 °C) and resuspend oocysts in PBS.
3. Count oocysts using a hemocytometer as mentioned in previous protocols.
4. Incubate overnight in 4% paraformaldehyde in PBS $1 \times 10^6$ oocysts in each chamber of an imaging eight-chamber slide and let dry at room temperature (for information, incubation at room temperature may promote excystation in some oocysts) (*see* **Note 14**).
5. The next day, wash slide three times in PBS (500 μL) for 5 min each wash.
6. Block nonspecific binding by incubation with 5% goat serum in PBS (200 μL) for at least 1 h.
7. Wash the slide once in PBS (500 μL) for 1 min.
8. Incubate overnight at 4 °C or for 1 h at room temperature with anti-p23 polyclonal antibodies (1:25 dilution, final concentration of 16 μg/mL) in 1% BSA in PBS (200 μL) or slides were incubated in 1% BSA in PBS (200 μL, negative control).
9. Wash the slide three times in PBS (500 μL) for 5 min each wash.
10. Incubate with Alexa Fluor 647 conjugate goat anti-rabbit secondary antibody at a final concentration of 2 μg/mL in 1% BSA in PBS (200 μL) for 1 h at room temperature.
11. Wash the slide three times in PBS (500 μL) for 5 min each wash.

12. Incubate with Alexa Fluor 488 conjugate monoclonal IgG mouse anti-*Cryptosporidium* clone BEL 0126 at a final concentration of 2 μg/mL in 1% BSA in PBS (200 μL) for 1 h at room temperature.
13. Wash the slide three times in PBS (500 μL) for 5 min each wash.
14. Apply 500 μL of PBS to each chamber with one drop of DAPI stain. Incubate with DAPI staining for 15 min.
15. Acquire image with confocal microscope such as Zeiss LSM780-NLO laser scanning confocal with IR-OPO laser microscope (*see* Fig. 7).

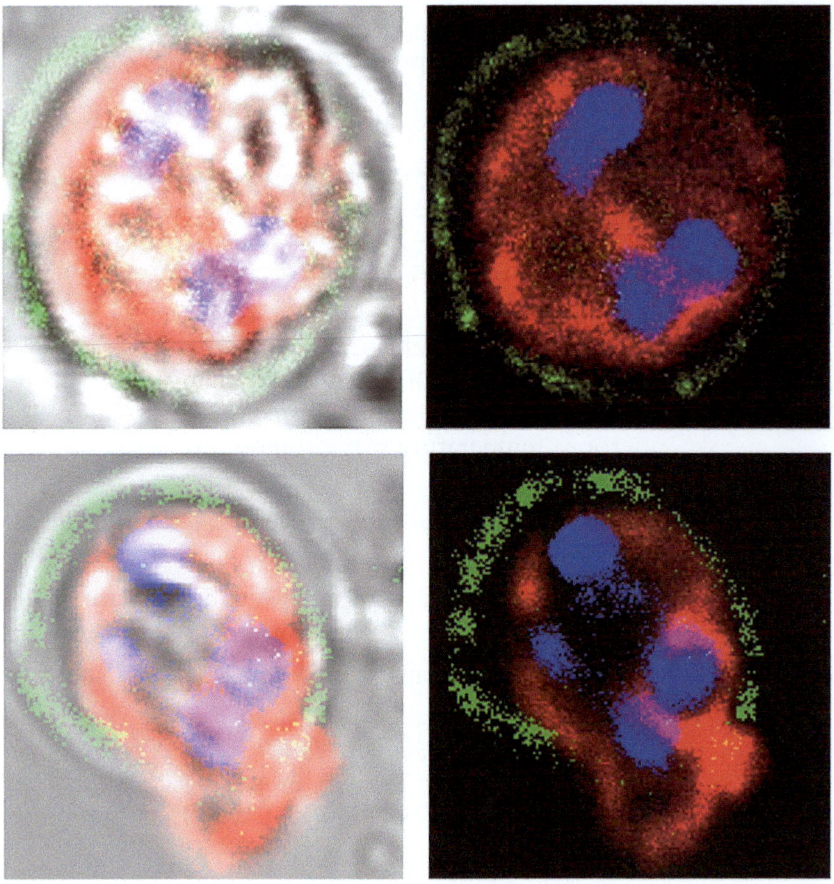

**Fig. 7** Confocal microscopy showing four panels of *C. parvum* oocysts close to or during excystation. Nuclei of sporozoites are side-by-side two-by-two stained with DAPI shown in blue. Oocyst wall is stained with Alexa Fluor 488 conjugate monoclonal IgG mouse anti-*Cryptosporidium* clone BEL 0126 shown in green. Shown in red, *C. parvum* p23 protein is stained with Alexa Fluor 647 conjugate goat anti-rabbit secondary antibody recognizing rabbit anti-p23 polyclonal antibodies. Staining with DAPI identifies all sporozoites even the ones outside oocysts and anti-*C. parvum* wall antibodies stain all oocysts, even "empty shells." In contrast, there will be no p23 protein to stain if the sample is only made of "empty shells"

### 3.14 Oocyst Genomic DNA Extraction and Purification

The extraction of genomic DNA from *C. parvum* oocysts presents a unique challenge due to their robust outer wall [65]. The freeze–thaw cycle technique has emerged as an efficient approach, capitalizing on the physical stress induced by extreme temperature fluctuations to disrupt the oocyst wall and release genetic material [66]. This method offers several advantages over chemical or enzymatic lysis, including reduced risk of introducing PCR inhibitors, minimal DNA degradation, and cost-effectiveness [67]. The mechanism relies on the expansion and contraction of water within the oocysts as temperatures alternate between freezing and thawing, eventually causing the oocyst wall to rupture [67]. Studies have shown that the number of freeze–thaw cycles, and the freezing temperature can significantly impact DNA yield and quality [66, 68]. Some research also suggests that moderate temperatures (e.g., −20 °C) may be as effective as more extreme conditions for specific sample types [68].

1. Vortex thoroughly oocyst sample kept in sterile, cold distilled water at 4 °C.
2. Transfer 250 μL of oocysts purified from stool or intestine in 1.7-mL microcentrifuge tubes.
3. Freeze the oocyst sample in liquid nitrogen for 2 min.
4. Briefly dip the bottom of the microcentrifuge tube in room temperature water to warm up the sample.
5. Thaw the oocyst sample in a 56 °C water bath for 2 min.
6. Briefly dip the bottom of the microcentrifuge tube in room temperature water to cool down the sample.
7. Repeat **steps 3–6** for a total of 10 freeze–thaw cycles.
8. Extract DNA from the oocyst sample using the Qiagen QIAamp DNA Mini Kit.
9. Quantify the DNA using the NanoDrop reader and store it at −20 °C suspended in 200 μL for qPCR analysis outlined below.

### 3.15 Quantification of Oocysts by qPCR Analysis

Quantitative PCR has become the gold standard for the detection and quantification of *Cryptosporidium* spp. in clinical and environmental samples due to its high sensitivity, specificity, and ability to quantify parasite load [69–71]. While several targets have been used for *C. parvum* detection and species differentiation, including the COWP (*Cryptosporidium* oocyst wall protein) gene, and the gp60 gene, which is particularly useful for subtyping [72, 73], our protocol focuses on the HSP70 gene, because this method showed high efficacy and specificity [17]. We are now using the Qiagen Rotor-Gene Q 5plex machine because of its versatility and its higher sensitivity (*see* **Note 15**). Compared to other machines, any tube can be used in the Rotor-Gene machine instead of using expensive capillaries [17, 74]. Also, supplies for the method with capillaries will soon be discontinued.

1. Prepare mastermix for qPCR reaction (multiply volume of reagents × number of samples):
   - 7.75 μL of PCR-grade water
   - 2 μL of 10 μM forward primer (5′-AACTTTAGCTCCAG TTGAGAAAGTACTC-3′)
   - 2 μL of 10 μM reverse primer (5′-CATGGCTCTTTACCG TTAAAGAATTCC-3′)
   - 2 μL of 2 μM 5′ 6-carboxyfluorescein reporter dye (FAM) probe with a 3′ TMR quencher dye (5′-AATACGTGTAG AACCACCAACCAATACAACATC-3′)
   - 6.25 μL of 4X mastermix from Qiagen QuantiNova Multiplex PCR Kit.
2. Transfer 20 μL of mastermix in Qiagen 200 μL PCR-grade tubes.
3. Add 5 μL of DNA extracted from oocyst sample to mastermix in 200-μL PCR grade tubes.
4. Run qPCR in Qiagen Rotor-Gene Q 5plex using settings optimized from previous protocols [14, 75]: 95 °C for 5 min, 46 cycles at 95 °C for 5 s, and 63 °C for 10 s.
5. In parallel, establish standard curve from known concentrations of oocysts (*see* Fig. 8): Infection of mice with *C. parvum* oocysts, wash oocysts from stock kept in 2.5% $K_2Cr_2O_7$ at 4 °C with PBS 1X, count oocysts using a hemocytometer and calculate concentration of oocysts.

   Prepare serial dilution of oocysts → series of samples with known concentrations of oocysts.

   Freeze–thaw, extract DNA and perform qPCR on series of samples with known concentrations of oocysts and determine their threshold cycle values (Ct values).
6. Extrapolate parasite burden of mouse sample by comparing Ct values with the standard curve made from Ct values of samples with known oocyst concentrations.
7. The threshold should be in the exponential phase of amplification.

### 3.16 Quantification of Oocysts by Histopathologic Analysis of Ileum

Histopathological examination is a crucial method for studying *C. parvum* infection in animal models. This technique allows for direct visualization of the parasites within the host tissue, primarily the ileum, and provides valuable insights into the extent of tissue damage and host–parasite interactions [76, 77]. Unlike quantitative methods such as qPCR or flow cytometry, histopathology offers a unique perspective on the spatial distribution of parasites and the associated inflammatory response [78, 79]. The histopathologic method offers a visual representation of the parasite burden

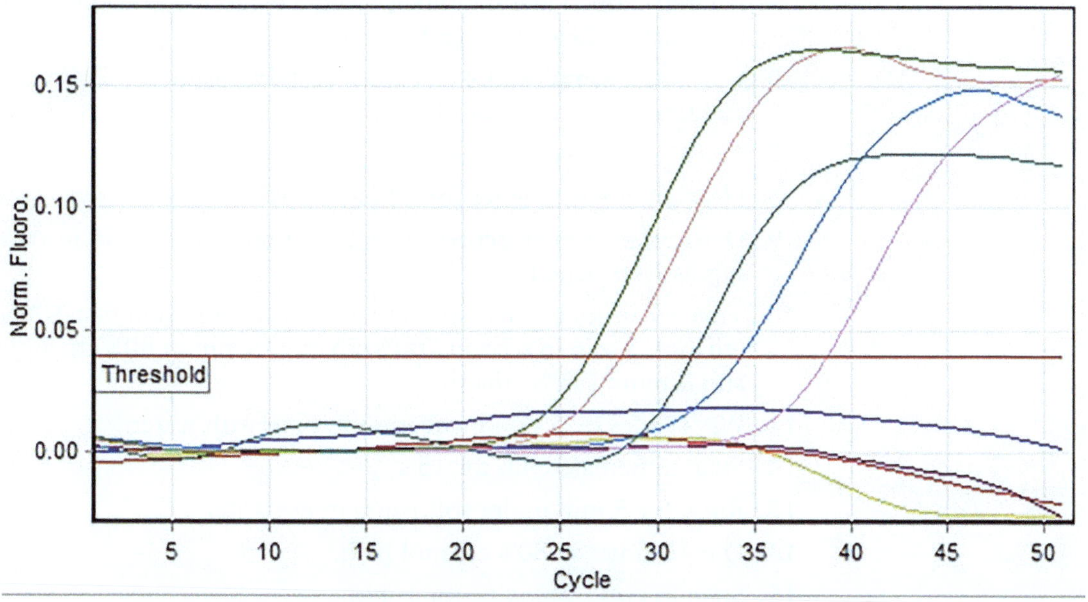


**Fig. 8** Representation of Ct values of known concentrations of *C. parvum* oocysts. The graph shows fluorescence over 50 cycles, with multiple colored curves representing different samples. A red horizontal line labeled 'Threshold' indicates a specific fluorescence level. The graph illustrates the amplification, with most curves rising sharply after cycle 25, indicating significant changes in fluorescence levels. From right to left, line represents sample of $10^3$ oocysts in light purple, Ct value = 38.76; $10^4$ oocysts in light blue, Ct value = 34.31; $10^5$ oocysts in dark green, Ct value = 31.81; $10^6$ oocysts in pink, Ct value = 28.17; $10^7$ oocysts in light green, Ct value = 26.54. Lines representing sample of $10^2$ oocysts and $10^1$ oocysts are dark purple and dark blue, respectively. The yellow line represents negative control (DNA from a human *Cryptosporidium*-negative patient) and the red line, no template control (no DNA)

in the ileum while providing only a semiquantitative value. We routinely use a custom scoring system in which grade 1 = absence of parasite, grade 2 = occasional presence of parasites, grade 3 = abundance of parasites without clusters, grade 4 = abundance of parasites with occasional clusters, and grade 5 = abundance of parasites with presence of multiple clusters. If a sample exceeds the preceding grade without meeting the criteria of the next grade, 0.5 is added to its score.

1. Collect 1.5 cm of ileum (distal part of the small intestine) at time of necropsy.
2. Store ileum in a histology cassette in 10% formalin for a minimum of 72 h.
3. Transfer histology cassettes in 70% ethanol just before processing to wash 10% formalin.
4. Transfer histology cassettes to a tissue processor such as Leica ASP300S tissue processor.

5. Paraffin-embed ileum samples using a paraffin embedding station such as Leica EG1150 station.
6. Cut ileum samples in 4-μm-thick sections using a microtome and transfer sections on slides.
7. Air dry microscope slides for at least 48 h.
8. Heat microscope slides for 30 min at 60 °C.
9. Deparaffinize microscope slides by 10-min incubation in two successive xylene baths.
10. Transfer slides to successive ethanol baths (5 min in first 100% ethanol, 5 min in second 100% ethanol, 3 min in 80% ethanol, and 3 min in 70% ethanol).
11. Wash slides by 5 min incubation in bath with water.
12. Stain in hematoxylin for 15 s.
13. Rinse for 5 min under soft running tap water.
14. Dip 10 times in 80% ethanol bath.
15. Stain in eosin-phloxine for 1 min.
16. Wash for 5 min in bath containing eosine-phloxine:100% ethanol in a 1:50 ratio.
17. Rinse slides with squeeze bottle filled with 100% ethanol.
18. Transfer slides for 5 min into 100% ethanol bath, then for 5 min into first xylene bath, then for 5 min into second xylene bath.
19. Wipe excess of xylene and apply two drops of mounting media and a coverslip.
20. Squeeze coverslip and microscope slide together to remove bubbles.
21. Use a light microscope to identify *C. parvum* parasites in ileum sections (*see* Fig. 9 and **Note 16**).

## 4 Notes

1. *C. parvum* oocysts are immediately infectious, measuring 4–6 μm in diameter [84]. The median infectious dose for humans is 132 oocysts in healthy adult volunteers [31, 80, 81]. Due to the zoonotic potential of this parasite, it would be advisable that immunocompromised people and pregnant women avoid contact with infectious *C. parvum* oocysts. Few cases of laboratory-acquired cryptosporidiosis have been reported, emphasizing the need for strict adherence to safety protocols [82].

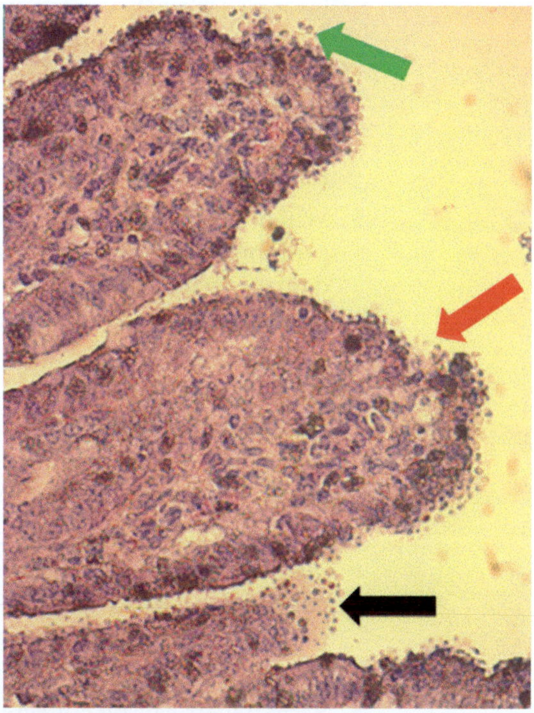

**Fig. 9** Histopathologic representation at 400× magnification of ileum from an infected mouse showing *C. parvum* parasites (green arrow), cellular damage/ loss of membrane integrity (red arrow), and clumps of *C. parvum* parasites (black arrow). This histopathological representation provides crucial insights into the parasite's interaction with host tissue, the extent of infection, and the resulting tissue damage

2. For Sheather's solution preparation, add 454 g of sucrose to 355 mL of distilled water. Dissolve slowly over gentle heat. Cool and add 6.0 mL full-strength phenol as a preservative. Store at room temperature.
3. To reduce the number of "empty shells" and improve the quality of a stock of parasites, *C. parvum* oocysts should be stored at 4 °C and, during experiments, reagents should be cold, and centrifugations performed at 4 °C to avoid excystation which releases sporozoites from oocysts.
4. To maintain *C. parvum* oocysts, we infect 6- to 8-week-old IFNγR-KO mice with 3000 oocysts stored in 2.5% $K_2Cr_2O_7$ at 4 °C. Oocysts are then purified from intestines of infected mice as per the protocol presented above and stored in 2.5% $K_2Cr_2O_7$ at 4 °C as we experienced a loss of infectivity of attenuated stains of *C. parvum* oocysts when kept in sterile cold distilled water at 4 °C. However, when conducting a study with mice to evaluate treatment efficacy of a compound, oocysts purified from stool or intestine of infected mice can

be kept in sterile cold distilled water at 4 °C instead of 2.5% $K_2Cr_2O_7$. If stored in $K_2Cr_2O_7$, wash the $K_2Cr_2O_7$ from the samples, or it may interfere with readout technique (i.e., reduce sensitivity of the protocol of flow cytometry using staining antibody) to quantify oocysts in stool or intestine samples of mice of different groups.

5. When inoculating mice, be sure that the grip is secure and prohibits the mouse from moving its head in order to prevent damage to internal tissues during gavage. The mouse should be immobilized, but not so much that it shows signs of distress.
6. During oral gavage using a 20G bulb gavage needle, no resistance should be felt in mice older than 6 weeks old. If any resistance is felt during insertion of the gavage needle or if the mouse at any time begins to gasp, gently remove the needle and again try to find the esophagus.
7. It is possible for reflux to occur during oral gavage with proper insertion of the gavage needle. Adjusting the volume and viscosity of the solution can help to limit reflux and ensure accurate dosing. Regular training and adherence to animal welfare guidelines are crucial for proper oral gavage technique.
8. Fecal pellets can be collected by placing mice in cages or other containers for a brief period of time. Another way is to pick the mice up (mice usually defecate instantly), and stool can be collected into a tube.
9. To prepare gradients microcentrifuge tubes, put the tip of the glass pipet at the very bottom of the tube when adding the 1: 2 sucrose solution, beneath the 1:4 sucrose solution, and make sure the interface rises.
10. Several factors can influence the quality and efficiency of oocyst purification. "Empty shells" are oocysts that have released their sporozoites. They can affect quantification and should be considered when analyzing results. Viability staining or excystation assays can help differentiate between viable and nonviable oocysts. Also, given the small size and mass of *Cryptosporidium* oocysts, they can sometimes fail to sediment properly. The protocol's centrifugation speeds and times have been optimized for oocyst recovery, but slight adjustments may be necessary depending on the specific centrifuge model used.
11. This unstained method to count oocysts by flow cytometry was proven efficient to differentiate oocysts from debris in numerous experiments in our laboratory and to be as effective as the protocol using staining antibodies with the additional advantage of being less expensive [17, 60]. Obviously, the sine qua non condition is to start with a purified population of oocysts containing rare intestinal debris which is an objective easily achievable with protocols presented above.

12. The choice between room temperature and 4 °C incubation for paraformaldehyde fixation involves a trade-off. On one hand, paraformaldehyde fully inactivates oocysts at room temperature. However, if oocysts are incubated overnight at room temperature, some excystation may occur. One can argue that all samples are exposed to the same conditions. Thus, samples can be compared with one another. On the other hand, incubation of oocysts overnight at 4 °C inhibits excystation, but paraformaldehyde loses some efficacy at 4 °C and doesn't inactivate 100% of oocysts. Therefore, some bleach should be passed into the flow cytometer for cleaning after acquisition, and some bleach will need to be added to the waste tank of the flow cytometer.
13. This protocol can be used to identify *C. parvum* oocysts in a sample. To do so, it is vital to take into consideration that some oocysts are lost during washes. Therefore, while this technique provides valuable insights, qPCR methods remain the most sensitive to assess the presence or absence of *C. parvum* oocysts in samples [83]. As for the anti-p23 polyclonal antibodies (which allow the localization of p23 proteins), antibodies specific for other *C. parvum* proteins will allow the localization of these proteins.
14. When using this protocol for protein localization, it's recommended to start with $1 \times 10^6$ oocysts to account for losses during processing. This applies not only to p23 but also to other *C. parvum* proteins of interest. Both polyclonal antibodies (obtained following the preliminary steps presented) and monoclonal antibodies can be used for protein localization studies.
15. It is crucial to note that the cycling conditions reported in our protocol are specific to the Qiagen Rotor-Gene Q 5plex machine. When using different systems, researchers should perform optimization steps, including standard curve, to determine the optimal annealing temperatures for their specific primer sets and machines. This optimization is essential to ensure accurate diagnosis and avoid potential false positive or negative results. Proper optimization and adherence to manufacturer guidelines will help maintain the high sensitivity and specificity of the assay.
16. *C. parvum* oocysts from different strains are indistinguishable by microscopy, but more virulent strains such as the Iowa strain can induce more tissue damage as well as melena. When interpreting histopathological results, consider that the parasite burden may vary along the intestinal tract, and multiple sections may need to be examined for a comprehensive assessment. Additionally, the timing of sample collection relative to the course of infection can significantly impact the observed parasite burden and tissue damage.

## Competing Interest Statement

The authors have no conflict of interests to declare that are relevant to the content of this chapter.

### References

1. Li T, Liu H, Jiang N, Wang Y, Wang Y, Zhang J, Shen Y, Cao J (2021) Comparative proteomics reveals *Cryptosporidium parvum* manipulation of the host cell molecular expression and immune response. PLoS Negl Trop Dis 15(11):e0009949. https://doi.org/10.1371/journal.pntd.0009949
2. Putignani L, Menichella D (2010) Global distribution, public health and clinical impact of the protozoan pathogen *cryptosporidium*. Interdiscip Perspect Infect Dis 2010. https://doi.org/10.1155/2010/753512
3. Painter JE, Hlavsa MC, Collier SA, Xiao L, Yoder JS, Centers for Disease Control and Prevention (2015) Cryptosporidiosis surveillance – United States, 2011–2012. MMWR Suppl 64(3):1–14
4. Collier SA, Deng L, Adam EA, Benedict KM, Beshearse EM, Blackstock AJ, Bruce BB, Derado G, Edens C, Fullerton KE, Gargano JW, Geissler AL, Hall AJ, Havelaar AH, Hill VR, Hoekstra RM, Reddy SC, Scallan E, Stokes EK, Yoder JS, Beach MJ (2021) Estimate of burden and direct healthcare cost of infectious waterborne disease in the United States. Emerg Infect Dis 27(1):140–149. https://doi.org/10.3201/eid2701.190676
5. Hlavsa MC, Cikesh BL, Roberts VA, Kahler AM, Vigar M, Hilborn ED, Wade TJ, Roellig DM, Murphy JL, Xiao L, Yates KM, Kunz JM, Arduino MJ, Reddy SC, Fullerton KE, Cooley LA, Beach MJ, Hill VR, Yoder JS (2018) Outbreaks associated with treated recreational water – United States, 2000–2014. MMWR 67(19):547–551. https://doi.org/10.15585/mmwr.mm6719a3
6. Kotloff KL, Nataro JP, Blackwelder WC, Nasrin D, Farag TH, Panchalingam S, Wu Y, Sow SO, Sur D, Breiman RF, Faruque AS, Zaidi AK, Saha D, Alonso PL, Tamboura B, Sanogo D, Onwuchekwa U, Manna B, Ramamurthy T, Kanungo S, Ochieng JB, Omore R, Oundo JO, Hossain A, Das SK, Ahmed S, Qureshi S, Quadri F, Adegbola RA, Antonio M, Hossain MJ, Akinsola A, Mandomando I, Nhampossa T, Acacio S, Biswas K, O'Reilly CE, Mintz ED, Berkeley LY, Muhsen K, Sommerfelt H, Robins-Browne RM, Levine MM (2013) Burden and aetiology of diarrhoeal disease in infants and young children in developing countries (the Global Enteric Multicenter Study, GEMS): a prospective, case-control study. Lancet 382(9888):209–222. https://doi.org/10.1016/S0140-6736(13)60844-2
7. Shirley DA, Moonah SN, Kotloff KL (2012) Burden of disease from cryptosporidiosis. Curr Opin Infect Dis 25(5):555–563. https://doi.org/10.1097/QCO.0b013e328357e569
8. Khalil IA, Troeger C, Rao PC, Blacker BF, Brown A, Brewer TG, Colombara DV, De Hostos EL, Engmann C, Guerrant RL, Haque R, Houpt ER, Kang G, Korpe PS, Kotloff KL, Lima AAM, Petri WA Jr, Platts-Mills JA, Shoultz DA, Forouzanfar MH, Hay SI, Reiner RC Jr, Mokdad AH (2018) Morbidity, mortality, and long-term consequences associated with diarrhoea from *Cryptosporidium* infection in children younger than 5 years: a meta-analyses study. Lancet Glob Health 6(7): e758–e768. https://doi.org/10.1016/S2214-109X(18)30283-3
9. Khalil S, Mirdha BR, Panda A, Singh Y, Makharia G, Paul J (2017) *Cryptosporidium* species subtypes and associated clinical manifestations in Indian patients. Gastroenterol Hepatol Bed Bench 10(4):311–318
10. Sow SO, Muhsen K, Nasrin D, Blackwelder WC, Wu Y, Farag TH, Panchalingam S, Sur D, Zaidi AK, Faruque AS, Saha D, Adegbola R, Alonso PL, Breiman RF, Bassat Q, Tamboura B, Sanogo D, Onwuchekwa U, Manna B, Ramamurthy T, Kanungo S, Ahmed S, Qureshi S, Quadri F, Hossain A, Das SK, Antonio M, Hossain MJ, Mandomando I, Nhampossa T, Acacio S, Omore R, Oundo JO, Ochieng JB, Mintz ED, O'Reilly CE, Berkeley LY, Livio S, Tennant SM, Sommerfelt H, Nataro JP, Ziv-Baran T, Robins-Browne RM, Mishcherkin V, Zhang J, Liu J, Houpt ER, Kotloff KL, Levine MM (2016) The burden of *Cryptosporidium* diarrheal disease among children < 24 months of age in moderate/high mortality regions of Sub-Saharan Africa and South Asia, utilizing data from the Global

Enteric Multicenter Study (GEMS). PLoS Negl Trop Dis 10(5):e0004729. https://doi.org/10.1371/journal.pntd.0004729

11. Khan SM, Witola WH (2023) Past, current, and potential treatments for cryptosporidiosis in humans and farm animals: a comprehensive review. Front Cell Infect Microbiol 13: 1115522. https://doi.org/10.3389/fcimb.2023.1115522
12. Jakobi V, Petry F (2008) Humoral immune response in IL-12 and IFN-gamma deficient mice after infection with *Cryptosporidium parvum*. Parasite Immunol 30(3):151–161
13. Ndao M, Nath-Chowdhury M, Sajid M, Marcus V, Mashiyama ST, Sakanari J, Chow E, Mackey Z, Land KM, Jacobson MP, Kalyanaraman C, McKerrow JH, Arrowood MJ, Caffrey CR (2013) A cysteine protease inhibitor rescues mice from a lethal *Cryptosporidium parvum* infection. Antimicrob Agents Chemother 57(12):6063–6073. https://doi.org/10.1128/AAC.00734-13
14. Sonzogni-Desautels K, Renteria Flores A, Vasquez Camargo F, Di Lenardo TZ, Mikhail A, Arrowood MJ, Fortin A, Ndao M (2015) Oleylphosphocholine (OlPC) arrests *Cryptosporidium parvum* growth in vitro and prevents lethal infection in interferon gamma receptor knock-out mice. Front Microbiol 6:973. https://doi.org/10.3389/fmicb.2015.00973
15. Tessema TS, Schwamb B, Lochner M, Forster I, Jakobi V, Petry F (2009) Dynamics of gut mucosal and systemic Th1/Th2 cytokine responses in interferon-gamma and interleukin-12p40 knock out mice during primary and challenge *Cryptosporidium parvum* infection. Immunobiology 214(6):454–466
16. von Oettingen J, Nath-Chowdhury M, Ward BJ, Rodloff AC, Arrowood MJ, Ndao M (2008) High-yield amplification of *Cryptosporidium parvum* in interferon gamma receptor knockout mice. Parasitology 135(10): 1151–1156
17. Sonzogni-Desautels K, Mead JR, Ndao M (2020) Mouse models for use in *Cryptosporidium* infection studies and quantification of parasite burden using flow cytometry, qPCR, and histopathology. Methods Mol Biol 2052:229–251. https://doi.org/10.1007/978-1-4939-9748-0_14
18. Basha S, Surendran N, Pichichero M (2014) Immune responses in neonates. Expert Rev Clin Immunol 10(9):1171–1184. https://doi.org/10.1586/1744666X.2014.942288
19. Delaunay A, Gargala G, Li X, Favennec L, Ballet JJ (2000) Quantitative flow cytometric evaluation of maximal *Cryptosporidium parvum* oocyst infectivity in a neonate mouse model. Appl Environ Microbiol 66(10):4315–4317
20. Downey AS, Chong CR, Graczyk TK, Sullivan DJ (2008) Efficacy of pyrvinium pamoate against *Cryptosporidium parvum* infection in vitro and in a neonatal mouse model. Antimicrob Agents Chemother 52(9):3106–3112. https://doi.org/10.1128/AAC.00207-08
21. Blagburn BL, Sundermann CA, Lindsay DS, Hall JE, Tidwell RR (1991) Inhibition of *Cryptosporidium parvum* in neonatal Hsd:(ICR)BR Swiss mice by polyether ionophores and aromatic amidines. Antimicrob Agents Chemother 35:1520–1523
22. Sherwood D, Angus KW, Snodgrass DR, Tzipori S (1982) Experimental cryptosporidiosis in laboratory mice. Infect Immun 38:471–475
23. VanDussen KL, Funkhouser-Jones LJ, Akey ME, Schaefer DA, Ackman K, Riggs MW, Stappenbeck TS, Sibley LD (2020) Neonatal mouse gut metabolites influence *Cryptosporidium* parvum infection in intestinal epithelial cells. mBio 11(6). https://doi.org/10.1128/mBio.02582-20
24. Mead JR, Arrowood MJ, Sidwell RW, Healey MC (1991) Chronic *Cryptosporidium parvum* infections in congenitally immunodeficient SCID and nude mice. J Infect Dis 163(6): 1297–1304
25. Theodos CM, Sullivan KL, Griffiths JK, Tzipori S (1997) Profiles of healing and nonhealing *Cryptosporidium parvum* infection in C57BL/6 mice with functional B and T lymphocytes: the extent of gamma interferon modulation determines the outcome of infection. Infect Immun 65(11):4761–4769
26. You X, Mead JR (1998) Characterization of experimental *Cryptosporidium parvum* infection in IFN-gamma knockout mice. Parasitology 117(6):525–531
27. Campbell LD, Stewart JN, Mead JR (2002) Susceptibility to *Cryptosporidium parvum* infections in cytokine- and chemokine-receptor knockout mice. J Parasitol 88(5):1014–1016
28. Urban JF Jr, Fayer R, Chen SJ, Gause WC, Gately MK, Finkelman FD (1996) IL-12 protects immunocompetent and immunodeficient neonatal mice against infection with *Cryptosporidium parvum*. J Immunol 156(1):263–268
29. Lemieux MW, Sonzogni-Desautels K, Ndao M (2017) Lessons learned from protective immune responses to optimize vaccines against cryptosporidiosis. Pathogens 7(1). https://doi.org/10.3390/pathogens7010002
30. Love MS, Beasley FC, Jumani RS, Wright TM, Chatterjee AK, Huston CD, Schultz PG, McNamara CW (2017) A high-throughput

phenotypic screen identifies clofazimine as a potential treatment for cryptosporidiosis. PLoS Negl Trop Dis 11(2):e0005373. https://doi.org/10.1371/journal.pntd.0005373

31. Zambriski JA, Nydam DV, Wilcox ZJ, Bowman DD, Mohammed HO, Liotta JL (2013) *Cryptosporidium parvum*: determination of $ID_{50}$ and the dose-response relationship in experimentally challenged dairy calves. Vet Parasitol 197(1–2):104–112. https://doi.org/10.1016/j.vetpar.2013.04.022
32. Zambriski JA, Nydam DV, Bowman DD, Bellosa ML, Burton AJ, Linden TC, Liotta JL, Ollivett TL, Tondello-Martins L, Mohammed HO (2013) Description of fecal shedding of *Cryptosporidium parvum* oocysts in experimentally challenged dairy calves. Parasitol Res 112(3):1247–1254. https://doi.org/10.1007/s00436-012-3258-2
33. Heine J, Moon HW, Woodmansee DB (1984) Persistent *cryptosporidium* infection in congenitally athymic (nude) mice. Infect Immun 43: 856–859
34. McDonald V, Deer R, Uni S, Iseki M, Bancroft GJ (1992) Immune responses to *Cryptosporidium muris* and *Cryptosporidium parvum* in adult immunocompetent or immunocompromised (nude and SCID) mice. Infect Immun 60(8):3325–3331
35. Jumani RS, Bessoff K, Love MS, Miller P, Stebbins EE, Teixeira JE, Campbell MA, Meyers MJ, Zambriski JA, Nunez V, Woods AK, McNamara CW, Huston CD (2018) A novel piperazine-based drug lead for cryptosporidiosis from the medicines for malaria venture open-access malaria box. Antimicrob Agents Chemother 62(4). https://doi.org/10.1128/AAC.01505-17
36. Bhalchandra S, Cardenas D, Ward HD (2018) Recent breakthroughs and ongoing limitations in *Cryptosporidium* research. F1000Res 7. https://doi.org/10.12688/f1000research.15333.1
37. Heo I, Dutta D, Schaefer DA, Iakobachvili N, Artegiani B, Sachs N, Boonekamp KE, Bowden G, Hendrickx APA, Willems RJL, Peters PJ, Riggs MW, O'Connor R, Clevers H (2018) Modelling *Cryptosporidium* infection in human small intestinal and lung organoids. Nat Microbiol 3(7):814–823. https://doi.org/10.1038/s41564-018-0177-8
38. Haserick JR, Klein JA, Costello CE, Samuelson J (2017) *Cryptosporidium parvum* vaccine candidates are incompletely modified with O-linked-N-acetylgalactosamine or contain N-terminal N-myristate and S-palmitate. PLoS One 12(8):e0182395. https://doi.org/10.1371/journal.pone.0182395
39. Dayao DA, Jaskiewcz J, Lee S, Oliveira BC, Sheoran A, Widmer G, Tzipori S (2022) Development of two mouse models for vaccine evaluation against cryptosporidiosis. Infect Immun 90(7):e0012722. https://doi.org/10.1128/iai.00127-22
40. Greigert V, Saraav I, Son J, Dayao D, Antia A, Tzipori S, Witola WH, Stappenbeck TS, Ding S, Sibley LD (2023) *Cryptosporidium* infection of human small intestinal epithelial cells induces type III interferon and impairs infectivity of Rotavirus. bioRxiv. https://doi.org/10.1101/2023.08.30.555581
41. Arrowood MJ, Hurd MR, Mead JR (1995) A new method for evaluating experimental cryptosporidial parasite loads using immunofluorescent flow cytometry. J Parasitol 81(3): 404–409
42. Jonkman J, Brown CM, Wright GD, Anderson KI, North AJ (2020) Tutorial: guidance for quantitative confocal microscopy. Nat Protoc 15(5):1585–1611. https://doi.org/10.1038/s41596-020-0313-9
43. Coppens I, Ward HD (2023) *Cryptosporidium* secretome sheds new light on host-parasite interactions. Trends Parasitol 39(7):499–500. https://doi.org/10.1016/j.pt.2023.04.008
44. Wyatt CR, Brackett EJ, Mason PH, Savidge J, Perryman LE (2000) Excretion patterns of mucosally delivered antibodies to p23 in *Cryptosporidium parvum* infected calves. Vet Immunol Immunopathol 76(3–4):309–317
45. Wyatt CR, Lindahl S, Austin K, Kapil S, Branch J (2005) Response of T lymphocytes from previously infected calves to recombinant *Cryptosporidium parvum* P23 vaccine antigen. J Parasitol 91(5):1239–1242. https://doi.org/10.1645/GE-3446RN.1
46. Perryman LE, Jasmer DP, Riggs MW, Bohnet SG, McGuire TC, Arrowood MJ (1996) A cloned gene of *Cryptosporidium parvum* encodes neutralization-sensitive epitopes. Mol Biochem Parasitol 80(2):137–147
47. Enriquez FJ, Riggs MW (1998) Role of immunoglobulin A monoclonal antibodies against P23 in controlling murine *Cryptosporidium parvum* infection. Infect Immun 66(9): 4469–4473
48. Ranucci L, Muller HM, La Rosa G, Reckmann I, Morales MA, Spano F, Pozio E, Crisanti A (1993) Characterization and immunolocalization of a *cryptosporidium* protein containing repeated amino acid motifs. Infect Immun 61(6):2347–2356

49. Yu JR, O'Hara SP, Lin JL, Dailey ME, Cain G, Lin JL (2002) A common oocyst surface antigen of *Cryptosporidium* recognized by monoclonal antibodies. Parasitol Res 88(5): 412–420. https://doi.org/10.1007/s00436-001-0586-z
50. Costa D, Soulieux L, Razakandrainibe R, Basmaciyan L, Gargala G, Valot S, Dalle F, Favennec L (2021) Comparative performance of eight PCR methods to detect *Cryptosporidium* species. Pathogens 10(6). https://doi.org/10.3390/pathogens10060647
51. Hulverson MA, Vinayak S, Choi R, Schaefer DA, Castellanos-Gonzalez A, Vidadala RSR, Brooks CF, Herbert GT, Betzer DP, Whitman GR, Sparks HN, Arnold SLM, Rivas KL, Barrett LK, White AC Jr, Maly DJ, Riggs MW, Striepen B, Van Voorhis WC, Ojo KK (2017) Bumped-kinase inhibitors for cryptosporidiosis therapy. J Infect Dis 215(8):1275–1284. https://doi.org/10.1093/infdis/jix120
52. Delling C, Holzhausen I, Daugschies A, Lendner M (2016) Inactivation of *Cryptosporidium parvum* under laboratory conditions. Parasitol Res 115(2):863–866. https://doi.org/10.1007/s00436-015-4813-4
53. Fujino T, Matsui T, Kobayashi F, Haruki K, Yoshino Y, Kajima J, Tsuji M (2002) The effect of heating against *Cryptosporidium* oocysts. J Vet Med Sci 64(3):199–200
54. Barbee SL, Weber DJ, Sobsey MD, Rutala WA (1999) Inactivation of *Cryptosporidium parvum* oocyst infectivity by disinfection and sterilization processes. Gastrointest Endosc 49(5): 605–611
55. Anderson BC (1985) Moist heat inactivation of *cryptosporidium* sp. Am J Public Health 75: 1433–1434
56. Rekha KM, Puttalakshmamma GC, D'Souza PE (2016) Comparison of different diagnostic techniques for the detection of cryptosporidiosis in bovines. Vet World 9(2):211–215. https://doi.org/10.14202/vetworld.2016.211-215
57. Chen F, Huang K, Qin S, Zhao Y, Pan C (2007) Comparison of viability and infectivity of *Cryptosporidium parvum* oocysts stored in potassium dichromate solution and chlorinated tap water. Vet Parasitol 150(1–2):13–17. https://doi.org/10.1016/j.vetpar.2007.09.001
58. Yang S, Healey MC, Du C (1996) Infectivity of preserved *Cryptosporidium parvum* oocysts for immunosuppressed adult mice. FEMS Immunol Med Microbiol 13(2):141–145. https://doi.org/10.1016/0928-8244(95)00096-8
59. Barbosa J, Costa-de-Oliveira S, Rodrigues AG, Pina-Vaz C (2008) Optimization of a flow cytometry protocol for detection and viability assessment of *Giardia lamblia*. Travel Med Infect Dis 6(4):234–239. https://doi.org/10.1016/j.tmaid.2008.01.004
60. Sonzogni-Desautels K, Di Lenardo TZ, Renteria AE, Gascon MA, Geary TG, Ndao M (2019) A protocol to count *Cryptosporidium* oocysts by flow cytometry without antibody staining. PLoS Negl Trop Dis 13(3): e0007259. https://doi.org/10.1371/journal.pntd.0007259
61. Melicherova J, Hofmannova L, Valigurova A (2018) Response of cell lines to actual and simulated inoculation with *Cryptosporidium proliferans*. Eur J Protistol 62:101–121. https://doi.org/10.1016/j.ejop.2017.12.003
62. Hasheminasab SS, Conejeros I, Velásquez ZD, Borggrefe T, Gartner U, Kamena F, Taubert A, Hermosilla C (2022) ATP purinergic receptor P2X1-dependent suicidal NETosis induced by *Cryptosporidium parvum* under physioxia conditions. Biology (Basel) 11:3. https://doi.org/10.3390/biology11030442
63. Lamisere H, Bhalchandra S, Kane AV, Zeng XL, Mo D, Adams W, Estes MK, Ward HD (2022) Differential response to the course of *Cryptosporidium parvum* infection and its impact on epithelial integrity in differentiated versus undifferentiated human intestinal enteroids. Infect Immun 90(11):e0039722. https://doi.org/10.1128/iai.00397-22
64. Omidian Z, Ebrahimzadeh E, Shahbazi P, Asghari Z, Shayan P (2014) Application of recombinant *Cryptosporidium parvum* P23 for isolation and prevention. Parasitol Res 113(1):229–237. https://doi.org/10.1007/s00436-013-3648-0
65. Mead JR (2014) Prospects for immunotherapy and vaccines against *Cryptosporidium*. Hum Vaccin Immunother 10(6):1505
66. Sharma S, Mann R, Kumar S, Mishra N, Srivastava B, Valecha N, Anvikar AR (2019) A simple and cost-effective freeze-thaw based method for *plasmodium* DNA extraction from dried blood spot. Iran J Parasitol 14(1):29–40
67. Jiang J, Alderisio KA, Singh A, Xiao L (2005) Development of procedures for direct extraction of *Cryptosporidium* DNA from water concentrates and for relief of PCR inhibitors. Appl Environ Microbiol 71(3):1135–1141. https://doi.org/10.1128/AEM.71.3.1135-1141.2005
68. Xin Y, Xie J, Nan B, Tang C, Xiao Y, Wu Q, Lin Y, Zhang X, Shen H (2021) Freeze-thaw pretreatment can improve efficiency of bacterial DNA extraction from meconium. Front Microbiol 12:753688. https://doi.org/10.3389/fmicb.2021.753688
69. O'Leary JK, Sleator RD, Lucey B (2021) *Cryptosporidium* spp. diagnosis and research in the

21(st) century. Food Waterborne Parasitol 24: e00131. https://doi.org/10.1016/j.fawpar.2021.e00131

70. Robinson G, Elwin K, Jones M, Chalmers RM (2023) A comparison of qPCR and microscopy for the detection and enumeration of *Cryptosporidium* oocysts from drinking water. J Med Microbiol 72(6). https://doi.org/10.1099/jmm.0.001715
71. Verweij JJ, Stensvold CR (2014) Molecular testing for clinical diagnosis and epidemiological investigations of intestinal parasitic infections. Clin Microbiol Rev 27(2):371–418. https://doi.org/10.1128/CMR.00122-13
72. Mary C, Chapey E, Dutoit E, Guyot K, Hasseine L, Jeddi F, Menotti J, Paraud C, Pomares C, Rabodonirina M, Rieux A, Derouin F, Network ACN (2013) Multicentric evaluation of a new real-time PCR assay for quantification of *Cryptosporidium* spp. and identification of *Cryptosporidium parvum* and *Cryptosporidium hominis*. J Clin Microbiol 51(8):2556–2563. https://doi.org/10.1128/JCM.03458-12
73. Hadfield SJ, Robinson G, Elwin K, Chalmers RM (2011) Detection and differentiation of *Cryptosporidium* spp. in human clinical samples by use of real-time PCR. J Clin Microbiol 49(3):918–924. https://doi.org/10.1128/JCM.01733-10
74. Nath-Chowdhury M, Sangaralingam M, Bastien P, Ravel C, Pratlong F, Mendez J, Libman M, Ndao M (2016) Real-time PCR using FRET technology for Old World cutaneous leishmaniasis species differentiation. Parasit Vectors 9:255. https://doi.org/10.1186/s13071-016-1531-4
75. Shahiduzzaman M, Dyachenko V, Keidel J, Schmaschke R, Daugschies A (2010) Combination of cell culture and quantitative PCR (cc-qPCR) to assess disinfectants efficacy on cryptosporidium oocysts under standardized conditions. Vet Parasitol 167(1):43–49. https://doi.org/10.1016/j.vetpar.2009.09.042
76. Enemark HL, Bille-Hansen V, Lind P, Heegaard PM, Vigre H, Ahrens P, Thamsborg SM (2003) Pathogenicity of *Cryptosporidium parvum*–evaluation of an animal infection model. Vet Parasitol 113(1):35–57. https://doi.org/10.1016/s0304-4017(03)00034-7
77. Gullicksrud JA, Sateriale A, Engiles JB, Gibson AR, Shaw S, Hutchins ZA, Martin L, Christian DA, Taylor GA, Yamamoto M, Beiting DP, Striepen B, Hunter CA (2022) Enterocyte-innate lymphoid cell crosstalk drives early IFN-gamma-mediated control of *cryptosporidium*. Mucosal Immunol 15(2):362–372. https://doi.org/10.1038/s41385-021-00468-6
78. Sawant M, Benamrouz-Vanneste S, Mouray A, Bouquet P, Gantois N, Creusy C, Duval E, Mihalache A, Gosset P, Chabe M, Hot D, Viscogliosi E, Certad G (2021) Persistent *Cryptosporidium parvum* infection leads to the development of the tumor microenvironment in an experimental mouse model: results of a microarray approach. Microorganisms 9(12). https://doi.org/10.3390/microorganisms9122569
79. Ogbuigwe P, Roberts JM, Knox MA, Heiser A, Pita A, Haack NA, Garcia-Ramirez JC, Velathanthiri N, Biggs PJ, French NP, Hayman DTS (2023) A novel, stain-free, natural autofluorescent signal, Sig M, identified from cytometric and transcriptomic analysis of infectivity of *Cryptosporidium hominis* and *Cryptosporidium parvum*. Front Cell Infect Microbiol 13: 1178576. https://doi.org/10.3389/fcimb.2023.1178576
80. Guerrant RL (1997) Cryptosporidiosis: an emerging, highly infectious threat. Emerg Infect Dis 3(1):51–57. https://doi.org/10.3201/eid0301.970106
81. Dupont HL, Chappell CL, Sterling CR, Okhuysen PC, Rose JB, Jakubowski W (1995) The infectivity of *Cryptosporidium parvum* in healthy volunteers. N Engl J Med 332(13):855–859
82. Bouzid M, Hunter PR, Chalmers RM, Tyler KM (2013) *Cryptosporidium* pathogenicity and virulence. Clin Microbiol Rev 26(1):115–134. https://doi.org/10.1128/CMR.00076-12
83. Checkley W, White AC Jr, Jaganath D, Arrowood MJ, Chalmers RM, Chen XM, Fayer R, Griffiths JK, Guerrant RL, Hedstrom L, Huston CD, Kotloff KL, Kang G, Mead JR, Miller M, Petri WA Jr, Priest JW, Roos DS, Striepen B, Thompson RC, Ward HD, Van Voorhis WA, Xiao L, Zhu G, Houpt ER (2015) A review of the global burden, novel diagnostics, therapeutics, and vaccine targets for *Cryptosporidium*. Lancet Infect Dis 15(1): 85–94. https://doi.org/10.1016/S1473-3099(14)70772-8
84. Natapol, Pumipuntu Supawadee, Piratae (2018) Cryptosporidiosis: A zoonotic disease concern Veterinary World 11(5) 681-686 10.14202/vetworld. https://doi.org/10.14202/vetworld.2018.681-686

# Chapter 15

# Calf Clinical Model of Cryptosporidiosis for Efficacy Evaluation of Therapeutics

**Michael W. Riggs and Deborah A. Schaefer**

## Abstract

Cryptosporidiosis, caused by the apicomplexan parasite *Cryptosporidium parvum*, is a moderate-to-severe diarrheal disease, now recognized as one of the leading causes of morbidity and mortality in livestock globally and in humans living in resource-limited parts of the world, particularly those with AIDS or who are malnourished. This recognition has fueled efforts for the discovery of effective therapeutics. While recent progress in drug discovery has been encouraging, there are presently no acceptably effective parasite-specific drugs for the disease. The urgent need for new drug discovery or drug repurposing has also increased the need for refined animal models of clinical disease for therapeutic efficacy evaluation. Here, we describe an acute model of cryptosporidiosis using newborn calves to evaluate well-defined clinical and parasitological parameter outcomes, including effect on diarrhea severity and duration, oocyst numbers produced, and multiple measures of clinical health. The model is highly reproducible and provides unequivocal direct measures of treatment efficacy on diarrhea severity and parasite replication.

**Key words** *Cryptosporidium*, Cryptosporidiosis, Bovine calf model, Therapeutic evaluation, Oocyst quantitation, Diarrhea quantitation, Clinical evaluation

## 1 Introduction

*Cryptosporidium* species are among the leading causative agents of diarrhea in humans and agriculturally important livestock species, particularly calves, throughout the world [1–3]. Despite this, there are currently no approved vaccines and no parasite-specific drugs to treat the disease in either humans or livestock in the USA. The only Food and Drug Administration-approved drug for treating cryptosporidiosis in humans is nitazoxanide [4]. However, the efficacy of nitazoxanide against *Cryptosporidium* in immunocompetent humans is marginal, and the drug has no treatment benefit against the parasite in AIDS patients [5–7]. The clear need for more effective parasite-specific drugs has increased the need for refined, reliable animal models of cryptosporidiosis for therapeutic efficacy evaluation. Several therapeutic candidates have shown efficacy

Jan R. Mead and Michael J. Arrowood (eds.), *Cryptosporidium: Methods and Protocols*, Methods in Molecular Biology, vol. 2978, https://doi.org/10.1007/978-1-0716-4824-7_15,

against cryptosporidiosis when tested in the neonatal calf model [8–10].

Neonatal calves can provide an excellent acute clinical model for both human and bovine cryptosporidiosis. During the 10–14-day-old age period most often used, neonatal calves are functional monogastrics and remain so up to 4 weeks of age [11, 12]. Clinical symptoms and diarrhea severity and duration following experimental infection in calves closely mirror what is observed in humans [11, 13]. Further, the anatomic distribution of infection in the GI tract and lesions are essentially the same as those observed in humans [11, 13]. Given these observations supporting the validity of calves as a useful clinical model most closely approximating human cryptosporidiosis, multiple methodologies and experimental designs for evaluation of therapeutics in the model have been published. The most commonly used calf models differ mainly in (1) frequency of fecal collection, (2) type of calf housing used (elevated stalls or box stalls/pens), (3) whether total daily fecal volume excreted is collected or a ~10–15 mL fecal sample is obtained per rectum once or twice daily, (4) methods used for determining diarrhea severity (quantitative or semi-quantitative), (5) methods used for evaluating oocyst shedding (quantitative, e.g., qPCR or hemocytometer enumeration of the total number of oocysts shed in total daily fecal collections, or various semi-quantitative measures, e.g., acid fast stain or immunofluorescence assay on fecal sample smears and microscopic scoring, or qPCR on fecal samples collected every 12–24 h), (6) trial duration, and (7) *C. parvum* isolate used (field vs characterized) [14–19]. Total daily fecal collection is required to unequivocally determine diarrhea severity and the total number of oocysts produced, and it requires housing in elevated stalls [14–16]; these measurements are not possible with box stall or pen housing in which a ~10–15 mL fecal sample is collected daily per rectum [17–19]. Model designs vary between investigator and institutional factors (e.g., housing facilities and equipment availability, investigator preferences, staffing resources and abilities, budgetary constraints) and can influence the accuracy of results for efficacy parameters. Therefore, the experimental design used in a particular calf model study should be considered when interpreting efficacy results of a given therapeutic and when making comparisons between studies. For example, in one calf model of cryptosporidiosis, treating calves with nitazoxanide reduced the duration of *C. parvum* oocyst shedding and diarrhea severity [17]. However, in another calf model, nitazoxanide had no therapeutic or prophylactic efficacy against clinical disease caused by *C. parvum*, and in some *C. parvum*-infected calves, nitazoxanide treatment actually prolonged and exacerbated diarrhea [18]. In a calf model similar to the preceding, paromomycin demonstrated prophylactic efficacy against *C. parvum* and provided proof of concept, but the drug dosage required was determined to be cost-prohibitive for veterinary use [19]. While there is no universally applicable design for the calf model given geographic, institutional, and laboratory constraints, we have used the acute model described here to consistently replicate data demonstrating therapeutic efficacy for multiple bumped kinase inhibitor (BKI) drugs for further study, as well as to consistently replicate data demonstrating lower, clinically less satisfactory efficacy for other BKIs to eliminate from further study ( [9, 10, 14, 20], M.W. Riggs and D.A. Schaefer, unpublished).

The model described here is quite labor-intensive and demanding on personnel compared to some other calf models for evaluation of therapeutics and requires housing calves in elevated stalls for total fecal collection. However, the increased labor and demands are justified because the method yields the most accurate efficacy data using defined, quantitative outcome parameters. Determining the total daily fecal volume and consistency for each calf provides an objective and unequivocal direct measure of diarrhea severity. Total daily fecal collection is also the preferred practice in human challenge studies to evaluate the efficacy of candidate anti-cryptosporidial preparations [21]. In addition, total daily fecal collection allows accurate determination of the total number of oocysts produced per day as an unequivocal measure of treatment effect on parasite replication, and it also avoids potential error resulting from diurnal variations in oocyst shedding that we have observed in this model [14, 22]. Reduction of oocyst shedding is an important treatment outcome when considering the epidemiologic importance of the environmental parasite burden and exposure risk to both humans and livestock [23]. By 8 or 9 days post-infection (DPI), most untreated control calves are beginning to resolve diarrhea in this model [10, 14]. Therefore, we end the experiment at 10 DPI to conclude initial efficacy determinations.

We previously attempted total fecal collection from calves using strap-on commercial fecal collection bags designed for small ruminant metabolic studies [24]. This allowed individual calves to be housed in larger box stalls. However, fecal collection bags invariably leaked or became detached and failed completely, despite emptying up to three times daily. Emptying the bags also required manual restraint of calves and resulted in visible distress. Given all of the above, the housing of calves in elevated stalls in this acute model design is thus scientifically and ethically justified and was approved by the Institutional Animal Care and Use Committee (IACUC). The method presented here is completed in the following major sequential steps or procedures to evaluate therapeutics in this calf model of cryptosporidiosis: (1) procure isolated *Cryptosporidium parvum* oocysts and age-matched (within 12 h) newborn pathogen-free Holstein bull calves for therapeutic efficacy evaluation; (2) manage husbandry of calves housed in elevated stalls, modified to accurately evaluate clinical health parameters, diarrhea severity, parasite production, and urine production; (3) infect calves with *C. parvum* at 36–48 h of age and administer candidate therapeutic agents per the appropriate dosage and treatment regimen commencing at 2 DPI; (4) collect and process feces to determine total daily fecal volumes and total daily oocyst counts; (5) complete terminal procedures and euthanize calves at 10 DPI; and (6) analyze data for significant treatment effects on clinical health, total fecal volume (diarrhea severity), total oocysts produced (parasite replication), total urine volume (an additional index of hydration), and body weight gain or loss.

## 2 Materials

### 2.1 Preparation of Cryptosporidium Parvum Oocysts

1. Fully equipped BSL-2 laboratory.
2. Peracetic acid.
3. Taurocholic acid.
4. Trypsin.
5. Phosphate-buffered saline (PBS).
6. Phase contrast and immunofluorescence microscopes.
7. Hemocytometer.

### 2.2 Newborn Calves for Therapeutic Efficacy Evaluation

1. Cargo van with climate control for heating and cooling.
2. Rubber boots and coveralls.
3. Roll of 10 mil plastic sheeting, ~185 cm wide.
4. Disinfectant (quaternary ammonium-based for animal facilities).
5. Weight scale for calves.
6. Wheelbarrow.
7. Anti-*Escherichia coli* and coronavirus passive antibody for calves (First Defense, ImmuCell, Portland, ME).
8. Attenuated bovine rotavirus and coronavirus vaccine (Calf-Guard®, Zoetis, Kalamazoo, MI).
9. Clean blankets and thick terry cloth cotton towels.
10. Newborn calf record sheets.
11. Surgical scissors to trim umbilical cords.
12. Ear tags and applicator.
13. 7% iodine scrub.
14. 10% iodine solution.
15. Colostrum replacer (bovine IgG colostrum replacer, Land O'Lakes, Shoreview, MN).

16. Disinfected 6 liter buckets.
17. Whisk.
18. Plastic feeding bottles with rubber nipples (2 liters).
19. Esophageal tube feeder and disposable 50 cc blunt-ended catheter-tip dosing syringes.
20. Hand soap and disposable disinfectant wipes.
21. Calf transport crates (crates designed for 90–125 pound dogs work well—one calf per kennel).
22. Autoclaved straw.

### 2.3 Modified Elevated Calf Stalls, Calf Husbandry

Coordinate with the local Institutional Animal Care and Use Committee (or equivalent) regarding housing requirements for calves.

1. Elevated calf stalls (~152 cm L × ~122 cm H × ~53.5 cm W; Wenke Mfg., Pender, NE).
2. Large rectangular plastic bins (~58.5 cm × ~43 cm × ~15 cm; ~26 liter) (4 per calf stall—3 for fecal collection and 1 for urine collection).
3. Reinforced waterproof industrial-strength adhesive tape.
4. Roll of 6 mil plastic sheeting, ~122 cm wide.
5. Large clamp binder clips to hold fecal collection bins together.
6. Wall mirrors (~100 cm × 100 cm).
7. Serum refractometer.

### 2.4 Calf Infection and Treatment

1. Isolated peracetic acid-disinfected *C. parvum* oocysts.
2. Candidate therapeutic in vehicle and vehicle alone.
3. *C. parvum* oocyst-specific monoclonal antibody(ies) and fluoresceinated secondary antibody for indirect immunofluorescence fecal assay.
4. Disposable 50 cc blunt-ended catheter-tip dosing syringes.
5. Disposable Tyvek® suits (various sizes) with elastic-cinched ankles and wrists.
6. Disposable plastic booties and plastic boot covers (all to be worn over shoes); alternatively, rubber boots.
7. Surgical masks.
8. Goggles and protective glasses.
9. Assorted disposable nitrile gloves.
10. Hairnet bonnets.
11. Fecal specimen cups (150 mL urinalysis cups) with lids.
12. Labels for fecal specimen cups.
13. Rectal thermometer.

14. Disinfectant (quaternary ammonium-based for animal facilities).
15. Mop bucket and mop.
16. Plastic scraper.
17. Assorted syringes and needles.
18. Appropriate blood collection tubes.
19. Flunixin meglumine.
20. Injectable selenium and vitamin E supplement (~1 mg selenium, 50 mg (68 USP units) vitamin E per mL).
21. Oral electrolytes for calves (or equivalent).
22. Milk replacer (crude protein ~22%, crude fat ~20%) and nonfat dry milk (crude protein ~34%).

### 2.5 Fecal Processing

1. Personal protective equipment listed in Subheading 2.4.
2. Assorted size graduated cylinders (4, 2, 1 liters, and 500 mL).
3. Buckets (4 liters).
4. Commercial blender (4 liter capacity).
5. Paper towels.
6. Plastic ladle and spatula.
7. Disinfectant (quaternary ammonium-based for animal facilities).
8. Pressure washer for cleaning post-trial.

### 2.6 Fecal Oocyst Quantitation

1. Real-time PCR (qPCR) system (ABI StepOne Plus®; Applied Biosystems, Grand Island, NY).
2. Real-time PCR (qPCR) reagents.
3. Fecal oocyst DNA extraction reagents and buffers (QIAamp® Fast DNA Stool Mini Kit; Qiagen, Gaithersburg, MD).
4. DNA quantification method (NanoDrop®; NanoDrop Technologies, Wilmington, DE).
5. Primers (we use the *C. parvum* oocyst wall protein gene (COWP) (GenBank no. AF248743) as the target for primers and selected a 151 bp sequence from the COWP gene).
6. SYBR Green (or equivalent nucleic acid stain).
7. Bovine serum albumin.

### 2.7 Terminal Procedures and Calf Euthanasia

1. Xylazine.
2. Ketamine.
3. Appropriate blood collection tubes.

4. Euthanasia solution (390 mg pentobarbital sodium/mL, 50 mg phenytoin sodium/mL, or equivalent).
5. Weight scale.

## 3 Methods

For use in the following methods for evaluation of therapeutics in a bovine clinical model of cryptosporidiosis, calves must be maintained in climate-controlled Animal Biosafety Level 2 (ABSL-2) biocontainment facilities in accordance with the *Biosafety in Microbiological and Biomedical Laboratories* guidelines and *Guide for the Care and Use of Laboratory Animals.* Approval by the Institutional Animal Care and Use Committee (IACUC) is required, and the methods must be performed in compliance with the guidelines in the Animal Welfare Act and the *Guide for the Care and Use of Agricultural Animals in Research and Teaching* (Federation of Animal Science Societies), preferably in facilities fully accredited by the Association for Assessment and Accreditation of Laboratory Animal Care International. The model described here has been established and operative in the authors' laboratory for two decades [14–16]. Based on the large number of unsolicited outside requests we receive for testing of candidate anti-cryptosporidials in this model from industry and established *Cryptosporidium* investigators at other institutions, we presume that the model is preferred practice. However, the model should only be used for further evaluation of lead candidate therapeutics or prophylactics that have demonstrated highly significant anti-cryptosporidial activity *in vitro* and in at least one less sentient species rodent model such as neonatal mice. While neonatal mice are the lowest species suitable for initial *in vivo* evaluation of therapeutics against *Cryptosporidium parvum*, the infection is asymptomatic. However, reduction of infection levels in the neonatal mouse model has positive predictive value for therapeutic efficacy against *C. parvum* in clinical models of disease, including calves [14, 16, 25, 26].

### 3.1 Procurement and Preparation of Cryptosporidium parvum Oocysts for Calf Infection

1. Decide on the *C. parvum* isolate to be used, and obtain purified oocysts from a source production laboratory. We routinely use the Iowa *C. parvum* isolate [27, 28] because it is well characterized and is maintained by monthly propagation in newborn *Cryptosporidium*-free Holstein bull calves in our laboratory [29–31]. Other established laboratories and commercial sources exist for obtaining purified *C. parvum* oocysts in the numbers required for calf trials including Bunch Grass Farm (Deary, ID; **HYPERLINK "mailto: bunchgrassfarm@gmail.com" bunchgrassfarm@gmail.com**) or hollow fiber cultured *C. parvum* oocysts

[32]. Alternatively, use the method described here to propagate oocysts in calves and isolate oocysts from feces using published methods [22] for calf infection. All procedures involving *C. parvum* must be performed at Biosafety Level-2 (BSL-2) in an approved BSL-2 laboratory.

2. Immediately prior to administration to calves, disinfect oocysts with 0.5% peracetic acid (4 °C) to inactivate any possible contaminant bacterial or viral pathogens [27, 33] (*see* **Note 1**). Peracetic acid is the preferred disinfectant, as treatment of oocysts with diluted sodium hypochlorite (bleach) has been shown to promote spontaneous excystation at 37 °C [29, 33]. Use of hypochlorite-disinfected oocysts could potentially result in excystation in the abomasum of experimental calves and in sporozoite killing by acidic conditions, effectively reducing the intended infectious dose [29]. In contrast, peracetic acid disinfection does not promote spontaneous excystation. All disinfection steps are performed at 4 °C. Centrifuge oocysts previously stored in saline/antibiotic solution (8000× *g*, 7 min, 4 °C), remove saline/antibiotic solution, resuspend the oocyst pellet in PBS (4 °C), and wash oocysts with PBS twice. Then, resuspend the pelleted oocysts thoroughly in 1–10 mL 0.5% peracetic acid (stock peracetic acid diluted in DI water to 0.5% vol/vol). Incubate resuspended oocysts at 4 °C for 3 min, inverting once each min. The 3 min timing period starts immediately upon the addition of peracetic acid. After the 3 min treatment, immediately pellet the oocysts (8000× *g*, 7 min, 4 °C), remove peracetic acid, and wash 4X with PBS as described above. Keep oocysts at 4 °C at all times post-peracetic acid treatment until immediately prior to inoculation of calves.

3. Following disinfection, determine percent oocyst excystation as an indirect viability index, and adjust the challenge dose if necessary (*see* **Note 2**). To determine percent excystation, resuspend a disinfected oocyst aliquot (~$10^6$ oocysts) in 250 μL excystation media (1.5% taurocholic acid, 0.5% trypsin in PBS, pH 7.2–7.4), and place in a shaking water bath (1.5 h, 37 °C). At the end of the incubation, count the number of intact, unexcysted oocysts and oocyst shells (fully/partially excysted oocysts) using a hemocytometer and phase-contrast microscopy. Divide the number of oocyst shells by the total number of oocyst shells plus intact oocysts, and multiply by 100 to determine percent excystation. Excystation is typically >95% for oocysts stored properly for up to 30 days following isolation and disinfection.

### 3.2 Procurement and Enrollment of Neonatal Calves for Therapeutic Efficacy Evaluation

1. Trained veterinary staff obtain age-matched (born within a 12 h period) Holstein bull calves from a closed-herd USDA-licensed dairy production unit, preferably the same source unit for all studies. Cohort size will vary depending on the investigator's facility capacity for ABSL-2 calf isolation and the source dairy's average daily calving numbers. The minimum cohort size desired is six calves (three treated, three control) (*see* **Note 3**). With two such cohorts completed, data for six treated and six control calves are obtained, allowing potential exclusion of up to one calf per group without compromising statistical analyses in most cases. Occasionally, such exclusions are required because of inadvertent coinfection with one or more other enteropathogens, need for euthanasia because of disease severity before completion of the study, or other confounding variables. Trained laboratory staff manually assist the birthing of calves, delivering them onto a 10 mil plastic sheet barrier to minimize possible exposure to other enteropathogens or *Cryptosporidium* isolates or species (*see* **Note 4**). Because newborn calves are wet and difficult to hold securely, use clean, thick cotton terrycloth towels to grasp them during delivery. Calves from cows which experience dystocia or prolonged calving or twin calves are not accepted. Weigh and perform a physical exam on each calf, including careful examination for congenital anomalies such as imperforate anus, cleft palate, and limb abnormalities. Calves weighing ≥38 kg and having normal physical exam findings are the most suitable for use in the method described here; calves weighing <38 kg are excluded (*see* **Note 5**). Place acceptable calves into a previously disinfected wheelbarrow for transport to a pre-disinfected designated concrete area. Disinfect the wheelbarrow between uses for each calf. Wash any body parts, especially the muzzle and head region, with 7% iodine scrub and warm water if necessary to remove any exogenous material, and then dry with a clean towel. Clean both sides of one ear with 7% iodine scrub and rinse, and then apply a numbered ear tag to each enrolled calf to serve as its identifying number. Record the date and time of birth. Dip the navel in 10% iodine solution (not scrub), again after colostrum replacer is fed, again on arrival at the ABSL-2 facility, and again 24 h after arrival. Orally administer to each calf First Defense (anti-*Escherichia coli* and coronavirus antibody; ImmuCell, Portland, ME) and Calf-Guard vaccine (attenuated bovine rotavirus and coronavirus; Zoetis, Kalamazoo, MI) within 20 min of birth and 1 h prior to feeding colostrum replacer per manufacturer's instructions. Commencing 1 h after administration of the preceding antibodies and vaccine, feed enrolled calves a commercial colostrum replacer from a single lot using a nipple bottle (*see* **Note 6**), continuing until 6 h after birth at twice the manufacturer's recommended calf

dose, reconstituted in 2 liters of water (*see* **Notes** 7 and **8**). Record the actual time and volume of colostrum replacer intake hourly to confirm that the optimal full dosing is achieved within 6 h after birth. If a calf will not voluntarily consume the full amount of colostrum replacer within 6 h of birth, attempt to nipple bottle-feed the remainder within 9 h of birth (*see* **Note 9**). If a calf will not voluntarily consume the full amount of colostrum replacer within 9 h of birth, carefully feed the remainder using an esophageal tube feeder or disposable 50 cc blunt-ended catheter-tip dosing syringes [34] (*see* **Note 10**).

2. Place calves into disinfected transport crates (crates for 90–125 pound dogs) with autoclaved straw bedding for transport by cargo van to ABSL-2 facilities. Disinfect calf rooms, and allow them to air-dry for at least 3 days prior to and after each experiment to inactivate any residual *Cryptosporidium* or other microbial agents. Upon arrival, assign calves to either treatment or control groups using a randomization method, such as Microsoft Excel Random Number Generation Tool, and then place calves into individual elevated calf stalls (Wenke Manufacturing, Pender, NE), equipped with coated slat flooring material (9 gauge flat expanded metal with rubberized coating and diamond-shaped openings) (*see* **Note 11**). House treatment and control groups in separate ABSL-2 rooms. Facilities will vary between institutions, but we house two to three calves per room (~3 meters wide by ~5 meters long), allowing placement of elevated stalls so that the calves face each other for environmental enrichment but remain at distances that do not allow direct contact between them. In instances when only a single calf is housed per room, place a mirror at eye level on the wall in front of the calf for environmental enrichment. Salt blocks and brushes are attached to the side of the stalls using zip ties for nutritional supplementation and environmental enrichment, respectively.

3. Using a nipple bottle, feed calves with non-medicated milk replacer (crude protein ~22% [bovine whey and milk-derived], crude fat ~20%) twice daily commencing at 12 h of age (volume equivalent to 10% of birth weight total per day divided into 2 feedings, 12 h apart through 5 days of age, increased by 25% at 6 days of age and increased again by 25% at 8 days of age) until the end of the experiment at 10 DPI. Nonfat dry milk (crude protein ~34%) can be substituted starting at 2 DPI if isolating oocysts (*see* **Note 12**). Allow 25–30 min per feeding for calves to consume the entire volume of milk replacer. Record the actual quantity of milk replacer consumed at each feeding. If a calf will not voluntarily consume any milk replacer, carefully feed the calculated volume of replacer using an

esophageal tube feeder or by disposable 50 cc blunt-ended catheter-tip dosing syringes [34] (*see* **Note 13**). Clean and disinfect all nipple bottles, buckets, and utensils used for replacer reconstitution and feeding after each use.

4. At 24 h of age, administer vitamin E/selenium supplement (3 mL/45 kg) subcutaneously. Collect a blood sample by venipuncture (~10 mL), isolate the serum fraction, and test serum to verify adequate passive transfer of colostral IgG using a rapid method such as serum refractometry to semi-quantitatively assess serum IgG levels [35, 36] (*see* **Note 14**).
5. At 36 h of age, perform another physical examination on each calf prior to final clearance and admittance into the study. Remove any unhealthy calves from the study (exclusion criteria: fever >103.0 °F, diarrhea, respiratory distress, inappetence, or inability to rise on own). Prior to experimental infection, examine fecal smears from each calf for the potential presence of *Cryptosporidium* oocysts by immunofluorescence assay (IFA) [37]. Remove any calves from the study that are positive for inadvertent field-acquired *Cryptosporidium* infection.

### *3.3 Infection of Calves with* C. parvum *and Administration of Candidate Therapeutic Agents*

1. Choose an optimal time for personnel to infect calves such that, at the time of inoculation, no calves in the cohort are younger than 36 h or older than 48 h of age. Following acclimation, at the chosen time between 36 and 48 h of age (defined as DPI 0), orally administer disinfected oocysts (*see* Subheading 3.1) to calves as follows. While the calf is standing, allow each calf to suckle milk replacer (250 mL) from a nipple bottle to close the esophageal (gastric) groove and thus bypass the rumen for delivery of the inoculum directly into the abomasum [34]. Immediately thereafter, orally administer to each calf $5 \times 10^7$ viable, peracetic acid-disinfected *C. parvum* oocysts previously enumerated using a hemocytometer and prepared as follows. Prior to use, keep the oocyst inoculum for each calf in ~5 mL PBS at 4 °C in a 15 mL centrifuge tube. Immediately prior to inoculation, allow the inoculum to gradually reach room temperature over a ~ 10 min period to avoid thermal shock, and then vortex the tube to resuspend oocysts. Immediately draw up the inoculum into a 50 cc blunt-ended catheter-tip dosing syringe, equipped with a 15 cm Nalgene tube (~2 mm inside diameter) to facilitate recovery of the entire inoculum. Rinse/vortex the oocyst tube once with 5 mL PBS, and draw up the rinse into the same 50 cc syringe. Finally, draw up ~20 mL milk replacer (at room temperature) into the syringe, and mix by inversion immediately prior to calf inoculation. For clarity, this is the final inoculum for one calf. For each additional calf, the inoculum is identically prepared using a new disposable dosing syringe. Administer the inoculum

immediately after closure of the esophageal (gastric) groove, as described above, by orally inserting the dosing syringe so that the tip is at the base of the tongue, while the calf is still in the suckling response. With the head elevated to optimize delivery of the entire inoculum, allow natural suckling of the syringe contents, facilitated by gradual manual expulsion. Administer the inoculum at a rate that does not allow any dripping from the calf's mouth (*see* **Note 15**). Reload the syringe once with ~25 mL milk replacer as a rinse, and repeat the administration procedure. Immediately thereafter feed each calf the calculated volume of milk replacer to meet daily dietary requirements as described above (Subheading 3.2, **step 3**) and to ensure that the full inoculum reaches the abomasum. Examine fecal samples for oocysts by IFA daily after infection to determine the pre-patent and patent periods.

2. Initial administration of a given therapeutic candidate begins on DPI 2, by which time untreated *C. parvum* infection is established and nearing completion of one round of its life cycle at the infection dose used [23]. The candidate therapeutic must be prepared according to the institution's Standard Operating Procedures for Preparation of Compounds for Use in Animals and must be approved by the IACUC prior to administration. The dosage and frequency of dosing for most candidate therapeutics in the calf model can be initially based on mg/kg extrapolation from the most efficacious dosage regimens identified in *C. parvum*-infected mice, allometric modelling in mice and rats, and ultimately pharmacokinetic/pharmacodynamic studies in uninfected and infected newborn calves [14]. Preliminary pharmacokinetics/absorption–distribution–metabolism–elimination studies may be performed on treated calves in a given trial to evaluate a candidate therapeutic, provided that additional stress is minimized and control (vehicle-only) calves are identically sampled. Of potential relevance to the desirable pharmacokinetic/pharmacodynamic properties of candidate chemotherapeutics, it has been shown that intestinal (rather than systemic) exposure is essential for the efficacy of bumped kinase inhibitor drugs against *C. parvum* in mouse models [38]. Treat control calves identically for the regimen selected with vehicle alone. Label the therapeutic (in vehicle) being evaluated and the vehicle alone by code to blind the treatment and control preparations. Just prior to each treatment, allow each calf to suckle milk replacer (~475 mL) from a nipple bottle to close the esophageal (gastric) groove. Then, feed each calf by nipple bottle the candidate therapeutic in vehicle, or vehicle alone, mixed well with milk replacer in a final volume of ~475 mL. Following each treatment, feed each calf the calculated volume of milk replacer to

meet daily dietary requirements, as described above (Subheading 3.2, **step 3**). Record all treatments with the therapeutic or vehicle alone. Make potable water available to all calves by bucket ad libitum for the entire experiment. At the earliest onset of diarrhea, make an oral electrolyte solution available to each calf by bucket for the remainder of the experiment. At each feeding record the quantity of water and oral electrolytes consumed from buckets since the previous feeding, clean both buckets, and fill each with fresh water or electrolytes.

### 3.4 Stall Modifications for Accurate Evaluation of Diarrhea Severity, Parasite Shedding, and Urine Production in Infected Calves

1. To ensure collection of all feces from individual calves for accurate total daily fecal volume quantitation and daily fecal oocyst count determinations, place three plastic fecal collection bins beneath the rear of each stall, held together by binder clips. Position fecal bins between the room wall and the rear part of the stall such that broad coverage is provided for all likely directions in which diarrhea may project (Fig. 1).
2. Prevent contamination of feces with urine by using only bull calves (birth weight ≥38 kg). Install a downward-angled 6 mil plastic sheet spanning the underside of the full width of the stall floor, extending forward from the front of fecal collection bins, to divert urine into a separate bin positioned beneath the front of each stall (Fig. 1). Adjust the position of the plastic sheet if needed based on calf size to ensure urine is diverted while feces pass freely into the fecal collection bins. Tightly secure the plastic sheet to avoid leakage with reinforced water-proof industrial-strength adhesive tape (Fig. 1) (*see* **Notes 5** and **16**). All collection bins, and urine diversion shields are most conveniently set up the day prior to calf arrival.

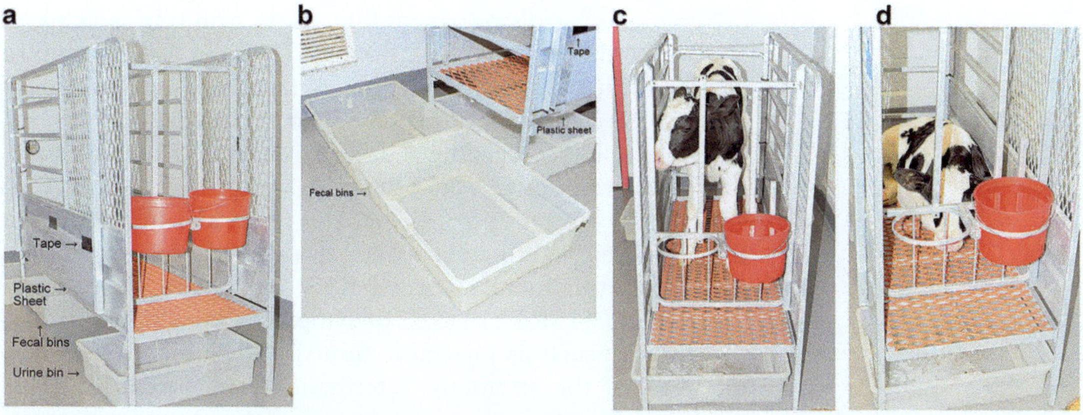


**Fig. 1** (**a**) Elevated calf stall depicting urine collection bin and plastic urine diversion sheet placement, viewed from front. (**b**) Stall depicting fecal collection bins and plastic urine diversion sheet placement, viewed from rear. (**c**) and (**d**) Newborn 41 kg Holstein bull calf in stall ~1 h following arrival at ABSL-2 facility, prior to infection

3. Loose stools or diarrhea typically commence between 48 and 72 h PI in this model. Determine total fecal volume and fecal oocyst counts daily for each calf from 3 DPI until the end of the experiment as described in Subheading 3.6. Use extreme care and adequate personal protective equipment to prevent zoonotic infection (*see* **Notes 17** and **18**).

### 3.5 Evaluation of Clinical Health, Diarrhea, and Oocyst Shedding in Infected Calves

1. Once the study has commenced, collect diarrheic fecal samples from each calf (~30 mL/calf) and have them examined for the presence of enteropathogenic bacteria (K99+ *E. coli*, *Clostridium* spp., and *Salmonella* spp.). Perform these tests on fecal samples collected after two consecutive episodes of diarrhea. Immediately remove from the study and exclude from analyses any calves found positive for other enteropathogens which would confound interpretation of the experimental results. Collect additional fecal samples from all calves at the completion of the experiment, and reexamine for the presence of the enteropathogens above. Exclude any positive calves from analyses (*see* **Note 19**).
2. Commencing 3 DPI, for the duration of the experiment, have trained veterinary clinical staff assess clinical health parameters on each calf twice daily (every 12 h) and assign an objective numerical score for each parameter. "Blind" personnel involved in the study as to the identity (treatment or control) of calves (single-blind study). Assign and record individual scores to each calf for (1) fecal consistency, (2) willingness to rise, (3) stance when up in stall, (4) appetite, (5) attitude, and (6) hydration status [39] (*see* **Note 20**). Determine the mean daily clinical health score using individual parameter scores for each calf. Determine rectal temperature once daily. Contact the attending veterinarian if any of the following are observed: (1) pain and/or distress (grinding teeth, grunting, kicking at abdomen, abnormal vocalizations); (2) severe watery profuse diarrhea or hemorrhagic diarrhea; (3) moderate or severe dehydration; (4) high fever (>103.5 °F) or hypothermia (<96 °F); (5) weight loss exceeding 10%; (6) depressed mental state, minimally responsive, or otherwise appearing in extremis; (7) inability to stand; or (8) inability to suckle. In such cases, the attending veterinarian makes all decisions on appropriate supportive treatment(s) or whether euthanasia is required. Treat fever (≥103.5 °F) and/or symptoms of pain with palliative therapy such as injectable flunixin meglumine under the direction of the attending veterinarian (*see* **Note 21**). Treat moderate dehydration at its very earliest onset with parenteral rehydration under the direction of the attending veterinarian to prevent progression to severe dehydration (Table 1). Record

**Table 1**
**Clinical health parameter evaluation and scoring**

| Score[a] | Fecal consistency | Willingness to rise | Stance when up | Appetite | Attitude | Hydration status[b] |
|---|---|---|---|---|---|---|
| 1 | Normal, formed stool | Normal, eager to rise or already up in stall | Normal, stable | Ravenous, consumes full ration, wants more | Normal, bright, alert, responsive | Normal, euhydrated, asymptomatic |
| 2 | Loose but not diarrheic | Willing to rise on own but needs encouragement or prodding | Somewhat unstable | Normal, consumes full ration, reaches satiety | Somewhat depressed, dull | Mild dehydration (~5%). Enophthalmos <3 mm. Skin tent duration <5 s. CRT 3–4 s |
| 3 | Moderately diarrheic | Reluctant to rise but able with assistance | Clearly unstable, loses balance or sways | Decreased, but will suckle if coaxed | Clearly depressed, minimally responsive | Moderate dehydration (~6–8%). Skin tent duration 5–6 s. Enophthalmos 3–4 mm. CRT >4 s |
| 4 | Severely diarrheic, watery and profuse | Unwilling or unable to rise on own, requires lifting | Unable to stand on own if lifted | Anorectic, will not suckle | | Severe dehydration (~9–12%). Enophthalmos 5–7 mm. Skin tent duration >6–>8 s. CRT >5 s |

[a]Use consensus assessment by two trained veterinary clinical staff to determine score for each parameter
[b]Determine hydration status based on lateral mid-cervical skin turgor (duration of skin tenting), severity of enophthalmos (distance between globe and palpebral conjunctiva of lower eyelid) [39], and gingival capillary refill time (CRT)

the use of any concomitant medications and parenteral fluid therapy in the records for each calf.

3. As an additional index of hydration status, measure total daily urine output for each calf 3–10 DPI. Based on estimated volume in the urine collection bin for each calf, pour all urine into an appropriately sized graduated cylinder to determine the actual volume of both AM and PM collections separately. Combine the volumes to determine the total daily urine volume for each calf, and record the values. Disinfect and rinse the urine collection bin, and then dry and place the bin back under the stall, positioned as previously.
4. On 3–10 DPI, measure and evaluate total daily fecal volume (mL) and consistency for each calf to accurately determine diarrhea severity and oocyst numbers produced daily (*see* **Note 22**). When caring for calves at AM and PM time points, complete fecal collection and assessment of clinical health parameters prior to feeding (*see* **Note 23**). Calves typically rise and stand prior to feeding, which facilitates fecal collection and clinical assessment (*see* **Note 24**). Measure the total daily fecal volume (combined AM and PM collections) just prior to the PM feeding as described in Subheading 3.6. Thoroughly clean and mop calf room floors with a quaternary ammonium-based disinfectant after procedures are completed each AM and PM.
5. Accurately quantify the total number of oocysts shed per day (3–10 DPI) from each calf as follows. Using a commercial-grade blender, thoroughly blend the total daily fecal collection from each calf to achieve homogeneity and to ensure the release of oocysts trapped in any clumps of excess mucus (as detailed in Subheading 3.6, **step 3**). Once homogenized, collect a 50 mL sample, representative of the total volume, and store at 4 °C for oocyst quantitation as described below (Subheading 3.8). Transport fecal samples from the ABSL-2 facility to the BSL-2 laboratory for analyses in a closed container labelled with a biohazard sticker.

### 3.6 Collection and Processing of Feces

1. Using a scraper, collect any feces on the floor that may have missed the collection bins, and place into the corresponding calf fecal bins. Use a plastic spatula to gently scrape any feces on the plastic urine diversion sheets taking care that no tears are created.
2. Based on the estimated volume in the three collection pans for each calf, pour all feces into an appropriately sized graduated cylinder to determine the actual volume of both AM and PM collections, separately. Just prior to the PM feeding, combine the AM and PM fecal collections to determine the total daily fecal volume for each calf, and record this value. Disinfect and

rinse out fecal collection bins. Dry the bins and place them back under the stall, positioned as previously or adjusted if needed.

3. Blend to homogeneity all feces collected over successive 24 h periods from each calf using a commercial blender. Blend on the low-speed setting for 5 min with the blender lid locked on securely. Place the blender on the floor undisturbed for ~5 min before unlocking and removing the lid to reduce aerosol risks. If the total daily fecal volume for any calf is near or greater than the 4 liter blender capacity, split the volume in two and blend each separately as described above. After blending, thoroughly mix the two using two 4 liter buckets, pouring back and forth at least five times. Immediately after completing the blending/ mixing steps, collect a representative ~50 mL fecal sample from the mid-point of the container using a ladle, and place it into a fecal specimen container (urinalysis cup) pre-labelled with date and calf number. The blending and mixing processes create aerosols. Take extreme precautions to avoid zoonotic infection.
4. Wash the blender thoroughly with hot water and dish detergent, and then disinfect and rinse well between processing feces from each calf.

### 3.7 Terminal Procedures and Euthanasia (See Note 25)

1. At 10 DPI, following the AM clinical health, diarrhea, and oocyst shedding evaluations described above (Subheading 3.5), sedate each calf by intramuscular injection of xylazine (3 mg/kg) combined with ketamine (100 mg/calf). Following sedation, collect blood samples from each treated and control calf into appropriate tubes for CBCs and serum chemistry panels to evaluate clinical pathology and screen for potential toxicity of the test therapeutic.
2. Euthanize each sedated calf by intracardiac injection of euthanasia solution (1 mL/10 lbs.; 1 mL/4.5 kg). This method (Subheading 3.7, **step 1** and **2**) is in accordance with the AVMA guidelines on euthanasia and euthanasia recommendations approved by the local IACUC. If clinically indicated before 10 DPI, as determined by the attending veterinarian, use the same euthanasia procedure.
3. Following euthanasia, weigh each calf for comparison to birth weight to compute average body weight gain or loss for the trial.

### 3.8 Determination of Daily Fecal Oocyst Counts from Infected Calves

Determine total daily fecal oocyst counts by real-time PCR (qPCR) as described previously [14, 40]. This step is typically performed following completion of a calf trial.

1. For each fecal sample, extract oocyst DNA using the QIAamp® Fast DNA Stool Mini Kit (Qiagen, Gaithersburg, MD). Add InhibitEx buffer, incubate at 95 °C (5 min), and then perform

5 freeze–thaw cycles. Quantify total DNA in the sample using NanoDrop® (NanoDrop Technologies, Wilmington, DE). We used the *C. parvum* oocyst wall protein gene (COWP) (GenBank no. AF248743) as the target for primers (COWPF, 5′--CAAATTGATACCGTTTGTCCTTCTG-3′; COWPR, 5′-AGCTGAATTAGAATCGACATGCC-3′) and selected a 151 bp sequence from the COWP gene. For each qPCR, prepare a 25-μL reaction containing 400 nM (final concentration) for each primer, 12.5 μL SYBR Green or similar nucleic acid stain, and 0.1 ng/μL bovine serum albumin. Add genomic DNA (2 μL), and perform the PCR in a real-time PCR system. Use cycling conditions of 95 °C for 10 min, followed by 50 cycles of 95 °C for 15 s and 60 °C for 1 min, with a final extension at 72 °C for 1 min. Run each sample in triplicate. Concurrently run a control with no template, and verify it to be negative. To determine the total number of oocysts shed per day, multiply the oocyst count per /mL—derived by comparison to standards run concurrently during the qPCR—by the total daily fecal volume (mL).

### 3.9 Statistical Analyses

For statistical analyses examine all calves given the same treatment (therapeutic candidate or vehicle-alone control) as one group for the evaluation of all outcome parameter means. Examine clinical health scores, daily and total fecal volumes, daily and total oocyst counts, daily and total urine production, and body weight gain/loss for statistically significant differences using Student's *t*-test and ANOVA or another preferred statistically valid method. Additional or more powerful statistical analyses are usually not necessary to validate clinically acceptable efficacy of candidate therapeutics.

## 4 Notes

1. Peracetic acid is highly toxic and explosive. Follow all precautions stated in the MSDS.
2. Use percent excystation to adjust the number of oocysts in the inoculum to deliver $5 \times 10^7$ viable oocysts/calf. For example, if percent excystation is 94%, adjust the oocyst inoculum dose/calf from $5 \times 10^7$ to $5.3 \times 10^7$.
3. Occasionally the desired number of age-matched calves for a cohort cannot be obtained (such as 6 calves total to provide 3 calves each for the treatment and control groups). If only three to five age-matched calves are obtained, the cohort can proceed, as long as one calf is a control (vehicle alone).
4. Dairy facilities vary in their willingness to allow laboratory personnel to manually assist in the delivery of calves on-site because of biosecurity and/or liability concerns for potential

personal injury. In such cases, accept only calves that dairy personnel have manually assisted delivery onto the plastic sheet barrier, then placed into a disinfected wheelbarrow for transport to the designated disinfected concrete area, where laboratory personnel determine acceptability and further process acceptable calves. If there is any inadvertent minimal soiling of otherwise acceptable calves, calves are thoroughly cleaned with 7% iodine scrub and towel dried.

5. Calves have room to rise/lie down and move/maneuver within the stall and may be able to turn completely around and reverse their position in the stall. This is most likely to occur if smaller calves <32 kg are used, during the first few days while acclimating to stalls, prior to infection and onset of diarrhea. If calves turn themselves around in the stall such that their posterior is at the front of the stall, reposition them properly. In such cases most of the feces excreted during the previous 12 h period will have ended up in the urine collection bin. If the feces are diarrheic, they may be mixed with urine, rendering the collection unsuitable for evaluation, and this should be noted in the record as such. However, we have rarely (~1 in 50 calves) observed calves turning around in the stall when larger calves ≥38 kg are used, and this is the primary reason calves <38 kg are excluded. Never use the front head-gate bars to headlock a calf for more than ~5 min, if at all. The headlock is intended strictly to facilitate brief restraint for blood draws and other brief procedures.

6. Some calves suckle so intensely that a vacuum may form in the bottle, eventually precluding additional ingestion. Only use nipple bottles that incorporate a ~1 mm diameter air-release valve port at the nipple base to allow air to enter the bottle as the calf is nursing, thus precluding vacuum formation. If necessary, massage or press next to the valve port to release the seal and any vacuum. After feeding is completed, verify that the valve is not plugged due to inadequate cleaning and is working properly.

7. Commercial colostrum replacer powder does not go into solution easily. To thoroughly reconstitute and solubilize colostrum replacer, gradually add the specified amount of powder per calf (twice the manufacturer's recommended dose or 2 packets totaling 940 grams powder) to a 6 liter bucket containing 2 liters of clean tap water at 110–120 °F using a manual whisk. It is critical to whisk the preparation during reconstitution until no visible clumps remain. Following consumption of the full dose of colostrum replacer, add ~250 mL clean warm water to the same nipple bottle, swirl vigorously to rinse residual colostrum replacer, and feed this to the calf to ensure complete dosing.

8. Bovine colostrum contains variable titers of antibody reactive with *C. parvum*, acquired from natural exposure or previous infection. However, no significant *neutralizing* antibody is present following previous infection, as normal colostrum-fed calves are readily infected with *C. parvum*. Nevertheless, while expensive, commercial colostrum replacer is used to eliminate this variable entirely. A commercial colostrum replacer, such as the one we use, has been formulated to replace maternal colostrum and remove both lot-to-lot and inherent variability found in source dairy colostrum. Frozen whole colostrum acquired from most source dairies varies substantially in antibody content, quality, and microbial contamination load and should be avoided.
9. Calves are born agammaglobulinemic. They require colostral antibodies following birth for passive transfer of immunity against the major enteropathogenic bacteria and viruses responsible for neonatal diarrhea and mortality for which commercially available maternal vaccines exist. The newborn calf's ability to absorb ingested colostral antibody decreases linearly over the first 24 h of age during the process of gut closure. By 12 h of age, the reduction is substantial, and by 24 h of age, the gut is essentially closed to absorption of colostral antibody and other macromolecules. For this reason, ensure that calves suckle the majority of the recommended colostrum replacer dose, preferably within 6 h after birth but no later than 9 h after birth. Feed any colostrum replacer remaining at 9 h after birth by esophageal intubation.
10. Feeding colostrum replacer by esophageal tube should be used as a last resort due to the fact that intubating (unlike suckling) does not stimulate the esophageal (gastric) groove closure reflex and thus delivers replacer into the rumen instead of the abomasum [34]. Alternatively, milk replacer could be fed using a disposable 50 cc blunt-ended catheter-tip dosing syringe to deliver the milk to the back of the throat and allow it to trickle down, initiating the swallowing mechanism. This takes more time to feed the calf, but it is less invasive.
11. Adjust the width of the stall's front head-gate bars to allow only the head and neck to have free access through the bars. This will allow calves to easily access water and/or electrolyte buckets while also preventing them from jumping through the front of the stall and getting trapped or injured. This potential problem is uncommon and typically occurs only if smaller calves <32 kg are used and the head-gate bars are improperly positioned. We have never observed this problem when using calves weighing ≥38 kg with the head-gate bars positioned properly. Other than bruising acquired by two small 30–32 kg calves jumping through the front of the stall, we

have also never had a calf acquire any injury being housed in elevated stalls. Nevertheless, after assembling newly purchased stalls, it is important to file down any sharp or jagged edges on the framing or exposed hardware. Periodically inspect stalls for the same, and ensure that all hardware fasteners are securely tightened to maintain stall stability. Periodically inspect the rubberized flooring for wear and deterioration from use of harsh disinfectants, and replace when necessary.

12. Similar to colostrum replacer, milk replacer powder does not go into solution easily. Use warm (110–120 °F) water and a whisk to thoroughly reconstitute replacer until all clumps are fully dissolved. Nonfat dry milk can be used to feed animals when trying to isolate oocysts from feces. A decrease in the fat content of the milk eases oocyst isolation.

13. Feed milk replacer via an esophageal tube as a last-resort measure for calves that will not consume any of the calculated replacer volume within a ~25–30 min feeding period. The suckle response resulting in esophageal (gastric) groove closure is voided by tube feeding, delivering milk replacer into the rumen. Additionally, in our experience, calves which are tube fed more than ~3–4 times are reluctant to feed from a nipple bottle subsequently. Alternatively, milk replacer could be fed using a disposable 50 cc blunt-ended catheter-tip dosing syringe to deliver the milk to the back of the throat and allow it to trickle down, initiating the swallowing mechanism. This takes more time to feed the calf, but it is less invasive.

14. Failure of passive transfer (FPT) of maternal immunoglobulins is rare if commercial colostrum replacer (at twice the label dose) is suckled within 6–9 h of birth. However, if FPT occurs, but the calf is otherwise healthy at 24 h of age, treat it with an intravenous transfusion of commercially available adult bovine plasma (Lake Immunogenics, Ontario, NY) administered under the supervision of a veterinarian.

15. We have never experienced a calf not receiving the entire oocyst inoculum using the method described, but it could potentially happen (leakage, expectoration). If this occurs, record the estimated volume of inoculum not received, and note that data from the calf may not be suitable for use in the final analyses.

16. It is potentially possible that some urine may enter the fecal collection bins, voiding accurate fecal volume determination. This possibility is largely eliminated if the plastic diversion sheet is stretched tightly against the bottom at the rear third of the stall and properly secured with reinforced water-proof industrial-strength adhesive tape. This ensures that urine flows only in one direction, towards the front of the stall and into the

urine collection bin. This is also a secondary reason calves <38 kg are excluded, as smaller calves can move back and forth in the stall to a greater extent, making accurate placement of the plastic diversion sheet difficult, especially for calves <32 kg.

17. Calves infected with *C. parvum* in this model shed extremely large numbers of oocysts, especially the untreated controls. A single experimentally infected newborn calf may shed $>10^{12}$ oocysts in feces during the peak days of diarrhea [14, 16]. Thus, personnel caring for calves are at high risk of zoonotic infection unless appropriate precautions are taken and adequate personal protective equipment is used while in calf rooms, including at a minimum: (1) disposable double gloves; (2) mid-calf-high plastic or rubber boots worn over disposable plastic shoe covers; (3) Tyvek coveralls with elastic-cinched ankle and wrist cuffs; (4) surgical masks; (5) disposable hair bonnets, completely covering all hair (long hair must be tied back into a bun); and (6) goggles, safety glasses, or full face shields.

18. An institutional protocol should be in place for the evaluation of any calf personnel who develop nausea, vomiting, or diarrhea during or after a trial to determine the etiology and appropriate treatment. We use the University's Occupational Health Division for independent assessment of ill personnel and compliance with HIPPA. If *C. parvum* is identified as the etiologic agent, the experimental protocol should be carefully reviewed to determine where the breach leading to zoonotic transmission likely occurred so that it can be addressed. Despite the precautions described here, we have had at least two documented cases of zoonotic transmission.

19. Adhering closely to the methods described here for calf procurement, we have never had a calf inadvertently infected with a field isolate of *C. parvum* and have had only 8 calves in the last ~200 used inadvertently infected with field-acquired enteropathogenic bacteria and/or viruses.

20. We use a standardized daily AM/PM hard-copy form for each calf that lists all health evaluation parameters along with the scoring scale for each to record consensus scores; rectal temperature; amount of milk replacer, water, and/or oral electrolytes consumed; date and actual AM/PM times (h, min) when data were recorded; and calf ear tag number, with a signature box for each of the two attending staff involved (Daily Health Observation Record). Standardized hard-copy forms are also used for all other information described in the Methods section for which it is stated that records are made (Newborn Calf Record, *Cryptosporidium* Infection Record, Daily Fecal

Volume Record, Test-Therapeutic Dosing Record, Concomitant Treatment Record, and Condition(s) Requiring Examination by the Attending Veterinarian Record). We have also started using an online e-record system using Microsoft Excel through Box to ensure that all personnel have 24/7 remote access to clinical records.

21. The nonsteroidal anti-inflammatory drug flunixin meglumine should be used sparingly and only when absolutely needed. No more than three doses in total should be administered to a given experimental calf during a trial period to avoid nephrotoxicity and gastrointestinal toxicity. Toxicity is of greatest concern when administered to dehydrated calves.
22. Fecal dry matter percentages may optionally be determined on total daily fecal collections as an additional index of diarrhea severity. However, in the model presented here, we found previously that fecal dry matter percentages added little to fecal consistency scores and total daily fecal volumes in determining diarrhea severity, and we have since discontinued percent dry matter determinations.
23. Calves with diarrhea are prone to cutaneous scalding in the perianal and tail base regions. Use human infant sanitary wipes or warm wet paper towels to gently clean these regions as needed. If indicated, apply generic human infant diaper rash ointment liberally after cleaning the region. In such cases this should be done prior to feeding or taking rectal temperature, both of which stimulate defecation.
24. When calves are reluctant but able to stand for evaluation and feeding, rapidly and repeatedly stroke the hair coat from the hip region towards the shoulders. This is usually effective in getting the calf to stand. Alternatively, grasp the hip region, and gently rock the calf back and forth laterally to stimulate initiation of the standing motion.
25. The drugs used for sedation and euthanasia are federally controlled (scheduled) substances. A DEA license is required to obtain these. If the investigator does not have a DEA license, obtain the drugs through the attending veterinarian, and comply with all recordkeeping and other requirements.

## Acknowledgments

This work was supported by USDA NIFA Animal Health Project numbers ARZT-5704180-A50-121 and ARZT-5704210-A50-133, Bill and Melinda Gates Foundation Contract 44418, and NIH grants R01 AI 111341, R01 HD 080670, and R01 AI 155412, as well as USDA NIFA grant 2014-67015-22106.

**Competing Interest Statement** The authors have no conflicts of interest to declare that are relevant to the content of this chapter.

## References


1. Prabakaran TM, Weible LJ, Champlain JD, Jiang RY, Biondi K, Weil AA, Van Voorhis WC, Ojo KK (2023) The gut wrenching effects of cryptosporidiosis and giardiasis in children. Microorganisms 11:1–17. https://doi.org/10.3390/microorganisms11092323
2. Brunauer M, Roch FF, Conrady B (2021) Prevalence of worldwide neonatal calf diarrhoea caused by bovine rotavirus in combination with bovine coronavirus, *Escherichia coli* K99 and *Cryptosporidium* spp.: a meta-analysis. Animals 11:1014
3. Khalil IA, Troeger C, Rao PC, Blacker BF, Brown A, Brewer TG, Colombara DV, De Hostos EL, Engmann C, Guerrant RL, Haque R, Houpt ER, Kang G, Korpe PS, Kotloff KL, Lima AAM, Petri WA Jr, Platts-Mills JA, Shoultz DA, Forouzanfar MH, Hay SI, Reiner RC Jr, Mokdad AH (2018) Morbidity, mortality, and long-term consequences associated with diarrhoea from *Cryptosporidium* infection in children younger than 5 years: a meta-analyses study. Lancet 6:758–768
4. Stockdale HD, Spencer JA, Blagburn BL (2008) Prophylaxis and chemotherapy. In: Fayer R, Xiao L (eds) *Cryptosporidium* and cryptosporidiosis, 2nd edn. CRC Press, Florida
5. Cabada MM, White AC (2010) Treatment of cryptosporidiosis: do we know what we think we know? Curr Opin Infect Dis 23:494–499
6. Amadi B, Mwiya M, Sianongo S, Payne L, Watuka A, Katubulushi M, Kelly P (2009) High dose prolonged treatment with nitazoxanide is not effective for cryptosporidiosis in HIV positive Zambian children: a randomised controlled trial. BMC Infect Dis 9:195. https://doi.org/10.1186/1471-2334-9-195
7. Amadi B, Mwiya M, Musuku J, Watuka A, Sianongo S, Ayoub A, Kelly P (2002) Effect of nitazoxanide on morbidity and mortality in Zambian children with cryptosporidiosis: a randomised controlled trial. Lancet 360: 1375–1380. https://doi.org/10.1016/S0140-6736(02)11401-2
8. Van Voorhis WC, Hulverson MA, Choi R, Huang W, Arnold SLM, Schaefer DA, Betzer DP, Vidadala RSR, Lee S, Whitman GR, Barrett LK, Maly DJ, Riggs MW, Fan E, Kennedy TJ, Tzipori S, Doggett JS, Winzer P, Anghel N, Imhof D, Müller J, Hemphill A, Ferre I, Sanchez-Sanchez R, Ortega-Mora LM, Ojo KK (2021) One health therapeutics: target-based drug development for cryptosporidiosis and other apicomplexa diseases. Vet Parasitol 289:109336. https://doi.org/10.1016/j.vetpar.2020.109336. Epub 2020 Dec 15. PMID: 33418437
9. Choi R, Hulverson MA, Huang W, Vidadala RSR, Whitman GR, Barrett LK, Schaefer DA, Betzer DP, Riggs MW, Doggett JS, Hemphill A, Ortega-Mora LM, McCloskey MC, Arnold SLM, Hackman RC, Marsh KC, Lynch JJ, Freiberg GM, Leroy BE, Kempf DJ, Choy RKM, de Hostos EL, Maly DJ, Fan E, Ojo KK, Van Voorhis WC (2020) Bumped kinase inhibitors as therapy for apicomplexan parasitic diseases: lessons learned. Int J Parasitol 50(5):413–422. https://doi.org/10.1016/j.ijpara.2020.01.006. Epub 2020 Mar 26.PMID: 32224121
10. Hulverson MA, Choi R, Arnold SLM, Schaefer DA, Hemphill A, McCloskey MC, Betzer DP, Müller J, Vidadala RSR, Whitman GR, Rivas KL, Barrett LK, Hackman RC, Love MS, McNamara CW, Shaughnessy TK, Kondratiuk A, Kurnick M, Banfor PN, Lynch JJ, Freiberg GM, Kempf DJ, Maly DJ, Riggs MW, Ojo KK, Van Voorhis WC (2017) Advances in bumped kinase inhibitors for human and animal therapy for cryptosporidiosis. Int J Parasitol 47(12):753–763. https://doi.org/10.1016/j.ijpara.2017.08.006. Epub 2017 Sep 9.PMID: 28899690
11. Santin M, Trout JM (2008) Livestock. In: Fayer R, Xiao L (eds) *Cryptosporidium* and cryptosporidiosis, 2nd edn. CRC Press, Florida
12. Heinrichs AJ (2005) Rumen development in the dairy calf. Adv Dairy Technol 17:179–188
13. Cirle A, Guerrant RL (2008) Clinical disease and pathology. In: Fayer R, Xiao L (eds) *Cryptosporidium* and cryptosporidiosis, 2nd edn. CRC Press, Florida
14. Schaefer DA, Betzer DP, Smith KD, Millman ZG, Michalski HC, Menchaca SE, Zambriski JA, Ojo KK, Hulverson MA, Arnold SL, Rivas KL, Vidadala RS, Huang W, Barrett LK, Maly DJ, Fan E, Van Voorhis WC, Riggs MW (2016) Novel bumped kinase inhibitors are safe and effective therapeutics in the calf clinical model for cryptosporidiosis. J Infect Dis 15(214):

1856–1864. https://doi.org/10.1093/infdis/jiw488

15. Imboden M, Schaefer DA, Bremel RD, Homan EJ, Riggs MW (2012) Antibody fusions reduce onset of experimental *Cryptosporidium parvum* infection in calves. Vet Parasitol 188(1–2):41–47. https://doi.org/10.1016/j.vetpar.2012.02.014
16. Perryman LE, Kapil SJ, Jones ML, Hunt EL (1999) Protection of calves against cryptosporidiosis with immune bovine colostrum induced by a *Cryptosporidium parvum* recombinant protein. Vaccine 17:2142–2149
17. Ollivett TL, Nydam DV, Bowman DD, Zambriski JA, Bellosa ML, Linden TC, Divers TJ (2009) Effect of nitazoxanide on cryptosporidiosis in experimentally infected neonatal dairy calves. J Dairy Sci 92:1643–1648. https://doi.org/10.3168/jds.2008-1474
18. Schnyder ML, Kohler L, Hemphill A, Deplazes P (2009) Prophylactic and therapeutic efficacy of nitazoxanide against *Cryptosporidium parvum* in experimentally challenged neonatal calves. Vet Parasitol 160:149–154. https://doi.org/10.1016/j.vetpar.2008.10.094
19. Fayer R, Ellis W (1993) Paromomycin is effective as prophylaxis for cryptosporidiosis in dairy calves. J Parasitol 79:771–774
20. Hulverson MA, Choi R, Vidadala RSR, Whitman GR, Vidadala VN, Ojo KK, Barrett LK, Lynch JJ, Marsh K, Kempf DJ, Maly DJ, Van Voorhis WC (2021) Pyrrolopyrimidine bumped kinase inhibitors for treatment of cryptosporidiosis. ACS Infect Dis 7(5): 1200–1207. https://doi.org/10.1021/acsinfecdis.0c00803
21. Okhuysen PC, Chappell CL, Crabb J, Valdez LM, Douglass ET, DuPont HL (1998) Prophylactic effect of bovine anti-*Cryptosporidium* hyperimmune colostrum immunoglobulin in healthy volunteers challenged with *Cryptosporidium parvum*. Clin Infect Dis 26:1324–1329
22. Mead JR, Arrowood MJ (eds) (2020) *Cryptosporidium*: methods and protocols. Methods Mol Biol 2052:43. https://doi.org/10.1007/978-1-4939-9748-0_15
23. Fayer R (2008) The general biology of *Cryptosporidium*. In: Fayer R, Xiao L (eds) *Cryptosporidium* and cryptosporidiosis, 2nd edn. CRC Press, Florida
24. Riggs MW (1987) Dissertation, Washington State University
25. Schaefer DA, Auerbach-Dixon BA, Riggs MW (2000) Characterization and formulation of multiple epitope-specific neutralizing monoclonal antibodies for passive immunization against cryptosporidiosis. Infect Immun 68: 2608–2616. https://doi.org/10.1128/IAI.68.5.2608-2616.2000
26. Perryman LE, Jasmer DP, Riggs MW, Bohnet SG, McGuire TC, Arrowood MJ (1996) A cloned gene of *Cryptosporidium parvum* encodes neutralization-sensitive epitopes. Mol Biochem Parasitol 80:137–147
27. Heine J, Pohlenz JF, Moon HW, Woode GN (1984) Enteric lesions and diarrhea in gnotobiotic calves monoinfected with *Cryptosporidium species*. J Infect Dis 150:768–775
28. Abrahamsen MS, Templeton TJ, Enomoto S, Abrahante JE, Zhu G, Lancto CA, Deng M, Liu C, Widmer G, Tzipori S, Buck GA, Xu P, Bankier AT, Dear PH, Konfortov BA, Spriggs HF, Iyer L, Anantharaman V, Aravind L, Kapur V (2004) Complete genome sequence of the apicomplexan, *Cryptosporidium parvum*. Science 304:441–445
29. Riggs MW, Perryman LE (1987) Infectivity and neutralization of *Cryptosporidium parvum* sporozoites. Infect Immun 55:2081–2087
30. Riggs MW, Stone AL, Yount PA, Langer RC, Arrowood MJ, Bentley DL (1997) Protective monoclonal antibody defines a circumsporozoite-like glycoprotein exoantigen of *Cryptosporidium parvum* sporozoites and merozoites. J Immunol 158:1787–1795
31. Riggs MW, Schaefer DA (2020) Chapter 15: Calf clinical model of cryptosporidiosis for efficacy evaluation of therapeutics. In: Mead J, Arrowood M (eds) *Cryptosporidium*. Springer nature methods in molecular biology, vol 2052. Humana, New York, pp 253–282
32. Yarlett N, Morada M, Schaefer DA, Ackman K, Carranza E, de Paula Baptista R, Riggs MW, Kissinger JC (2024) Genomic and virulence analysis of *in vitro* cultured *Cryptosporidium parvum*. PLoS Pathog 28:20–34. https://doi.org/10.1371/journal.ppat.1011992
33. Riggs MW, Cama VA, Leary JR, Sterling CR (1994) Bovine antibody against *Cryptosporidium parvum* elicits a circumsporozoite precipitate-like reaction and has immunotherapeutic effect against persistent cryptosporidiosis in SCID mice. Infect Immun 62:1927–1939
34. Jones CM, Heinrichs AJ (2003) Feeding the newborn dairy calf. In: Agricultural research and cooperative extension publications. Pennsylvania State University. Available at: https://extension.psu.edu/feeding-the-newborn-dairy-calf
35. Morrill KM, Polo J, Lago A, Campbell J, Quigley J, Tyler H (2013) Estimate of serum immunoglobulin G concentration using

refractometry with or without caprylic acid fractionation. J Dairy Sci 96:4535–4541. https://doi.org/10.3168/jds.2012-5843

36. Hernandez D, Nydam DV, Godden SM, Bristol LS, Kryzer A, Ranum J, Schaefer D (2016) Brix refractometry in serum as a measure of failure of passive transfer compared to measured immunoglobulin G and total protein by refractometry in serum from dairy calves. Vet J 211:82–87
37. Riggs MW, Schaefer DA, Kapil SJ, Barley-Maloney L, Perryman LE (2002) Efficacy of monoclonal antibodies against defined antigens for passive immunotherapy of chronic gastrointestinal cryptosporidiosis. Antimicrob Agents Chemother 46:275–282. https://doi.org/10.1128/AAC.46.2.275-282.2002
38. Arnold SLM, Choi R, Hulverson MA, Schaefer DA, Vinayak S, Vidadala RSR, McCloskey MC, Whitman GR, Huang W, Barrett LK, Ojo KK, Fan E, Maly DJ, Riggs MW, Striepen B, Van Voorhis WC (2017) Necessity of bumped kinase inhibitor gastrointestinal exposure in treating *Cryptosporidium* infection. J Infect Dis 216:55–63
39. Constable PD, Walker PG, Morin DE, Foreman JH (1998) Clinical and laboratory assessment of hydration status of neonatal calves with diarrhea. J Am Vet Med Assoc 212:991–996
40. Guy RA, Payment P, Krull UJ, Horgen PA (2003) Real-time PCR for quantification of *Giardia* and *Cryptosporidium* in environmental water samples and sewage. Appl Environ Microbiol 69:5178–5185. https://doi.org/10.1128/AEM.69.9.5178-5185.2003

# Chapter 16

# Cell Culture Infectivity to Assess Chlorine Disinfection of *Cryptosporidium* Oocysts in Water

**Jennifer L. Murphy and Michael J. Arrowood**

## Abstract

This chapter provides a detailed protocol to assess disinfection efficacy of chlorine against *Cryptosporidium* oocysts including the core chlorine disinfection assay, the in vitro cell culture infectivity assay, and microscopy analysis and data interpretation.

**Key words** Cell culture infectivity, Chlorine, Disinfection, In vitro culture, *Cryptosporidium*, Oocysts

## 1 Introduction

*Cryptosporidium* is a leading cause of waterborne disease in both developing [1, 2] and industrialized [3–5] countries. Due to a robust outer oocyst wall, *Cryptosporidium* oocysts are extremely tolerant of many chemical disinfectants, including chlorine, one of the most widely used water treatment chemicals in the world. Laboratory experiments have assessed the efficacy of a number of chlorine-based disinfectants, including sodium hypochlorite [6–9], stabilized chlorine [10, 11], monochloramine [7, 12], mixed oxidants [9, 13], and chlorine dioxide [7, 14–18]. A variety of methods were used to assess the inactivation of *Cryptosporidium* in these disinfection studies, including animal infectivity [7, 9, 15, 18], in vitro excystation [12, 17], cell culture-quantitative polymerase chain reaction [13], and in vitro cell culture infectivity [8, 10, 11, 14, 16, 19]. This chapter provides a detailed protocol we have used to assess efficacy of chlorine against *Cryptosporidium* oocysts, including the core chlorine disinfection assay, the in vitro cell culture infectivity assay, and microscopy analysis and data

**Supplementary Information:** The online version contains supplementary material available at https://doi.org/10.1007/978-1-0716-4824-7_16.

Jan R. Mead and Michael J. Arrowood (eds.), *Cryptosporidium: Methods and Protocols*, Methods in Molecular Biology, vol. 2978, https://doi.org/10.1007/978-1-0716-4824-7_16,

interpretation. These methods can be used as a basis for designing studies to assess the efficacy of new or existing disinfectants, either alone or in combination. Also included in the online Electronic Supplementary Files are image analysis macro file examples and a MS Excel workbook file with macros that automates image analysis data importation and statistical analyses.

## 2 Materials

1. MDCK (NBL-2) (ATCC CCL-34) Madin–Darby canine kidney cells (*see* **Note 1**).
2. Water bath (37 °C).
3. 5% $CO_2$ incubators (37 °C) (one for routine cell passaging, one for infectivity assay).
4. DMEM growth medium: DMEM base (high glucose), 10% heat-inactivated fetal bovine serum, 1X sodium pyruvate, 1X L-glutamine, 1X MEM nonessential amino acids, 0.01 M HEPES buffer, 1X penicillin–streptomycin antibiotic mix, ~0.1% sodium bicarbonate.
5. UltraCULTURE™ synthetic culture medium (Lonza Bioscience, Walkersville, MD).
6. Vacuum aspirator flasks (2 or 4 liters) with vacuum flask traps (typically 1 liter or 500 mL) outfitted with disposable glass Pasteur pipets connected via silicone tubing.
7. Hydrophobic in-line filter for vacuum aspirator to protect vacuum system from contamination.
8. Micropipettes and tips, including aerosol-barrier tips.
9. Automatic pipet aid (line or battery powered).
10. 1, 5, 10, 50, and 100 mL disposable pipets.
11. 50 mL polypropylene conical tubes.
12. Phosphate-buffered saline (PBS): 0.01 M, pH 7.2.
13. PBS (0.01 M, pH 7.2) supplemented with 0.1% (w/v) bovine serum albumin (BSA).
14. PBS (0.01 M pH 7.2), 0.1% bovine serum albumin (BSA), supplemented with sodium thiosulfate at 50 mg/L per 1 mg/L free chlorine neutralization.
15. Trypsin 1X: ~0.25% w/v trypsin, ~0.5 mM EDTA in PBS for dissociating cells during routine culture.
16. Versene: ~0.5 mM EDTA in PBS (monolayer rinse before trypsinization).
17. T75 (75 $cm^2$) vented or non-vented tissue culture flasks.

18. SensoPlate™ 24-well flat-bottom microplates with lids (Greiner Bio-One North America, Inc.) with confluent MDCK cell monolayers (*see* Subheading 3.1).
19. Digital colorimeter or spectrophotometer for chlorine determination.
20. DPD 1 reagent (N,N-diethyl-p-phenylenediamine: tablets or pillows) for chlorine determination.
21. Timer.
22. pH meter.
23. Environmental chamber (or recirculating water bath) with multi-position magnetic stir plate.
24. Sodium hypochlorite solution: ~5.65% w/v (laboratory grade).
25. Chlorine demand-free (CDF) water (*see* **Note 2**).
26. Chlorine demand-free (CDF) glassware and magnetic stir bars (*see* **Note 3**).
27. *Cryptosporidium parvum* oocyst stock suspension (*see* **Note 4**).
28. Potassium dichromate: 2.5% w/v aqueous solution.
29. Hemocytometer or Neubauer chamber and coverslips.
30. Vortex mixer.
31. Sodium thiosulfate solution: 10% w/v in distilled water.
32. Refrigerated table top centrifuge (should reach 3200× *g* or more).
33. Refrigerated microcentrifuge (should reach 16,000× *g* or more).
34. Flame (ethanol or butane).
35. 1.5 mL microcentrifuge tubes, low-binding (*see* **Note 5**).
36. Refrigerator (4 °C).
37. Class II Biosafety Cabinet (BSC).
38. DMEM or UltraCULTURE base medium, plus 0.75% synthetic sodium taurocholate (to promote excystation).
39. Bouin's fixative solution.
40. 70% ethyl alcohol (v/v, prepared from 200 proof anhydrous non-denatured ethanol and deionized water).
41. *Cryptosporidium*-specific monoclonal antibody C3C3 (*see* **Note 6**).
42. 4′, 6-Diamidino-2-phenylindole (DAPI): 1000X frozen stock solution: 1 mg/mL in sterile distilled water, stored at −20 °C; working solution: dilute 1000X stock into PBS (0.01 M, pH 7.2) to yield 1.0–0.5 μg/mL (i.e., 1/1000 to 1/2000 dilution).

43. Platform shakers (rocking, rotary, and/or orbital wave motion).
44. Goat anti-mouse IgG (H + L) secondary antibody, conjugated with fluorescein isothiocyanate (FITC), cyanine (Cy3), or similar fluorochromes (e.g., Alexa Fluor 555), supplemented with 2 mM sodium azide.
45. Aluminum foil.
46. Coverslip mounting medium (PVA/DABCO) with a fluorescence anti-quenching agent (*see* **Note 7** *for recipe and preparation instructions*).
47. 15 mm #1 round glass coverslips.
48. Suction pen (vacuum sucking pen with silicone tips for manipulating coverslips).
49. Inverted microscope for routine cell culture applications.
50. Inverted epifluorescence microscope with digital camera (*see* **Note 8**).
51. Zeiss AxioVert 200 M microscope (or similar) (Carl Zeiss, Inc., Thornwood, NY).
52. Zeiss HRm digital camera (or similar) (Carl Zeiss, Inc., Thornwood, NY).
53. Zeiss AxioVision software (or similar software for capturing images) (Carl Zeiss, Inc., Thornwood, NY).
54. ImageJ software (for image analysis and parasite quantification [20]).
55. Microsoft Excel (for data analysis).

## 3 Methods (See Note 9)

### 3.1 MDCK Cell Maintenance (Twice Weekly Subculture)

1. Warm all reagents in a 37 °C water bath (or incubator) prior to splitting cells.
2. Aspirate media from T75 culture flask containing confluent MDCK cells (typically ~3.5 days old).
3. Rinse the cell monolayer with 10 mL PBS. Aspirate rinsate.
4. Flood cell monolayer with 7 mL Versene, cap flask, gently rock the flask to get complete coverage of the monolayer, and incubate at 37 °C for 10 min. Aspirate Versene.
5. Add 3 mL trypsin with EDTA. Gently rock the flask to get complete coverage of the monolayer.
6. Incubate the flask in a 5% $CO_2$ incubator at 37°C for up to 30 min or until monolayer visibly dissociates from the flask. It might be necessary to dislodge some cells using the tip of a pipet (triturate) or cell scraper.

7. Add 7 mL DMEM growth medium to cells.
8. Use a pipet to triturate cells ~10 times (or more to disperse cells into suspension).
9. Remove 9 mL of the cell/media suspension in the T75 flask. This will serve as the backup flask.
10. Transfer 1 mL of the cell–media suspension to a new T75 flask, and discard the remainder.
11. Add 15 mL of fresh DMEM growth medium to each flask.
12. Tighten (if vented flask) or loosely close (if non-vented flask) cap.
13. Label flasks with passage number and date.
14. Incubate flasks in a 5% $CO_2$ incubator at 37°C for ~3.5 days.

### 3.2 Microplate Preparation

1. During routine cell passaging, transfer sufficient volume of triturated cells (*see* Subheading 3.1, **step 8**) to a flask containing fresh DMEM growth medium (~1:100 dilution of cells) to prepare the necessary number of culture microplates (*see* **Note 10**). Swirl to mix.
2. Transfer 1.5 mL of this diluted cell suspension to each of 24 wells in a microplate. Place lid on microplate.
3. Incubate microplate in a 5% $CO_2$ incubator at 37°C for ~3.5 days (until confluent).
4. At ~3.5 days, aspirate media from wells.
5. Add 1.5 mL of fresh DMEM growth medium to each well.
6. Incubate the microplate in a 5% $CO_2$ incubator at 37°C for an additional day prior to performing the infectivity assay.

### 3.3 Titer C. parvum Oocyst Stock

1. Calculate the total number of oocysts necessary for the experimental run (all treatment and control flasks).
2. Wash out storage medium using sterile PBS (0.01 M, pH 7.2) and centrifugation (~16,000× *g* for 3 min).
3. Repeat **step 2** twice more, especially if the oocysts were previously stored in potassium dichromate.
4. Resuspend oocysts in 1 mL PBS (0.01 M, pH 7.2) by vortexing for 30 s.
5. Oocyst stock concentrations are verified using a hemocytometer (or other cell counting tool).

### 3.4 Disinfection Assay

1. Set the environmental chamber (or prepare a recirculating water bath) to the desired experimental temperature.
2. Label three CDF glass flasks each containing a CDF magnetic stir bar as experimental replicates for each chlorine concentration being tested (*see* **Note 11**). Label one additional CDF

glass flask containing a CDF magnetic stir bar as a control (no chlorine) flask.

3. Prepare stock volume of experimental water at the desired chlorine concentration (*see* **Note 12**).
4. Use a CDF graduated cylinder to add sterile CDF water to a large CDF beaker containing a CDF stir bar. Immediately return the foil cover to the beaker and keep it covered at all times.
5. Add the appropriate volume of 5.65% sodium hypochlorite to the CDF water to achieve the target chlorine concentration. Stir gently on magnetic stir plate.
6. Measure free chlorine concentration using standardized colorimeter (*see* **Note 13**). Record.
7. Measure pH using calibrated pH meter (*see* **Note 14**). Record.
8. Use a CDF graduated cylinder to transfer an equal volume of chlorinated water into each experimental flask and the control flask (*see* **Note 15**). Immediately return the foil cover to the flasks and keep them covered at all times.
9. Place all flasks on a magnetic stir plate in the environmental chamber (or in a recirculating water bath).
10. Measure the free chlorine concentration and pH for each flask to ensure that target experimental parameters are met (Time 0). Record.
11. Vortex oocyst stock for 30 s.
12. Add predetermined volume of oocyst stock to the first experimental flask for a final concentration of ~$10^5$ oocysts/mL. Start timer.
13. Repeat **step 12** for the two other experimental flasks.
14. Repeat **step 12** for the control flask. After 1 min, add sodium thiosulfate solution to the control flask at 50 mg/L per 1 mg/L free chlorine (*see* **Note 16**).
15. At set time points, measure free chlorine and other water parameters (e.g., pH) as required.
16. At set time points, transfer 4 mL, 40 mL, or 120 mL (in triplicate 40 mL volumes) from each flask into sterile 50 mL screw-cap tubes containing PBS–BSA and 10% sodium thiosulfate (*see* **Note 17**). Gently swirl and invert to mix, neutralizing chlorine.

### 3.5 Sample Concentration

1. Immediately after collection, centrifuge 50 mL tube(s) at ~3290× *g* for 10 min at 4 °C.
2. Carefully aspirate supernatant down to just above the pellet (~1.5 mL should remain in the tube). Be careful not to disturb the pellet (*see* **Note 18**).

3. Resuspend the pellet by gently triturating the remaining material at least five times with a micropipettor equipped with a new aerosol-barrier tip.
4. Transfer the resuspended pellet solution to a 1.5 mL microcentrifuge tube.
5. Centrifuge the microcentrifuge tube at ~16,000× *g* for 3 min.
6. Carefully aspirate the supernatant down to the 0.1 mL demarcation on the microcentrifuge tube. Be careful not to disturb the pellet.
7. Rinse the 50 mL tube with 1 mL PBS, and transfer rinsate to the corresponding 1.5 mL microcentrifuge tube containing the pellet.
8. Centrifuge the microcentrifuge tube at ~16,000× *g* for 3 min.
9. Carefully aspirate the supernatant down to the 0.1 mL demarcation on the microcentrifuge tube. Be careful not to disturb the pellet.
10. Resuspend the pellet up to 1 mL with 0.9 mL DMEM media containing 0.75% synthetic sodium taurocholate. Gently triturate suspension as you slowly add the DMEM to resuspend the pellet. Incubation in DMEM/taurocholate facilitates excystation.
11. Incubate tubes at ambient temperature for 10 min.

### 3.6 Cell Culture Infectivity Assay

1. During oocyst incubation in DMEM base medium or UltraCULTURE with 0.75% synthetic sodium taurocholate, aspirate media from MDCK monolayers in culture plate wells, and replace with 1 mL fresh, room temperature DMEM growth medium or UltraCULTURE.
2. Following incubation, invert the microcentrifuge tube containing the resuspended oocyst pellet three times.
3. Inoculate 150 μL of the resuspended pellet onto MDCK monolayers in duplicate microplate wells (*see* **Note 19**).
4. Inoculate 150 μL of the appropriate dilution of the *C. parvum* oocyst stock (Subheading 3.3, **step 5**) which will result in an inoculum of ~$10^4$ oocysts onto MDCK monolayers in duplicate microplate wells (positive oocyst control).
5. Inoculate 150 μL of DMEM media containing 0.75% synthetic sodium taurocholate onto MDCK monolayers in duplicate microplate wells (reagent blank/no parasites).
6. Incubate the microplate for 48–60 h at 37 °C under 5% $CO_2$. Ensure that control and treatment wells are incubated for the same time period.

7. After incubation, aspirate the media from the microplate wells (the media will have changed color from red to yellow/ orange).
8. Add and remove (aspirate) 1 mL sterile PBS to wash the remaining media from wells.
9. Add 0.5 mL Bouin's fixative solution to each well. Place microplate on a platform shaker, and gently rock for 30 min at ambient temperature (*see* **Note 20**).
10. After incubation, use a micropipettor to transfer Bouin's fixative solution into the appropriate waste container for disposal as hazardous chemical waste.
11. Add 0.5 mL 70% ethanol to each well. Place microplate on platform shaker, and gently rock for 10 min at ambient temperature. After incubation, transfer solution into hazardous waste container (as in **step 8**).
12. Repeat **step 9** four additional times. At this point, waste can be aspirated into regular aspiration flasks and does not need to go into a specialized waste container.
13. Add and remove (aspirate) 1 mL sterile 0.01 M PBS pH 7.2 to wash the remaining 70% ethanol from wells.
14. Add 1 mL PBS containing 0.1% BSA to each well, and incubate at ambient temperature for at least 30 min ("blocking" step). At this point slides can be stored overnight at 4 °C, if necessary.
15. Remove (aspirate) PBS containing 0.1% BSA from each well.
16. Add 350 μL working DAPI solution (prepared in PBS), and incubate for 10 min at room temperature on a platform shaker (with light-protective cover).
17. Remove (aspirate) DAPI solution from each well.
18. Add 0.25 mL appropriately diluted C3C3 monoclonal antibody (or alternative reagent) to each well. Place microplate on platform shaker, and gently rock for 60 min at ambient temperature (*see* **Note 21**).
19. After incubation, remove C3C3 monoclonal antibody from each well.
20. Add 1 mL sterile PBS to each well to wash unbound C3C3 antibody from wells. Place microplate on platform shaker, and gently rock for 10 min at ambient temperature. After incubation, remove PBS from each well by aspiration.
21. Repeat **step 18** two additional times.
22. [Skip to **step 24** if using a fluorochrome-labeled primary reagent to label developing parasites in host cells]. Add 0.25 mL FITC-goat anti-mouse IgG (H + L) antibody (or alternative reagent, as desired) to each well. Place microplate on

platform shaker (with light-protective cover), and gently rock for 60 min at ambient temperature.

23. After incubation, remove FITC-goat anti-mouse IgG (H + L) antibody from each well.
24. Add 1 mL sterile PBS to each well to wash unbound FITC-goat anti-mouse IgG (H + L) antibody from wells. Place microplate on platform shaker (with light-protective cover), and gently rock for 10 min at ambient temperature. After incubation, remove PBS from each well by aspiration.
25. Repeat **step 22** two additional times.
26. Add two to three drops of room temperature PVA/DABCO to each well.
27. Use suction pen tool to place 15 mm coverslip in well (*see* **Note 22**).
28. Aspirate excess PVA/DABCO from each well (*see* **Note 23**).
29. Cover microplate with foil and transfer it to the microscope for image capture. Alternatively, allow microplate to sit overnight at 4 °C before microscopic examination.

### 3.7 Microscopy and Data Analysis

1. If previously stored at 4 °C, remove culture plates from refrigerator, and allow to warm to ambient temperature (make sure there is no condensation on plate bottom).
2. Set up inverted epifluorescent microscope for use (*see* **Note 24**).
3. Place 24-well plate on microscope stage.
4. Run software image capture protocol/experiment to capture up to 64 images per culture well at 100X or 200X magnification (depending on the desired imaging coverage and analysis configuration). Capture images of parasite developing stages and, if desired, host cell nuclei (DAPI staining). The latter will facilitate characterizing parasite-to-host cell ratios and is also useful in assessing whether host cell monolayers vary significantly between wells (i.e., reveal differences in host cell growth or perhaps show signs of toxicity or bacterial or fungal infection).
5. Because most native image capture files are very large, it may be useful (for portability and to conserve storage space) to convert the files to the JPEG format. We find no differences in parasite quantification between JPEG images and uncompressed native image file formats. The ImageJ macro language has convenient tools for file format conversion.
6. Using relevant criteria (*see* **Note 25**), use ImageJ to count parasites and host cell nuclei for each frame (digital image) within a microscopical field (*see* **Note 26 and supplemental**

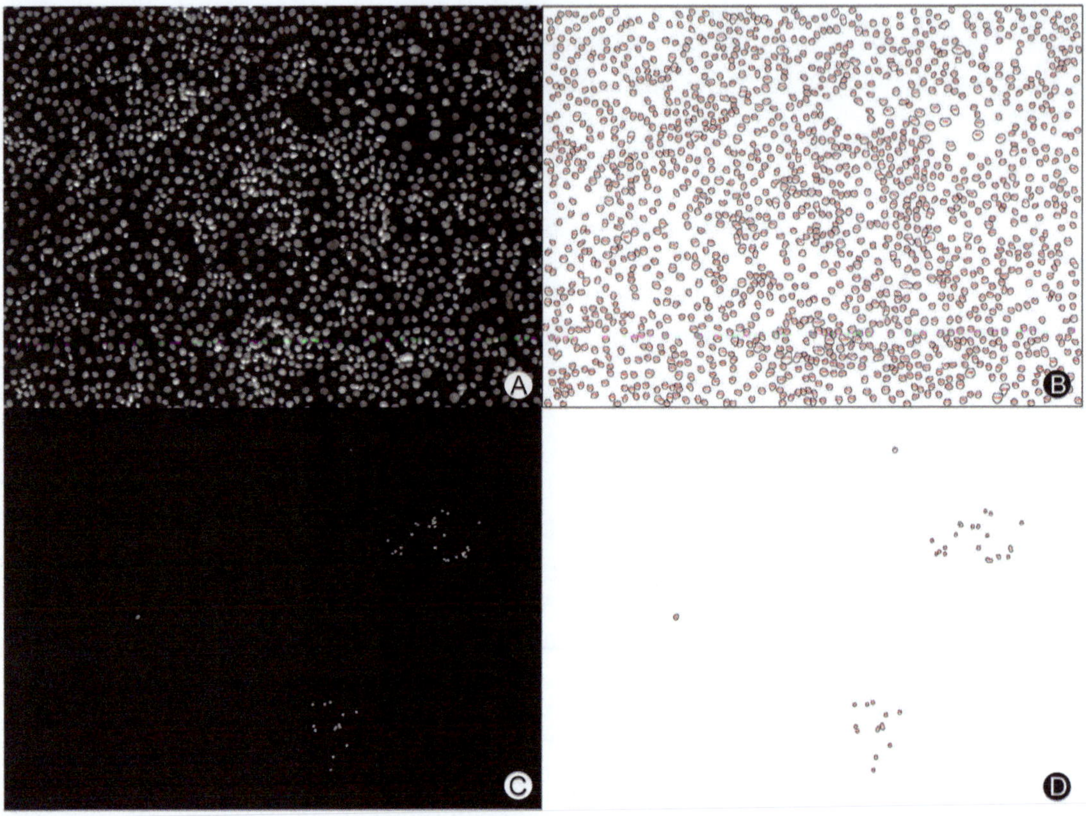


**Fig. 1** Representative images captured from an MDCK cell culture of chlorine-treated *C. parvum* oocysts and corresponding images showing ImageJ-quantified parasites or host cell nuclei (after DAPI staining). Panel A illustrates MDCK cell nuclei stained with DAPI, and panel B shows the ImageJ-identified nuclei (note nucleus perimeter/outlines and numbers corresponding to each nucleus in red font). Panel C illustrates the reduced numbers of *C. parvum* life cycle stages labeled with specific monoclonal antibody (C3C3), while panel D shows the corresponding ImageJ-identified parasites. Microscopical images captured at a total magnification of 100X. *See* Subheading 4 for details on ImageJ criteria and macro used to identify parasites under these assay conditions

**ImageJ example macro: "CountParasiteFolder_Universal_8-bit.txt"**). Note that derivative images are generated for each original image (Fig. 1, panels A and C) with the identified parasites (or host cell nuclei) indicated as numbered, schematic "ring" objects (Fig. 1, panels B and D). The ImageJ-generated data file (MS Excel *.xls format) tabulates the relevant data for each countable object (i.e., Area, X, Y, Perim., BX, BY, Width, Height, Major, Minor, Angle, Circ., Feret, FeretX, FeretY, FeretAngle, MinFeret, AR, Round, Solidity, XStart, YStart). The most useful data (for quantification purposes) includes the area (volume in pixels), perimeter, and circularity of each object. The ImageJ macro **"CountParasiteFolder_Universal_v4_8-bit.txt"** generates a summary data file (Well ID + "Summary") that consolidates the results for all of the images in each folder analyzed (corresponding to each culture well) (*see* **Note 27**). These tables are imported into Microsoft Excel to calculate the average number and size of developing parasite stages (or host cell nuclei) per culture well. The macro language included in MS Excel can automate importing data files and calculating descriptive statistics. See in Electronic Supplementary File an example MS Excel workbook file with macros **"Excel_Template_24_Well_10X_Parasite_Analysis_Blank.xlsm"** that automates data importation and statistical analysis.

7. Parasite quantification is used to calculate the $\log_{10}$ inactivation of oocysts in each condition over time as compared with the oocyst concentration in the control flasks. For a basic disinfection analysis, calculate $\log_{10}$ inactivation or percent reduction of parasites:

$$\text{Log Reduction} = \log_{10}\left(\frac{A}{B}\right)$$

Alternatively, calculate percent reduction using this formula:

$$\text{Percent Reduction} = \frac{(A - B) * 100}{A}$$

where
$A$ = parasite stage count in control wells
$B$ = parasite stage count in treatment wells

Adjust calculations based on any dilutions incorporated into well inocula values. For example, if a tenfold higher oocyst inocula was added to a culture well, the parasite count for that well would need to be decremented by a factor of 10 before calculating log or percent reduction values (*see* **Note 28**). Consider more elaborate analyses if accounting for disinfectant decay, etc. [14, 21, 22].

## 4 Notes

1. We routinely use MDCK cells for immunofluorescence-based assays of *C. parvum* growth primarily because they are robust, support cryptosporidial growth comparable to or exceeding that in HCT-8 cells, and grow "flat" in culture chambers. The flat monolayer topology facilitates image capture of developing parasites, as they are located in a narrow focal plane. HCT-8 cells, however, grow much more three-dimensionally ("hills and valleys"), complicating the autofocus of parasites during image capture-based microscopic analyses (this is more pronounced at microscope magnifications >100X). While recent advancements in cell culture environments facilitate

extended cultivation of *C. parvum*, the conditions described in this chapter are well suited to evaluating and enumerating the survival of disinfectant-treated oocysts.

2. Prepare a 0.01 M phosphate-buffered solution (pH 7.5) in reverse osmosis-treated water using dibasic sodium phosphate (anhydrous) and monobasic sodium phosphate (monohydrate). If different pH is desired, quantities of mono and dibasic sodium phosphate can be adjusted empirically. Add sufficient 5.65% sodium hypochlorite solution to achieve ≥5 mg/L free chlorine. Cover tightly with aluminum foil. After 2 days, ensure the solution contains at least 2 mg/L free chlorine; discard if not. Place uncovered container under an ultraviolet lamp to remove all remaining free chlorine. Ensure there is no detectable free chlorine in the solution, cover tightly with aluminum foil, and autoclave at 121 °C at 15 psi for 15 min. CDF water can be used for up to 2 days after preparation; otherwise, prepare fresh CDF water.
3. Add sufficient 5.65% sodium hypochlorite solution to deionized water in a large tank to achieve ≥10 mg/L free chlorine. Submerge glass beakers and Erlenmeyer flasks each containing magnetic stir bars and glass graduated cylinders. After ≥3 h, rinse thoroughly with CDF water, cover tightly with aluminum foil, and autoclave at 121 °C at 15 psi for 15 min.
4. *C. parvum* oocysts are available from commercial vendors or may be produced and isolated from experimentally infected animals, from naturally infected animals (e.g., calves), or from clinical stool samples collected from patients with cryptosporidiosis (*see* Chapter 22 for details on oocyst purification). If oocysts are derived from human samples, genotyping may be warranted to determine the species of *Cryptosporidium* present. In general, *C. parvum* is more readily cultured in vitro than *C. hominis* when using MDCK cells. For disinfection studies, oocysts are typically processed and stored in saline or PBS supplemented with antibiotics. Such oocysts may harbor contaminating bacteria that can overgrow control wells in cell culture studies. Consider using a short-term bleach exposure of control oocysts (1/10 dilution of commercial bleach for 5 min) to "sanitize" and provide a reference "zero" for comparison to experimentally treated oocysts. Oocysts stored in 2.5% aqueous potassium dichromate are likewise suitable for disinfection studies, although some investigators prefer antibiotic-stored oocysts. One advantage of using potassium dichromate-stored oocysts is that they are "sanitized" already and can be used directly in cultures after washing out the dichromate via centrifugation. While "freshly passaged" oocysts (≤4 months old) are preferred for disinfection studies, we have found that as long as the oocysts used are from the

same lot, results are consistent using oocysts up to 6 months old. Slight differences might be observed between antibiotic-stored and potassium dichromate-stored oocysts; these differences are typically less than observed in lot-to-lot variation.

5. Microcentrifuge tubes should be "low binding" or "low retention" to minimize losses of oocysts during centrifugation and other processing steps. Conventional polypropylene microcentrifuge tubes are not recommended. If tubes are labeled "RNase, DNase, and/or DNA free," they can typically be considered suitable for use without the need to sterilize by autoclave (just be sure to handle aseptically).
6. C3C3 monoclonal antibody can be used in a direct or indirect fluorescence assay format [19, 23]. If using the fluorochrome-labeled antibody, no secondary antibody step is necessary. If using the unlabeled C3C3 antibody, use an appropriate secondary antibody (e.g., fluorescein-conjugated goat anti-mouse IgG reagent). Alternatively, *Vicia villosa* lectin has been successfully used to label and identify the developing stages of *C. parvum* in cell cultures [24].
7. Antiquenching (anti-fading) long-term coverslip mounting medium for immunofluorescence assays. DABCO (1,4-diazabicyclo-[2.2.2]octane) is recommended as a fluorescence antiquenching agent due to its great solubility and chemical stability [25]. DABCO protects fluorochrome dyes against quenching by scavenging free radicals liberated by the excitation of the fluorochrome. The higher pH of this recipe typically enhances fluorescence intensity versus neutral pH recipes. Slides prepared with PVA/DABCO can be stored at 4 °C or −20 °C for years with little loss of fluorescence. Preparation: Mix 2.4 g polyvinyl alcohol (PVA, average molecular weight 30,000–70,000) gradually into 6 g glycerol. Mix carefully to avoid creating air bubbles. Add 6 mL of $H_2O$, mix, and allow to stand at room temperature for several hours. Add 12 mL Tris buffer (0.2 M, pH 8.5), mix, and heat to ~50 °C for 10 min with occasional mixing. When the PVA has dissolved, clarify by centrifugation at ~3000× *g* for 15 min. Decant supernatant (discard pellet, if any), and add DABCO to a final concentration of 2.5% (w/v). The mounting medium is viscous, so use a pipet to mix in the DABCO. Avoid introducing bubbles while mixing. When dissolved, decant aliquots of the "working" PVA/DABCO into "snap cap" polypropylene tubes (~4 mL/tube), and store at −20 °C (long-term storage) or at 4 °C (short-term storage, i.e., several months).
8. Ideally, the inverted epifluorescence microscope will have objectives suitable for fluorescence imaging at 10X and 20X. Epifluorescence light source (e.g., mercury vapor lamp or LED

source) should produce wavelengths needed for nuclear stain observation (DAPI) as well as for fluorescein, cyanine, or other fluorochromes used in the parasite and host cell labeling portion of the assay. Typical fluorescence filter sets will have the following or similar features: For DAPI observation, excitation filter ~340–380 nm; dichroic mirror ~400 nm; barrier filter ~420 nm. For fluorescein or Alexa Fluor 488 observation, excitation filter ~450–490 nm; dichroic mirror ~510 nm; barrier filter ~515–550 nm. For cyanine dye (e.g., Cy3) or TRITC or Alexa Fluor 555 observation, excitation filter ~530–560 nm; dichroic mirror ~565 nm; barrier filter ~580–660 nm. Digital camera should be suitable for automatic capture of fluorescent images. Users will need to calculate the frame size scale at the magnification of choice and determine how many frames/fields cover the area of the culture well. While we use 24-well plates, other investigators have used 96-, 384-, and 1536-well formats successfully.

9. Unless otherwise indicated, all steps are performed in an approved Class II Biosafety Cabinet.
10. Each volume of water sampled from experimental and control flasks (*see* **Note 17**) will be individually concentrated and inoculated onto cell monolayers in duplicate microplate wells. This number of wells, plus additional wells needed for infectivity assay controls (*see* Subheading 3.6, **steps 4** and **5**), will dictate the required number of culture microplates needed per experiment.
11. At least three flasks per chlorine concentration should be used. All experimental flasks should be of equal volume. Use flasks that will be no more than 2/3 full at maximum volume during the experiment.
12. If experiments will be conducted in triplicate, prepare at least five times the volume needed for each replicate. Experimental water should be prepared on the day of the experiment. A slightly higher concentration should be prepared to account for chlorine off-gassing when pouring chlorinated water into flasks. Preliminary experiments to determine dose volume and to assess chlorine decay during pouring steps should be conducted to determine target chlorine concentration in stock water.
13. If target chlorine concentrations are near the detection limit of the colorimeter, use CDF water to prepare dilutions for measurement within the working range of the colorimeter.
14. Preliminary experiments to assess pH changes due to chlorine addition should be conducted to ensure that the target pH is achieved after the addition of sodium hypochlorite.

15. After measuring free chlorine, proceed immediately to the following step because chlorine will begin to decay.
16. Sodium thiosulfate is added to stop hypochlorite oxidation and sanitization of control oocysts. Measure free chlorine in control flask right after sodium thiosulfate neutralization to verify it is zero. Note that this hypochlorite step assures that the control oocysts (stored in PBS/antibiotics) are free of residual (viable) bacterial contaminants, preventing potential bacterial infection of positive control wells. Also, the short hypochlorite exposure provides a slight enhancement of the excystation of the control oocysts, providing a better reference for the experimentally treated oocysts. The latter is why we recommend using the bleach treatment step for potassium dichromate-stored oocysts, too. If the bleach step is omitted, early time points in the chlorine treatment of oocysts can actually generate surprising infection results where treated oocysts yield higher infection levels compared to control oocysts.
17. The volume assayed is dependent on the efficacy of the disinfectant on oocyst inactivation. Smaller sample volumes are needed when a high number of infectious oocysts is expected (e.g., short contact time). Larger sample volumes are needed when a low number of infectious oocysts (survivors) is expected (e.g., long contact time) to increase the detection limit of the assay. Preliminary "range-finding" experiments may be necessary to help determine optimal sample volumes required at assay time points.
18. The pellet will likely not be visible to the naked eye. When aspirating, keep suction low, and keep the Pasteur pipet tip just below the surface of the supernatant.
19. Oocysts suspended in DMEM/taurocholate (or UltraCULTURE/taurocholate) can be directly inoculated into culture wells. The residual sodium taurocholate actually enhances cryptosporidial growth (keep final sodium taurocholate concentration at ≤0.1%).
20. Platform shaker should provide gentle rocking or orbital wave (variable tilt) motion to ensure the monolayer is continuously covered with reagent, but not agitated so aggressively that the monolayer detaches.
21. Alternatives to the use of unlabeled monoclonal antibody include using a fluorochrome-labeled version (e.g., C3C3-Cy3, 1/500 dilution, 250 μL/chamber cell) or fluorochrome-labeled lectin at an appropriate dilution. Red fluorescent dyes are typically preferred as host cell monolayers as they typically exhibit lower background fluorescence when observed using "red" filter sets compared to fluorescein or similar dyes.

22. Use suction pen tool with silicone tip to pick up 15 mm coverslip. Tilt 24-well plate at an angle that causes PVA/DABCO mounting media to pool to the lower side of the well. Carefully place the edge of the glass coverslip toward the base of the well to contact mounting medium. Carefully release the coverslip (release suction) to allow it to drop into place. Avoid introducing air bubbles under the coverslip. It may be useful to "tap" the coverslip with the tip of the aspirating Pasteur pipet to dislodge any bubbles trapped beneath the coverslip before aspirating excess mounting medium.
23. PVA/DABCO mounting medium will "gel" (immobilizing the coverslip), protect the labeled monolayer from drying, and protect fluorochromes from quenching (fading) under microscope lamp illumination.
24. Power up microscope and computer. Load microscopy software that controls the microscope and captures images using the digital camera. Load image capture protocol or experiment in software that will capture images. For the described protocol, we employed Zeiss AxioVision software (Carl Zeiss, Inc., Thornwood, NY), a Zeiss HRm digital camera, and a Zeiss AxioVert 200 M microscope to capture 64 images in an 8 × 8 "checkerboard" grid at 100X total magnification (10X objective) which is equivalent to approximately 22% of the well (monolayer) surface area. Ideally, the entire well should be captured for optimal sensitivity, but the subset of 64 images has been successful for us and covers a significant portion of the monolayer. Each microscope system will require optimization, and it is beyond the scope of this chapter to describe all alternatives.
25. Criteria for identifying and counting parasites and host cells include the following. Developing parasite life cycle stages can be selectively identified by parasite-specific monoclonal antibody reactivity (or by general labeling by lectin reactivity). Immediately after sporozoite invasion, early trophozoites are approximately 1–2 μm in diameter (normally round in morphology). While some sporozoites do not successfully invade host cells, they typically "round up" anyway even if they remain free in the culture medium or adhere to host cells (though they do not actually invade, grow, or proliferate). If fluorescing objects smaller than ~2 μm are excluded (nonviable sporozoites/trophozoites or debris), as well as objects much larger than ~6 μm (some parasite clumps and large debris), the quantification can be confidently limited to developing stages (late trophozoites, meronts, gamonts, and zygotes, depending on the time of analysis post-inoculation). Also keep in mind that there is little, if any, "recycling" of the asexual stages in MDCK or HCT-8 cells under conventional cell culture conditions.

This means that life cycle development progresses essentially directly through the life cycle, going from sporozoite to Type I meront, then to Type II meront, and finally to the gamont stages. This effectively limits the amplification potential to the numbers predicted by a direct path through the life cycle. Consequently, the developing stage numbers (which "peak" between 48 and 72 h post-inoculation) are predictive of the number of surviving oocysts after disinfection treatment.

26. Pre-process images using the ImageJ built-in commands "Make Binary," "Fill Holes," and "Watershed." The ImageJ command "Analyze Particles" is used to count parasites with the criteria described below (realizing the values indicated may need to be customized for the microscopy hardware, magnification, fluorescent labeling reagents, etc.). Based on our configuration (*see* Subheading 3) where imaging is performed at 100X total magnification, the following settings can be used as defaults: size min, 30; size max, 250 (the size range of 30–250 is a volumetric range of pixels that developing parasites occupy at this magnification); circularity min, 0.7; circularity max, 1 (the circularity range of 0.7–1 indicates the parasites are essentially round to slightly oval); brightness/contrast min, 15; brightness/contrast max, 255; threshold min, 25; and threshold max, 255. Depending on the host cells used in the culture assays, determine the appropriate conditions and values to identify DAPI-stained host cell nuclei (if calculating this data for determining host cell monolayer characteristics and to calculate parasite/host cell ratios). ImageJ has robust macro options for automating the image analysis process. See example in Electronic Supplementary File ImageJ macro: **"CountParasiteFolder_Universal_8-bit.txt" in the online supplement to this chapter**. Note that "Size min" and "Size max" can be adjusted to conform to the characteristics of the images captured by the microscope, digital camera, and fluorescence labeling results used by the respective investigators. The size parameters used above actually capture slightly larger objects than might be expected from uniformly dispersed parasites (i.e., closely adjacent parasites that cannot be split using the "Watershed" command may be counted as a single parasite); consequently, the total counts may be somewhat conservatively estimated depending on the parasite density in the image and the quality of the fluorescence labeling). Manually scanning image sets (raw images and the processed images) is warranted to determine if excessive over- or under-counts have occurred because of heavy parasite density in the images (tightly packed clusters of parasites can be undercounted because they may not be resolvable into individual parasites and may be excluded because the estimated size exceeds the maximum for inclusion

in the overall count). This result can also be gleaned by examining the average object (parasite) size in the data tables generated for the images. See in Electronic Supplementary File an example MS Excel workbook file with macros "**Excel_Template_24_Well_10X_Parasite_Analysis_Blank.xlsm**" that automates data importation and statistical analysis.

27. Several options are available for the analysis of the parasite and host cell nuclei data. Data for each well can be simply summed from the total number of frames captured (64) and used as a data point. Alternatively, the average number of parasites per frame can be calculated from the 64 frames and used as a data point. If an estimate of the total number of parasites per well is preferred, the estimate can be calculated using the following information (we are using the data from our microscope configuration in this example): total magnification (100X), well area (1.9 $cm^2$), digital camera frame dimension (1388 × 1040 pixels), total number of frames per well at 100X magnification (293), and subset of frames actually captured per well (64, representing ~21.8% of total area). The following calculation can be made: total number of parasites per well = (average number of parasites in the 64 frames captured) * 293. Alternatively, parasites per $mm^2$ can be derived using a variation on the formula described by Augustine [26] where stages/$mm^2$ = ((mean # stages per frame) / ($K \times$ PC)) where PC is the average percent confluence for each field and the constant $K$ represents the area in $mm^2$ for each video frame digitized at 100X (in our configuration $K$ = ~0.648).

28. After range-finding experiments, it may be useful to adjust the treated inoculum oocyst numbers introduced to test wells/monolayers. For example, if it is expected that a treatment will produce a 2–3 log reduction in oocyst viability (infectivity), the inoculum can be increased by 1–2 logs (or more) compared to the control. In this way, the numbers of developing parasites will occupy the optimal range for the assay. Too few survivors will result in parasite counts near the lower limit of detection, while too many survivors will lead to monolayer injury or parasite overcrowding, which can interfere with the identification of individual life cycle stages (reducing the sensitivity of the assay). We find that viable oocyst inocula should be in the following range for the 24-well plates described in this assay: ~$1 \times 10^2$ to $1 \times 10^5$ oocysts, with a preference for close to $1 \times 10^4$ for the best data.

## Acknowledgments

The use of trade names is for identification only and does not imply endorsement by the Public Health Service or by the US Department of Health and Human Services. The findings and conclusions in this report are those of the authors and do not necessarily represent the official position of the Centers for Disease Control and Prevention. Michael J. Arrowood is retired from service at the Centers for Disease Control and Prevention.
The authors would like to acknowledge the expert assistance of Long Ti Xie, who helped optimize the culture methods and parasite quantification in the immunofluorescence assays.

**Competing Interest Statement** The authors have no conflicts of interest to declare that are relevant to the content of this chapter.

## References

1. Snelling WJ, Lin Q, Moore JE, Millar BC, Tosini F, Pozio E, Dooley JS, Lowery CJ (2007) Proteomics analysis and protein expression during sporozoite excystation of *Cryptosporidium parvum* (Coccidia, Apicomplexa). Mol Cell Proteomics 6(2):346–355. https://doi.org/10.1074/mcp.M600372-MCP200
2. Sow SO, Muhsen K, Nasrin D, Blackwelder WC, Wu Y, Farag TH, Panchalingam S, Sur D, Zaidi AK, Faruque AS, Saha D, Adegbola R, Alonso PL, Breiman RF, Bassat Q, Tamboura B, Sanogo D, Onwuchekwa U, Manna B, Ramamurthy T, Kanungo S, Ahmed S, Qureshi S, Quadri F, Hossain A, Das SK, Antonio M, Hossain MJ, Mandomando I, Nhampossa T, Acacio S, Omore R, Oundo JO, Ochieng JB, Mintz ED, O'Reilly CE, Berkeley LY, Livio S, Tennant SM, Sommerfelt H, Nataro JP, Ziv-Baran T, Robins-Browne RM, Mishcherkin V, Zhang J, Liu J, Houpt ER, Kotloff KL, Levine MM (2016) The burden of *Cryptosporidium* diarrheal disease among children $<$ 24 months of age in moderate/high mortality regions of Sub-Saharan Africa and South Asia, utilizing data from the Global Enteric Multicenter Study (GEMS). PLoS Negl Trop Dis 10(5):e0004729. https://doi.org/10.1371/journal.pntd.0004729
3. MacKenzie WR, Hoxie NJ, Proctor ME, Gradus MS, Blair KA, Peterson DE, Kazmierczak JJ, Addiss DG, Fox KR, Rose JB et al (1994) A massive outbreak in Milwaukee of *Cryptosporidium* infection transmitted through the public water supply. N Engl J Med 331(3): 161–167
4. Shirley DA, Moonah SN, Kotloff KL (2012) Burden of disease from cryptosporidiosis. Curr Opin Infect Dis 25(5):555–563. https://doi.org/10.1097/QCO.0b013e328357e569
5. Hlavsa MC, Cikesh BL, Roberts VA, Kahler AM, Vigar M, Hilborn ED, Wade TJ, Roellig DM, Murphy JL, Xiao L, Yates KM, Kunz JM, Arduino MJ, Reddy SC, Fullerton KE, Cooley LA, Beach MJ, Hill VR, Yoder JS (2018) Outbreaks associated with treated recreational water – United States, 2000–2014. MMWR 67(19):547–551. https://doi.org/10.15585/mmwr.mm6719a3
6. Carpenter C, Fayer R, Trout J, Beach MJ (1999) Chlorine disinfection of recreational water for *Cryptosporidium parvum*. Emerg Infect Dis 5(4):579–584
7. Korich DG, Mead JR, Madore MS, Sinclair NA, Sterling CR (1990) Effects of ozone, chlorine dioxide, chlorine, monochloramine, on *Cryptosporidium parvum* oocyst viability. Appl Environ Microbiol 56:1423–1428
8. Shields JM, Hill VR, Arrowood MJ, Beach MJ (2008) Inactivation of *Cryptosporidium parvum* under chlorinated recreational water conditions. J Water Health 6(4):513–520
9. Venczel LV, Arrowood M, Hurd M, Sobsey MD (1997) Inactivation of *Cryptosporidium parvum* oocysts and *Clostridium perfringens* spores by a mixed-oxidant disinfectant and by free chlorine. Appl Environ Microbiol 63: 1598–1601

10. Murphy JL, Arrowood MJ, Lu X, Hlavsa MC, Beach MJ, Hill VR (2015) Effect of cyanuric acid on the inactivation of *Cryptosporidium parvum* under hyperchlorination conditions. Env Sci Technol 49(12):7348–7355. https://doi.org/10.1021/acs.est.5b00962
11. Shields JM, Arrowood MJ, Hill VR, Beach MJ (2009) The effect of cyanuric acid on the disinfection rate of *Cryptosporidium parvum* in 20-ppm free chlorine. J Water Health 7(1): 109–114
12. Driedger AM, Rennecker JL, Mariñas BJ (2001) Inactivation of *Cryptosporidium parvum* oocysts with ozone and monochloramine at low temperatures. Water Res 35(1):41–48
13. Keegan AR, Fanok S, Monis PT, Saint CP (2003) Cell culture-Taqman PCR assay for evaluation of *Cryptosporidium parvum* disinfection. Appl Environ Microbiol 69(5): 2505–2511
14. Murphy JL, Haas CN, Arrowood MJ, Hlavsa MC, Beach MJ, Hill VR (2014) Efficacy of chlorine dioxide tablets on inactivation of *Cryptosporidium* oocysts. Environ Sci Technol 48(10):5849–5856. https://doi.org/10.1021/es500644d
15. Peeters JE, Mazas EA, Masschelein WJ, Martinez de Maturana IV, Debacker E (1989) Effects of disinfection of drinking water with ozone or chlorine dioxide on survival of *Cryptosporidium parvum* oocysts. Appl Environ Microbiol 55:1519–1522
16. Chauret CP, Radziminski CZ, Lepuil M, Creason R, Andrews RC (2001) Chlorine dioxide inactivation of *Cryptosporidium parvum* oocysts and bacterial spore indicators. Appl Environ Microbiol 67(7):2993–3001. https://doi.org/10.1128/AEM.67.7.2993-3001.2001
17. Ruffell KM, Rennecker JL, Mariñas BJ (2000) Inactivation of *Cryptosporidium parvum* oocysts with chlorine dioxide. Water Res 34(3):868–876
18. Finch GR, Li H (1999) Inactivation of *Cryptosporidium* at 1°C using ozone or chlorine dioxide. Ozone Sci Eng 21(5):477–486. https://doi.org/10.1080/01919512.1999.10382886
19. Arrowood MJ (2002) In vitro cultivation of *Cryptosporidium* species. Clin Microbiol Rev 15(3):390–400
20. Schneider CA, Rasband WS, Eliceiri KW (2012) NIH image to ImageJ: 25 years of image analysis. Nat Methods 9(7):671–675
21. Haas CN, Joffe J (1994) Disinfection under dynamic conditions: modification of Hom's model for decay. Environ Sci Technol 28(7): 1367–1369. https://doi.org/10.1021/es00056a028
22. Haas CN, Kaymak B (2003) Effect of initial microbial density on inactivation of *Giardia muris* by ozone. Water Res 37(12): 2980–2988. https://doi.org/10.1016/S0043-1354(03)00112-X
23. Arrowood MJ, Mead JR, Xie L, You X (1996) In vitro anticryptosporidial activity of dinitroaniline herbicides. FEMS Microbiol Lett 136(3):245–249
24. Gut J, Nelson RG (1999) *Cryptosporidium parvum*: synchronized excystation in vitro and evaluation of sporozoite infectivity with a new lectin-based assay. J Eukaryot Microbiol 46(5): 56S–57S
25. Harlow E, Lane D (1988) Cell staining. In: Antibodies: a laboratory manual. Cold Spring Harbor Laboratory, Cold Spring Harbor/New York, p 418
26. Augustine PC (1980) Effects of polyions, $Ca^{++}$, and enzymes on penetration of cultured cells by *Eimeria meleagridis* sporozoites. J Parasitol 66:498–501

Chapter 17

# High-Content Screening for *Cryptosporidium* Drug Discovery

**Malina A. Bakowski, Melissa S. Love, and Case W. McNamara**


## Abstract

High-content screening (HCS) is a cell-based type of phenotypic screening that combines multiple simultaneous readouts with a high level of throughput. A particular benefit of this form of screening for drug discovery is the ability to perform the interrogation in a biologically relevant system. This approach has greatly advanced the field of drug discovery for cryptosporidiosis, a diarrheal disease caused by protozoan parasites of *Cryptosporidium* spp. These parasites are obligate intracellular parasites and cannot be cultured in vitro without the support of a host cell, limiting the options for potential assay readout. Here we describe an established 384- or 1536-well format high-content imaging (HCI) assay of *Cryptosporidium*-infected HCT-8 human ileocecal adenocarcinoma cells. This HCS assay is a powerful tool to assess large numbers of compounds to power drug discovery, as well as to phenotypically characterize known *Cryptosporidium*-active compounds.


**Key words** Phenotypic screening, High-throughput screening, High-content imaging, *Cryptosporidium*, Cryptosporidiosis, Anti-cryptosporidial, Drug discovery, Neglected tropical diseases, Infectious diseases

## 1 Introduction

Interest in developing novel anti-cryptosporidial drugs [1, 2] has led to advances in characterizing *Cryptosporidium* infections in vitro [3–5], including screening of diverse compound libraries. Screening efforts have included targeted screens of compounds with known mechanisms of action [6–10], compounds with known activity against related pathogens [11, 12], libraries of known bioactive compounds [13], natural products [14], and broader small-molecule compound libraries [15]. The largest screen published to date against *Cryptosporidium* spp. parasites included ~80 K compounds in total, and all published screening efforts to date have led to a diverse set of lead compounds in the pipeline, recently reviewed [15].

Jan R. Mead and Michael J. Arrowood (eds.), *Cryptosporidium: Methods and Protocols*, Methods in Molecular Biology, vol. 2978, https://doi.org/10.1007/978-1-0716-4824-7_17,

Here we describe a method of high-throughput screening (HTS) for *Cryptosporidium* spp. The compound treatment of *Cryptosporidium*-infected host cells is carried out in a microtiter plate and analyzed via high-content imaging (HCI). Specifically, HCT-8 host cells are seeded in microtiter plates 24 h prior to infection, compounds of interest are transferred into the wells, and then "activated" *Cryptosporidium* oocysts are added to allow for cell infection and subsequent intracellular parasite proliferation over a 48-h period. After 48 h, the cells are fixed and differentially stained with fluorescent markers to allow for enumeration of both HCT-8 cells and parasites. All steps are capable of automation to allow for maximum throughput. This method can be used for *C. parvum* or *C. hominis* parasites and is amenable to 384-well or 1536-well formats [6, 16].

HCI capabilities have also greatly advanced, with many systems allowing automated imaging of microtiter plates [17, 18]. An advantage to using HCI as the readout for anti-cryptosporidial HTS is that phenotypic effects on both host cells and parasites can be assessed concomitantly. Host cell effects can be used as a determination of cytotoxicity, and therefore compounds can be quickly triaged as either *Cryptosporidium*-specific or more generally cytotoxic. The assay described herein is designed to be flexible for phenotypic characterization of a wide range of molecules. Although we have capabilities of automation to increase throughput, the 384-well format is easily carried out "by hand" with multichannel pipettors and is amenable to most laboratory setups.

## 2 Materials

### 2.1 Media and Cell Culture Reagents

1. Propagation medium for HCT-8 host cells: RPMI 1640 medium with L-glutamine, 10% heat-inactivated (55 °C for 30 min) fetal bovine serum, 100 IU penicillin, and 100 μg/mL streptomycin. Combine all components and sterile filter through a 0.22 μm filter flask. Store at 4 °C.
2. Assay medium: phenol red-free RPMI 1640 with L-glutamine, 2% heat-inactivated horse serum, 100 IU penicillin, and 100 μg/mL streptomycin. Combine all components and sterile filter through a 0.22 μm filter flask. Store at 4 °C.
3. Dulbecco's Phosphate-Buffered Saline (DPBS).
4. TrypLE™ Dissociation Reagent (recombinant trypsin, EDTA) (Thermo Fisher Scientific, Inc., Norcross, GA).
5. T75 treated tissue culture flasks.
6. Cell scrapers.
7. 70 μm mesh filters.
8. C-Chip disposable hemocytometer.

### 2.2 *Cryptosporidium* Oocysts and Excystation Reagents

1. *C. parvum* oocysts, Iowa strain, propagated through neonatal Holstein calves, isolated, and stored at 4 °C in a solution with 0.01% Tween 20, 100 U of penicillin, and 100 μg of gentamicin per mL (*Cryptosporidium* Production Laboratory, School of Animal and Comparative Biomedical Sciences, University of Arizona). UGA1 Nluc [19], a nanoluciferase-expressing strain, is also available.
2. *C. parvum* oocysts, Iowa strain, propagated through neonatal calves, isolated, and stored at 4 °C in a solution with 0.01% Tween 20, 100 U of penicillin, and 100 μg of gentamicin per mL (Bunch Grass Farm, Deary, ID).
3. *C. hominis* oocysts, TU502 strain, propagated through neonatal gnotobiotic piglets, isolated, and stored at 4 °C in a solution with 0.01% Tween 20, 100 U of penicillin, and 100 μg of gentamicin per mL (Tzipori Laboratory, Cummings School of Veterinary Medicine, Tufts University).
4. 10 mM hydrochloric acid (HCl).
5. Excystation buffer: dissolve 10 mg of sodium taurocholate hydrate in 9 mL of DPBS with 0.9 mM $CaCl_2$ and 0.5 mM $MgCl_2$, sterile filter (*see* **Note 1**), giving a final concentration of 2 mM sodium taurocholate.
6. Microcentrifuge.
7. 37 °C water bath incubator.

### 2.3 *Cryptosporidium* Cell Assay

1. Low-evaporation metal lids or comparable plastic lids.
2. BioTek MultiFlo™ FX multi-mode dispenser (Agilent, Santa Clara, CA).
3. BioTek MultiFlo™ cassettes: 1 μL and 5 μL sizes.
4. Labcyte Echo® 550 acoustic transfer liquid handling system (Labcyte, Inc., San Jose, CA) (*see* **Note 2**).
5. Echo®-compatible 384-well source plates (*see* **Note 3**).
6. Dimethyl sulfoxide (DMSO) as a solvent and neutral control.
7. 2'-Deoxy-5-fluorouridine (FDU, Sigma-Aldrich Co., St. Louis, MO) and nitazoxanide (NTZ) as positive controls for *Cryptosporidium* growth inhibition; puromycin (Pur) as a positive control for HCT-8 host cell growth inhibition.
8. Multichannel electronic pipettor.
9. Tabletop centrifuge with microplate adapters.
10. Tissue culture incubator programmable for 37 °C with 5% $CO_2$.
11. Assay plates: 384-well black μClear bottom or 1536-well black μClear bottom (LoBase) microtiter plates (Greiner Bio-One, Monroe, NC).

### 2.4 High-Content Imaging

1. Microplate washer/dispenser (*see* **Note 4**).
2. 20% paraformaldehyde for cell fixation.
3. Wash buffer A: phosphate-buffered saline (PBS) with 0.05% sodium azide.
4. Permeabilization buffer: 0.25% triton X-100 in PBS.
5. Wash buffer B: PBS with 0.1% Tween 20 (PBS-T) and 0.05% sodium azide.
6. Blocking agent: SuperBlock™ T20 (PBS) Blocking Buffer (Thermo Fisher Scientific, Waltham, MA).
7. Visualization buffer: 10% SuperBlock™ T20 (PBS) Blocking Buffer diluted with PBS-T, 1 μg/mL fluorescein-labeled *Vicia villosa* lectin, and 3 μM 4′,6-diamidino-2-phenylindole dihydrochloride (DAPI). Protect from light.
8. Adhesive foil seals.
9. High-content imaging microscope: MetaXpress IXMC (Molecular Devices), CellInsight CX5 high-content screening platform (Thermo Fisher Scientific, Waltham, MA), or equivalent.

## 3 Methods

HTS workflow is typically as follows: first is a single-point primary screen, followed by reconfirmation at a single concentration in triplicate, and then dose-response reconfirmation in either duplicate or triplicate. Secondary or complementary assays may be employed for follow-up characterization, such as to assess the compound stage of action, mode of action, cidality, or rapidity of killing.

This cell-based screening assay takes place with multiple steps over multiple days. On day 1, host cell seeding occurs. On day 2, compounds of interest and controls are transferred onto the cells in a "wet" dispense with an acoustic liquid handler. Compound preparation depends on the step in the overall screening workflow. After compound transfer, *Cryptosporidium* oocysts are subjected to an excystation protocol and then dispensed onto the cells and left to infect and replicate within cells while remaining in the presence of compound for 48 h. On day 4, the infected cells are fixed and then labeled with fluorescent markers for imaging. This can be broken up into multiple steps if needed, and stopping points are indicated in the Notes. The final steps are imaging the microplates to obtain cell and parasite counts, followed by data analysis to determine inhibition and to identify either "hits" or the half-maximal effective concentration ($EC_{50}$) of a confirmed hit compound.

$$\text{S/N} = \frac{\text{mean signal} - \text{mean background}}{\text{standard deviation of background}}$$

$$\text{S/B} = \frac{\text{mean signal}}{\text{mean background}}$$

$$\text{SD} = \sqrt{\frac{\Sigma(x-\mu)^2}{n}}$$

$$\text{CV} = \frac{\text{SD}}{\mu}$$

$$\text{Z'-factor} = 1 - \frac{3(\text{SD}_p + \text{SD}_n)}{|\mu_p - \mu_n|}$$

**Fig. 1** Statistical parameters for high-throughput screening. Signal-to-noise ratio (S/N), signal-to-background ratio (S/B), standard deviation (SD), coefficient of variation (CV), and the Z'-factor, where $x$ = each value in the data set, $\mu$ = mean, $n$ = number of values in the data set, and ($\text{SD}_p$, $\mu_p$, $\text{SD}_n$, $\mu_n$) are the standard deviations and means of both the positive (p) and negative (n) controls [20]

Our compound preparation is geared toward the use of an acoustic liquid handler, and therefore all compound dilutions are made in compatible source plates. DMSO is the preferred solvent, although aqueous transfers are also possible. For primary screening of a compound library, compounds should be prepared at a single concentration, usually 2 mM or 10 mM stock. These same stocks can be used throughout the reconfirmation stages, though a resupply from the vendor or resynthesis of compounds (especially small molecules) is recommended as a final step to ensure the quality and purity of the active molecule. For dose-response reconfirmation, serial dilutions can be carried out ahead of time on the source plate, or all points can come from a single source well by programming the acoustic liquid handler and employing an intermediate plate as necessary.

Cell densities and any automation should be optimized within each laboratory setup. Statistical analyses and reproducibility using negative and positive controls are essential to ensure assay quality. Factors that should be examined include the signal-to-noise ratio (S/N), signal-to-background ratio (S/B), standard deviation (SD), coefficient of variation (CV), and the Z' (or RZ')-factor (Fig. 1) [20]. An assay with a Z'-factor of >0.5 is considered to be high-quality, though Z'-factors of 0.2–0.4 may be acceptable depending on the signal separation between the neutral and inhibitor controls. Automation of liquid-handling steps can greatly improve throughput for this assay, although it is not critical.

### 3.1 Host Cell Seeding

1. HCT-8 cells should be passaged twice per week with a 2:5 split ratio in T75 flasks (*see* **Note 5**).
2. Harvest cells 3–4 days after a passage (*see* **Note 6**) by first rinsing with 6 mL DPBS and then adding 3 mL TrypLE™.
3. Cells will not fully release with TrypLE™ alone. Use a cell scraper (can be used across multiple flasks) to release the cells from the bottom of the flask.
4. Add 7 mL of assay medium to collect the cells and transfer to a 50 mL conical tube.
5. Pellet cells in a benchtop centrifuge at 300× *g* for 5 min.
6. Aspirate the supernatant and resuspend cells in 10 mL assay medium.
7. Strain the cell suspension through a 70 μm mesh filter into a fresh 50 mL conical tube. Wash the filter once with an additional 5 mL of assay medium.
8. Dilute the cell suspension 1:10 in 100 μL, and determine cell concentration using a C-Chip hemocytometer.
9. Dilute the cell suspension to $6 \times 10^5$ cells/mL, preparing enough suspension for the number of assay plates required, plus an additional 30% for priming and loss.

   For 384-well plates: 25 μL per well or 9.6 mL per plate; 12.5 mL with dead volume. This is equivalent to 15,000 cells/well.

   For 1536-well plates: 5 μL per well or 7.68 mL per plate; 10 mL with dead volume. This is equivalent to 3000 cells/well.
10. Prepare the MultiFlo™ dispenser and 5 μL cassette for dispense. Assemble the cassette and prime the line with 70% ethanol to disinfect, followed by sterile water, assay medium, and the cell suspension last. If the volume and setup allow, a sterile stir bar may be added and the cell suspension placed on a stir plate to prevent the cells from settling and causing uneven quantities to be dispensed into the assay plates.
11. Dispense cells into assay plates at the volumes indicated above (**step 9**) for each microplate type.
12. Cover the plates with low-evaporation metal lids, and place them in the humidified 37 °C, 5% $CO_2$ incubator for 24 h.

### 3.2 Compound Preparation and Transfer

1. Prepare a 10 mM stock of compounds of interest and add to a 384-well Echo® Qualified microplate. This will serve as the source plate.
2. If testing in dose-response, an 8-point 1:3 serial dilution can be prepared on the source plate at this time.
3. 24 h after host cell seeding, remove plates from the 37 °C incubator (*see* **Note 7**).

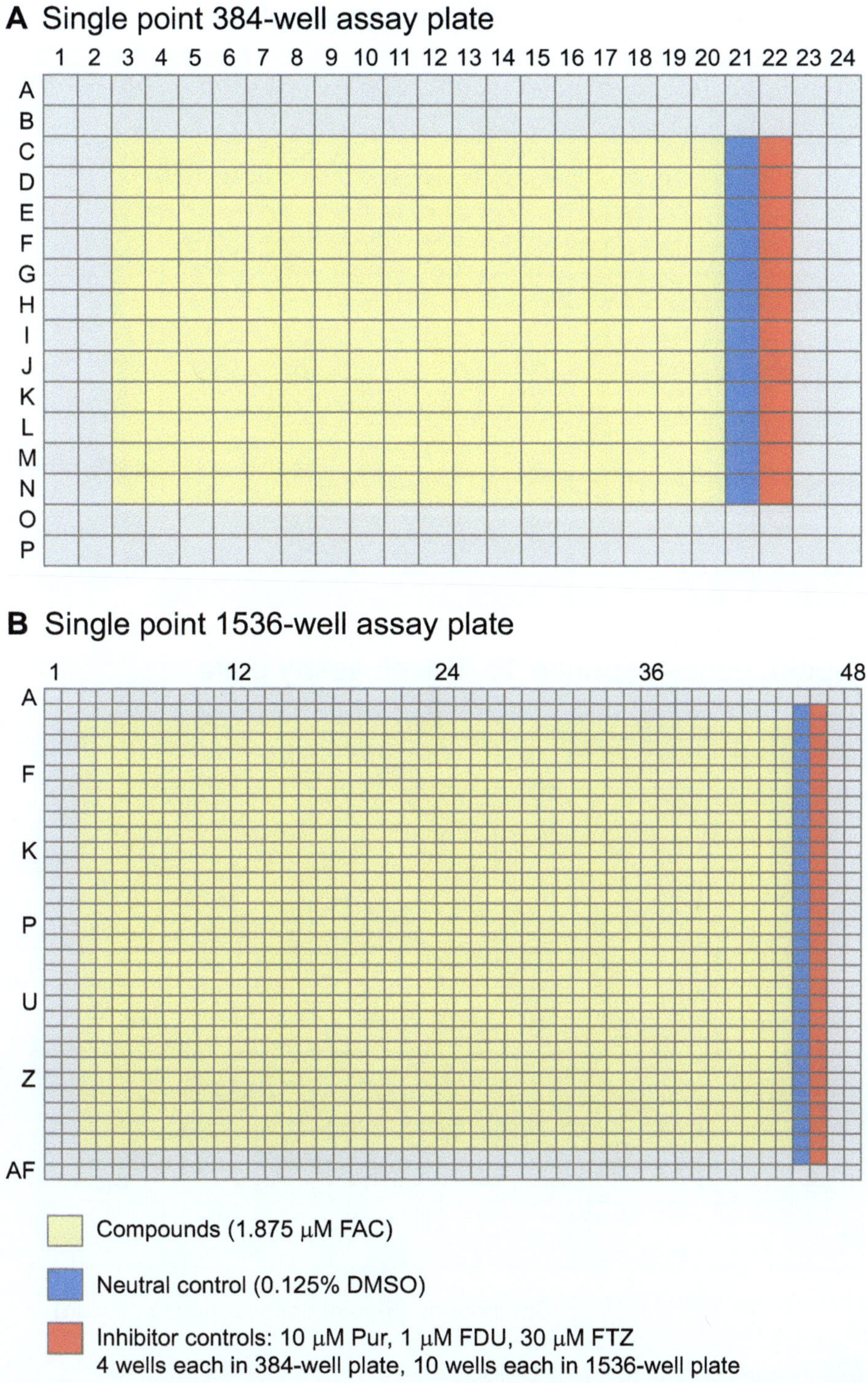


**Fig. 2** Plate layouts for (**a**) 384- and (**b**) 1536-well formats in single-point testing. These formats are used for both primary and triplicate reconfirmation steps. *FAC* final assay concentration; *Pur* puromycin; *FDU* 5-fluoro-2′-deoxyuridine; *NTZ* nitazoxanide. The values provided are the most often used in our *Cryptosporidium* screening assays

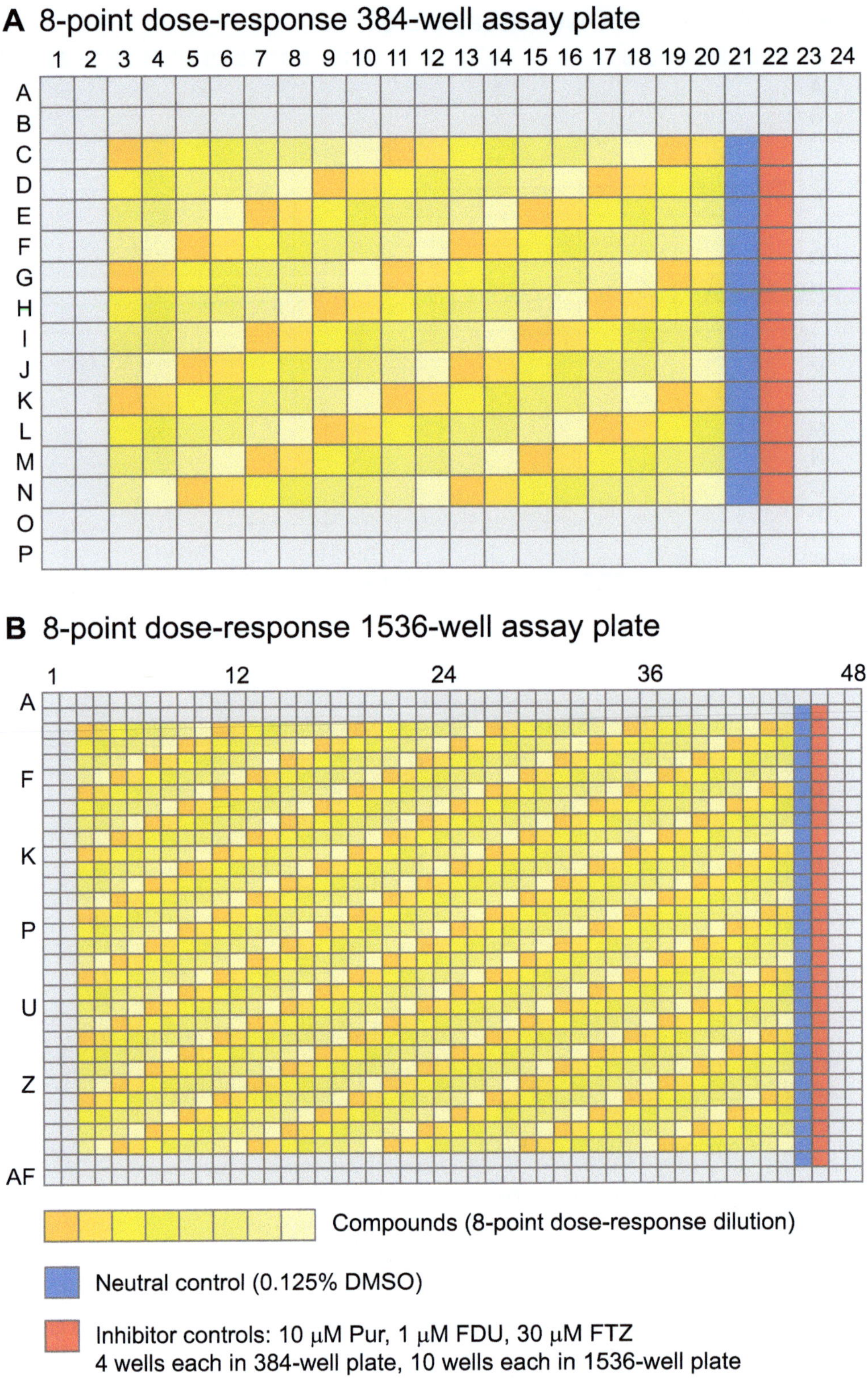


**Fig. 3** Plate layouts for (**a**) 384- and (**b**) 1536-well formats in dose-response testing. A 1:3 serial dilution is recommended for the dose-response titration. *Pur* puromycin; *FDU* 2′-deoxy-5-fluorouridine; *NTZ* nitazoxanide

4. Using the Echo®, transfer compounds of interest into the cell-seeded assay plates in either single-point (Fig. 2) or 8-point dose-response (Fig. 3).
5. Ensure that controls at the proper concentration, as well as DMSO in an equivalent volume (for negative or neutral control), are added to each assay plate.
6. After compound transfer, place the assay plates back into the 37 °C incubator until the infection step (Subheading 3.3).

### 3.3 *Cryptosporidium* Oocyst Excystation and Infection

1. Prewarm 1 mL of 10 mM HCl in a microcentrifuge tube to 37 °C in a water bath incubator for ≥10 min.
2. Calculate the volume of oocyst suspension needed to infect the number of plates seeded the day prior:

   *For 384-well plates*: 15 μL per well or 5.76 mL per plate; 7.5 mL with dead volume.

   *For 1536-well plates*: 3 μL per well or 4.61 mL per plate; 6 mL with dead volume.
3. Calculate the quantity of oocysts needed (*see* **Note 8**).
4. Remove the number of oocysts required from the stock, and transfer to the microcentrifuge tube containing prewarmed 10 mM HCl. If the volume of oocyst stock is >200 μL, separate into equal volumes across 2–3 microcentrifuge tubes and centrifuge at 16,000× *g* for 2 min. Remove the supernatant, and resuspend with the 10 mM HCl.
5. Incubate at 37 °C in a water bath incubator for 10 min.
6. Centrifuge at 16,000× *g* for 2 min. Carefully remove the supernatant.
7. Resuspend in enough excystation buffer to achieve 0.05 mM sodium taurocholate in the final dispense volume.
8. Incubate for 40–60 min in the excystation buffer at room temperature.
9. Dilute excysted oocysts with assay medium to an optimized density (*see* **Note 9**).
10. Remove 10 μL of oocyst suspension and count on a C-Chip disposable hemocytometer. Both unexcysted oocysts and sporozoites should be examined to determine the excystation ratio, though this is more qualitative than quantitative. For 100% excystation, there should be four sporozoites per empty oocyst shell. If excystation has failed, there will be no visible sporozoites. If excystation has initiated in the 40–60 min incubation, the dispensing can commence as oocysts will continue to excyst in the assay plates.
11. Prepare the MultiFlo™ dispenser and 1 μL cassette for dispense: assemble the cassette, and prime the line with 70%

ethanol to disinfect, followed by sterile water, assay medium, and the oocyst suspension last. If the volume and setup allow, a sterile stir bar may be added and the oocyst suspension placed on a stir plate to prevent the oocysts and sporozoites from settling and causing uneven quantities to be dispensed into the assay plates.

12. Dispense excysted oocysts into assay plates at the indicated volumes in **step 2** for each microplate type.
13. Centrifuge plates in a tabletop centrifuge with microplate adapters at 150× *g* for 3 min with soft deceleration.
14. Replace the low-evaporation metal lids onto each plate, and place in the 37 °C incubator for 48 h.

### 3.4 Post-assay Microplate Processing and Cell Staining

1. After 48 h, remove the assay plates from the incubator.
2. Using the microplate bottle valve washer (or BlueCatBio Washer together with MultiFlo™, *see* **Note 4**), add 2 μL (for 1536-well plates) or 10 μL (for 384-well plates) of 20% paraformaldehyde to each well to fix the cells (final well volume 10 μL in 1536-well plates; 50 μL in 384-well plates; final concentration 4% paraformaldehyde).
3. Incubate at room temperature for 15 min.
4. Aspirate the assay media/paraformaldehyde solution.
5. For all the remaining steps, use the following volumes:

   8 μL/well for 1536-well plates

   40 μL/well for 384-well plates
6. Wash 2X with wash buffer A (*see* **Note 10**).
7. Dispense permeabilization buffer.
8. Incubate at room temperature for 5–10 min. If possible, keep incubation ≤20 min.
9. Wash 2X with wash buffer B.
10. Dispense blocking agent.
11. Incubate at room temperature for 1 h (*see* **Note 11**).
12. Aspirate the blocking agent.
13. Dispense the visualization buffer.
14. Incubate at room temperature for 1 h.
15. Wash 2X with wash buffer B.
16. Seal each plate with adhesive foil seals and keep plate bottoms protected from light.

### 3.5 High-Content Imaging and Data Processing

1. Image plates with an available high-content imaging microscope. We use the Molecular Devices IXMC imager (10x objective, DAPI and FITC channels) and the MetaXpress Multi Wavelength Cell Scoring algorithm, which detects all nuclei (DAPI) and infected cells based on the identification of

FITC-stained area in each cell. Previously we successfully used the Cellomics instrument coupled with High-Content Screening analysis software: CellInsight CX5 system with HCS Studio 2.0, Spot Detector module, a 10X objective using two channels: excitation/emission (nm) 384/440 for DAPI (host nuclei) and 485/521 for FITC (*Cryptosporidium* parasites).

*For 384-well plates*, up to nine microscopic fields (802,511.39 $\mu m^2$ per field) can be captured.

*For 1536-well plates*, only one microscopic field per well can be captured.

2. DAPI-stained host cell nuclei are identified through Wavelength 1 Total Cell count of the MetaXpress Multi Wavelength Cell Scoring algorithm, while FITC-labeled *Cryptosporidium* intracellular parasites are assessed through Wavelength 2 thresholding, which yields the total count of *Cryptosporidium*-infected host cells as determined by the optimized protocol (Fig. 4).
3. Host cell and *Cryptosporidium*-infected cell counts are loaded into Genedata Screener analysis software (Version 21.0.4 Standard) (Danaher Corporation, USA).
4. Normalize *Cryptosporidium*-infected cell counts and HCT-8 total cell counts to neutral (DMSO) controls minus inhibitor controls. We used 1 μM FDU (thymidylate synthase inhibitor; F0503, Sigma-Aldrich, 5-Fluoro-2′-deoxyuridine) as the inhibitor control for *Cryptosporidium* proliferation and 10 μM Pur (Puromycin dihydrochloride, P7255) as the HCT-8 cytotoxicity control.
5. Correct for any obvious effects due to liquid-handling errors in specific wells or groups of wells, e.g., if there is a tip that failed to dispense evenly, the affected wells may need to be excluded from analysis.
6. Select hits based on an appropriate activity cutoff (*see* **Note 12**).
7. If testing in dose-response, fit curves to determine the $EC_{50}$ value of each compound. We used the Smart Fit option in the Genedata software.
8. Values for both host cells and *Cryptosporidium* parasites can be determined (*see* **Note 13**).

## 4 Notes

1. The excystation buffer should be prepared fresh before each assay.

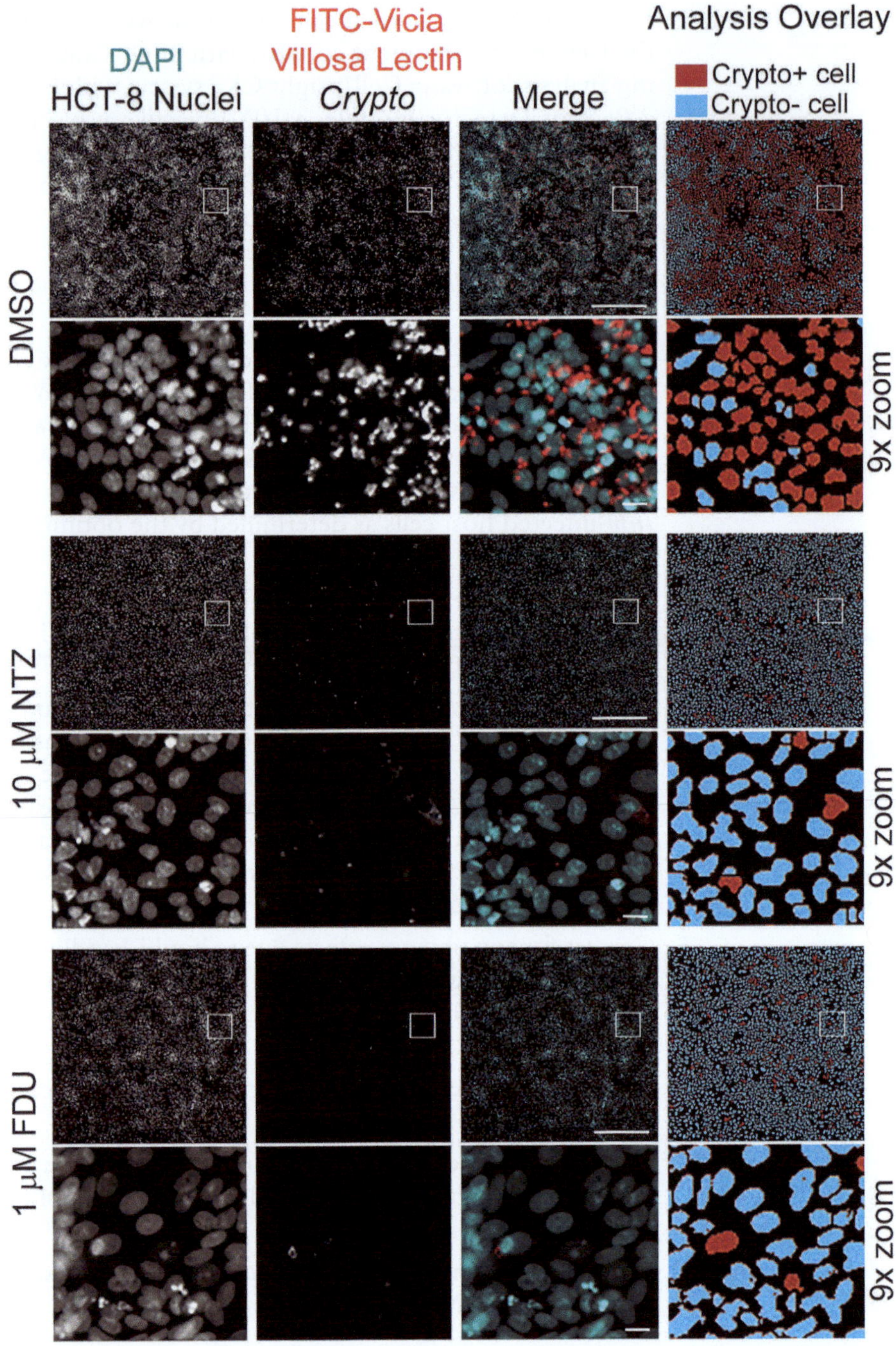


**Fig. 4** Screening images and software analysis. Representative images and software analysis from 1536-well high-content imaging on the MetaXpress IXMC platform. Images of wells either mock-treated with dimethyl sulfoxide (DMSO), 30 μM nitazoxanide (NTZ), or 1 μM FDU are shown. A merged image of the two fluorescent channels is artificially colored to show contrast: DAPI in cyan and FITC in red. A 200 μm and a 10 μm (9X zoomed-in image) scale bars are shown. Zoomed-in images from the inset white squares in each panel are also shown. The image overlays are generated by the Multi Wavelength Cell Scoring algorithm of the MetaXpress software to assess assay metrics: HCT-8 total cell count and *Cryptosporidium*-infected cell count.

2. There are several acoustic transfer liquid handlers on the market, but we are equipped with the Labcyte Echo® 555 system. This type of liquid handler is not essential for 384-well plates but is necessary for 1536-well plates, as pipetting "by hand" is difficult in this format.
3. The two types of Echo® qualified microplates include a polypropylene plate with a 15–65 μL working volume and a cyclic olefin copolymer plate with a 3–12 μL working volume.
4. We have used a "bottle-valve" type of dispenser/washer fitted with a 32-head aspirator for 1536-well formats. Our dispenser was custom made, but any 384- or 1536-well compatible washer will do. For example, recently we adapted the assay to use the centrifugal Blue®Washer (BlueCatBio NH, Inc., Lebanon, NH). If an automated instrument is not available, "flicking" plates is also an option. This involves holding a microplate upside down and flicking in a downward motion to expel the liquid out of the wells. For an infectious agent such as *Cryptosporidium*, care must be taken to contain the expelled liquid into a trap. Flicking works for both 1536- and 384-well formats.
5. We find that HCT-8 cells grow best in a 75 $cm^2$ flask (T75) with 20 mL of propagation medium and gentle splitting. This cell line maintains a consistent phenotype when left in high confluency and a 2:5 split ratio twice per week. We find that we have consistent infections from passages 30–45 (P30–P45), and therefore we keep frozen stocks of cells around P28 to quickly reach an "aged" passage number from thaw. Also of note is that the propagation medium contains fetal bovine serum, whereas the assay medium contains horse serum. The recommended serum for this cell line from the source (ATCC) is horse serum; however, we found that the cells did not grow consistently when using horse serum for *Cryptosporidium* infection, although these conditions gave better assay results.
6. Prior to setting up an assay, be sure to propagate the proper number of cells needed for seeding the quantity of microplates intended. In our hands, a single T75 flask contains roughly $4 \times 10^7$ HCT-8 cells when at 90–100% confluency.
7. This is a "wet" transfer, as opposed to transferring compounds into an empty assay plate, and is a time-sensitive protocol. Compound transfer should take ≤2 h, and plates should be kept at 37 °C whenever possible.

**Fig. 4** (continued) First, host cell nuclei are identified and counted based on the size and intensity of the DAPI signal. Next, the software identifies signal from the FITC-conjugated *Vicia villosa* lectin within the cell area based on size and intensity thresholding and computes the number of infected cells [16]

8. The oocyst density is calculated from the stock and not from the number of excysted sporozoites. This is because the excystation rate is variable across strains, species, and even assays. A quick check to ensure that excystation has initiated is necessary, but oocysts will continue to excyst within the assay plates. For the four lines that we have routinely used, these are the optimized densities:

   For University of Arizona *C. parvum* (Iowa) oocysts, use 3125 oocysts/well (MOI = 4).

   For University of Arizona *C. parvum* UGA1 Nluc (Iowa) oocysts, use 3125 oocysts/well (MOI = 4).

   For Bunch Grass Farm *C. parvum* (Iowa) oocysts, use 1500 oocysts/well (MOI = 2).

   For wild-type *C. parvum* (field isolate) oocysts, use 1500 oocysts/well (MOI = 2).

   For *C. hominis* (TU502) oocysts, use 3125 oocysts/well (MOI = 4).

9. The optimized densities are based on the starting number of oocysts as in **Note 8**. The densities for each commonly-used *Cryptosporidium* line in assay medium are as follows:

   For University of Arizona *C. parvum* (Iowa) oocysts, $1.04 \times 10^6$ oocysts/mL.

   For Bunch Grass Farm *C. parvum* (Iowa) oocysts, $5.0 \times 10^5$ oocysts/mL.

   For wild-type *C. parvum* (field isolate) oocysts, $5.0 \times 10^5$ oocysts/mL.

   For *C. hominis* (TU502) oocysts, $1.04 \times 10^6$ oocysts/mL.

10. This is the first stopping point. Plates can be kept at 4 °C until processing can continue.

11. This is the second stopping point. Plates can be kept at 4 °C until processing can continue.

12. We have found that due to high variability (1–3X across samples and negative controls), a high hit cutoff of 70% inhibition improved our assay reproducibility and reconfirmation rates.

13. We incorporated a cytotoxicity cutoff into our workflow for hit selection. Compounds that inhibited *Cryptosporidium* propagation by ≥70% but inhibited HCT-8 growth by ≥40% were excluded from our hit lists. These cutoffs were selected for our assay specifically and can be changed depending on the goals of the screen.

## Competing Interest Statement

The authors have no conflicts of interest to declare that are relevant to the content of this chapter.

### References


1. Kotloff KL, Nataro JP, Blackwelder WC, Nasrin D, Farag TH, Panchalingam S, Wu Y, Sow SO, Sur D, Breiman RF, Faruque AS, Zaidi AK, Saha D, Alonso PL, Tamboura B, Sanogo D, Onwuchekwa U, Manna B, Ramamurthy T, Kanungo S, Ochieng JB, Omore R, Oundo JO, Hossain A, Das SK, Ahmed S, Qureshi S, Quadri F, Adegbola RA, Antonio M, Hossain MJ, Akinsola A, Mandomando I, Nhampossa T, Acacio S, Biswas K, O'Reilly CE, Mintz ED, Berkeley LY, Muhsen K, Sommerfelt H, Robins-Browne RM, Levine MM (2013) Burden and aetiology of diarrhoeal disease in infants and young children in developing countries (the Global Enteric Multicenter Study, GEMS): a prospective, case-control study. Lancet 382(9888):209–222. https://doi.org/10.1016/S0140-6736(13)60844-2
2. Platts-Mills JA, Babji S, Bodhidatta L, Gratz J, Haque R, Havt A, McCormick BJ, McGrath M, Olortegui MP, Samie A, Shakoor S, Mondal D, Lima IF, Hariraju D, Rayamajhi BB, Qureshi S, Kabir F, Yori PP, Mufamadi B, Amour C, Carreon JD, Richard SA, Lang D, Bessong P, Mduma E, Ahmed T, Lima AA, Mason CJ, Zaidi AK, Bhutta ZA, Kosek M, Guerrant RL, Gottlieb M, Miller M, Kang G, Houpt ER (2015) Pathogen-specific burdens of community diarrhoea in developing countries: a multisite birth cohort study (MAL-ED). Lancet Glob Health 3(9):e564–e575. https://doi.org/10.1016/s2214-109x(15)00151-5
3. Arrowood MJ (2002) In vitro cultivation of *Cryptosporidium* species. Clin Microbiol Rev 15(3):390–400
4. Hijjawi NS, Meloni BP, Morgan UM, Thompson RC (2001) Complete development and long-term maintenance of *Cryptosporidium parvum* human and cattle genotypes in cell culture. Int J Parasitol 31(10):1048–1055
5. Upton SJ, Tilley M, Nesterenko MV, Brillhart DB (1994) A simple and reliable method of producing in vitro infections of *Cryptosporidium parvum* (Apicomplexa). FEMS Microbiol Lett 118(1–2):45–49
6. Bessoff K, Sateriale A, Lee KK, Huston CD (2013) Drug repurposing screen reveals FDA-approved inhibitors of human HMG-CoA reductase and isoprenoid synthesis that block *Cryptosporidium parvum* growth. Antimicrob Agents Chemother 57(4):1804–1814. https://doi.org/10.1128/AAC.02460-12
7. Murphy RC, Ojo KK, Larson ET, Castellanos-Gonzalez A, Perera BG, Keyloun KR, Kim JE, Bhandari JG, Muller NR, Verlinde CL, White AC Jr, Merritt EA, Van Voorhis WC, Maly DJ (2010) Discovery of potent and selective inhibitors of calcium-dependent protein kinase 1 (CDPK1) from *C. parvum* and *T. gondii*. ACS Med Chem Lett 1(7):331–335. https://doi.org/10.1021/ml100096t
8. Sharling L, Liu X, Gollapalli DR, Maurya SK, Hedstrom L, Striepen B (2010) A screening pipeline for antiparasitic agents targeting *Cryptosporidium* inosine monophosphate dehydrogenase. PLoS Negl Trop Dis 4(8):e794. https://doi.org/10.1371/journal.pntd.0000794
9. Umejiego NN, Gollapalli D, Sharling L, Volftsun A, Lu J, Benjamin NN, Stroupe AH, Riera TV, Striepen B, Hedstrom L (2008) Targeting a prokaryotic protein in a eukaryotic pathogen: identification of lead compounds against cryptosporidiosis. Chem Biol 15(1):70–77. https://doi.org/10.1016/j.chembiol.2007.12.010
10. Gorla SK, McNair NN, Yang G, Gao S, Hu M, Jala VR, Haribabu B, Striepen B, Cuny GD, Mead JR, Hedstrom L (2014) Validation of IMP dehydrogenase inhibitors in a mouse model of cryptosporidiosis. Antimicrob Agents Chemother 58(3):1603–1614. https://doi.org/10.1128/AAC.02075-13
11. Bessoff K, Spangenberg T, Foderaro JE, Jumani RS, Ward GE, Huston CD (2014) Identification of *Cryptosporidium parvum* active chemical series by repurposing the open access malaria box. Antimicrob Agents Chemother 58(5):2731–2739. https://doi.org/10.1128/AAC.02641-13
12. Manjunatha UH, Vinayak S, Zambriski JA, Chao AT, Sy T, Noble CG, Bonamy GMC, Kondreddi RR, Zou B, Gedeck P, Brooks CF, Herbert GT, Sateriale A, Tandel J, Noh S, Lakshminarayana SB, Lim SH, Goodman LB,

Bodenreider C, Feng G, Zhang L, Blasco F, Wagner J, Leong FJ, Striepen B, Diagana TT (2017) A *Cryptosporidium* PI(4)K inhibitor is a drug candidate for cryptosporidiosis. Nature 546(7658):376–380. https://doi.org/10.1038/nature22337

13. Chen JJ, Rateb ME, Love MS, Xu Z, Yang D, Zhu X, Huang Y, Zhao LX, Jiang Y, Duan Y, McNamara CW, Shen B (2018) Herbicidins from *Streptomyces* sp. CB01388 showing anti-*Cryptosporidium* activity. J Nat Prod 81(4): 791–797. https://doi.org/10.1021/acs.jnatprod.7b00850
14. Jin Z, Ma J, Zhu G, Zhang H (2019) Discovery of novel anti-cryptosporidial activities from natural products by in vitro high-throughput phenotypic screening. Front Microbiol 10: 1999. https://doi.org/10.3389/fmicb.2019.01999
15. Love MS, McNamara CW (2021) Phenotypic screening techniques for *Cryptosporidium* drug discovery. Expert Opin Drug Discov 16(1): 59–74. https://doi.org/10.1080/17460441.2020.1812577
16. Love MS, Beasley FC, Jumani RS, Wright TM, Chatterjee AK, Huston CD, Schultz PG, McNamara CW (2017) A high-throughput phenotypic screen identifies clofazimine as a potential treatment for cryptosporidiosis. PLoS Negl Trop Dis 11(2):e0005373. https://doi.org/10.1371/journal.pntd.0005373
17. Giuliano KA, Haskins JR, Taylor DL (2003) Advances in high content screening for drug discovery. Assay Drug Dev Technol 1(4): 565–577. https://doi.org/10.1089/154065803322302826
18. Ang ML, Pethe K (2016) Contribution of high-content imaging technologies to the development of anti-infective drugs. Cytometry A 89(8):755–760. https://doi.org/10.1002/cyto.a.22885
19. Vinayak S, Pawlowic MC, Sateriale A, Brooks CF, Studstill CJ, Bar-Peled Y, Cipriano MJ, Striepen B (2015) Genetic modification of the diarrhoeal pathogen *Cryptosporidium parvum*. Nature 523(7561):477–480. https://doi.org/10.1038/nature14651
20. Zhang JH, Chung TD, Oldenburg KR (1999) A simple statistical parameter for use in evaluation and validation of high throughput screening assays. J Biomol Screen 4(2):67–73. https://doi.org/10.1177/108705719900400206

# Chapter 18

# Phenotypic High-Throughput Screening and Drug Efficacy Test Against the Growth of *Cryptosporidium parvum* In Vitro by qRT-PCR

**Chenchen Wang, Dongqiang Wang, Haili Zhang, and Guan Zhu**

## Abstract

An effective method to quantify parasite loads is key to evaluating anti-cryptosporidial drug efficacy in vitro. However, high-throughput screening of drugs against *Cryptosporidium parvum* was previously infeasible using labor-intensive traditional assays. Here, we describe a simplified quantitative RT-PCR assay that directly uses cell lysates as templates, enabling high-throughput quantification of parasite loads. This assay can also be used to evaluate drug efficacy against the growth of *C. parvum* in vitro.

**Key words** *Cryptosporidium parvum*, In vitro cell culture, Real-time quantitative RT-PCR (qRT-PCR), Drug screening, Anti-cryptosporidial drug efficacy

## 1 Introduction

High-throughput screening (HTS) of compounds in vitro is essential in the drug discovery pipeline against infectious diseases, but it was previously impractical for the intracellular parasite *Cryptosporidium parvum* due to the lack of suitable assays. This changed with the development of three phenotypic assays suitable for HTS. The first assay is based on high-content imaging analysis ($Z' = 0.21–0.47$). In this method, cells are grown in 384-well, black-walled, clear-bottom microwell plates, and intracellular parasites are labeled with immunofluorescence-conjugated *Vicia villosa* lectin (VVL), which mainly labels the parasitophorous vacuole membrane (PVM). Selected microscopic fields are captured for batch image analysis with an ImageJ macro [1]. Our laboratory developed a second assay based on quantitative RT-PCR (qRT-PCR) detection of the parasite 18S rRNA [2]. In this assay, host cells are grown in regular 96-well cell culture plates for inoculation with *C. parvum* and drug treatment, and HTS is achieved by directly using cell lysates as templates for qRT-PCR in 384-well

Jan R. Mead and Michael J. Arrowood (eds.), *Cryptosporidium: Methods and Protocols*, Methods in Molecular Biology, vol. 2978, https://doi.org/10.1007/978-1-0716-4824-7_18,

plates. The third assay uses transgenic *C. parvum* that carries a nanoluciferase reporter for chemiluminescent detection [3]. The first two assays are applicable to all parasite strains, whereas the third one is restricted to the transgenic strain. This chapter provides a detailed protocol for the qRT-PCR-based assay.

Compared to qPCR, qRT-PCR has several advantages for assessing parasite loads in vitro: (1) The standard curves between cycle threshold ($C_T$) values and the logarithm of parasite loads produced by qRT-PCR are more linear than those produced by qPCR; (2) RNA levels are a better indication of cell viability than DNA and degrade more quickly after cell death; and (3) total RNA can be more easily extracted, enabling the direct use of cell lysates prepared with appropriate lysis buffers in a high-throughput format. The qRT-PCR assay avoids several issues faced when counting parasites manually or by high-content imaging analysis, including: (1) dead parasites killed by drugs may take time to degrade, leading to overcounting, as these dead but yet undegraded parasites are still counted; (2) mature meronts, which contain eight merozoites, are large in size but counted as one parasite, whereas when the merozoites are released, they might be counted individually; and (3) empty PVM after the release of merozoites can still be fluorescence-labeled and mistakenly counted, potentially leading to overcounting of parasite numbers.

In the qRT-PCR assay, host cell 18S rRNA is also detected for normalization. Combined with the use of multichannel pipettors and robotic pipetting, the qRT-PCR-based assay achieves excellent assay uniformity (i.e., <5% coefficient of variation [CV]); satisfactory intra-plate, inter-plate, and inter-day uniformity; high signal-to-noise ratios (SW = 7.4–44.1, $Z' = 0.73$–$0.87$); and a wide linear dynamic range (i.e., >150-fold linear range in detecting parasite loads). The procedure for the qRT-PCR assay for drug screening is highlighted in Fig. 1.

In addition to drug screening, the qRT-PCR assay can also be used for detecting parasite loads in other studies (e.g., the growth of *C. parvum* under various conditions) and adapted for drug screening against other intracellular pathogens using specific primers. When more sensitive detection and a better linear dynamic range are needed—such as when evaluating lower levels of parasite loads in the early stage of development—RNA samples can be prepared using traditional column-based purification procedures to achieve up to a 10,000-fold linear dynamic range [4].

## 2 Materials

### 2.1 Common Reagents and Kits

1. Ultrapure water.
2. Phosphate-buffered saline (PBS; pH 7.4).

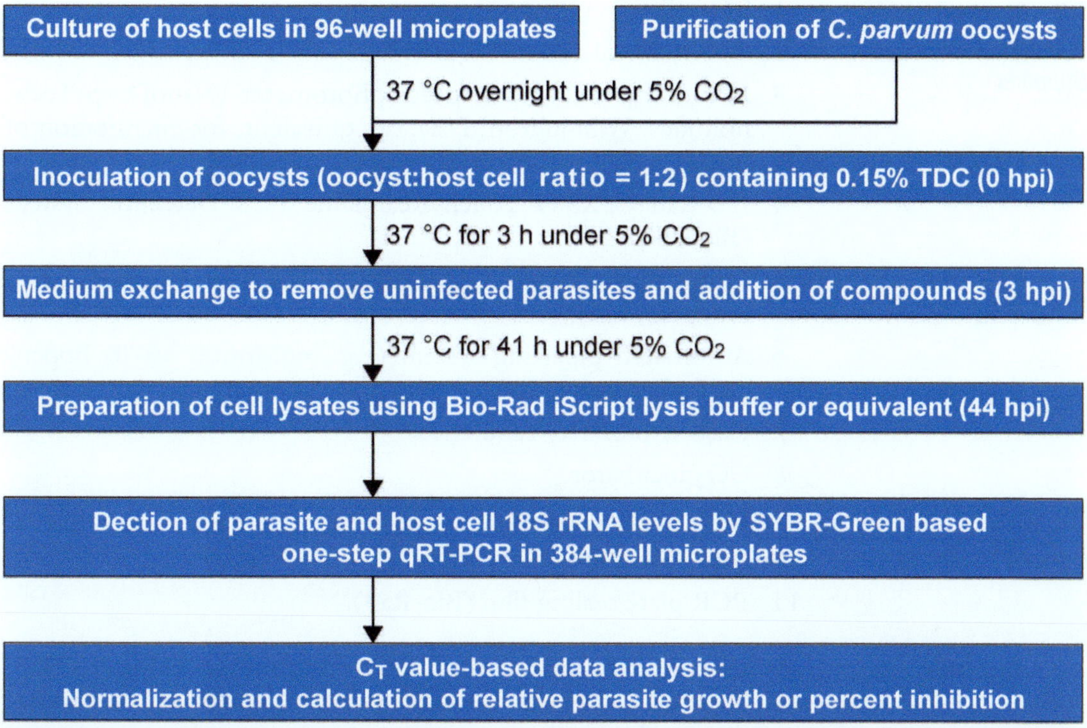


**Fig. 1** Illustration of the procedure for high-throughput screening of drugs or evaluating drug efficacy against *Cryptosporidium parvum* in vitro using qRT-PCR to detect the relative levels of parasite 18S rRNA (Cp18S), normalized with levels of host cell 18S rRNA (Hs18S)

3. Taurodeoxycholic acid (TDC).
4. Trypsin.
5. RPMI 1640 culture medium.
6. Fetal bovine serum (FBS).
7. 10X trypsin/EDTA cell detachment solution (0.5% trypsin and 0.2% EDTA) or a commercial alternative, such as Detachin™ cell detachment solution (Genlantis, San Diego, CA).
8. Percoll solution (Sigma-Aldrich, St. Louis, MO).
9. iScript RT-qPCR sample preparation reagent (Bio-Rad Laboratories, Hercules, CA). Similar extraction-free lysis buffers may also be used, such as the Invitrogen "Cells-to-CT 1-Step Power SYBR Green Kit" and "Cell-to-cDNA II Kit" (Thermo Fisher Scientific, Waltham, MA).
10. qScript One-Step SYBR Green qRT-PCR Kit (Quanta Biosciences, Beverly, MA), iTaq Universal SYBR® Green One-Step Kit (Bio-Rad), or other one-step qRT-PCR kits.

### 2.2 Common Equipment and Supplies

1. 96-well cell culture plates.
2. 384-well real-time PCR plates.
3. NanoDrop ND-1000 spectrophotometer (NanoDrop Technologies, Wilmington, DE), or equivalent, for quantitation of total RNA.
4. Bio-Rad CFX384 Touch Real-Time PCR Detection System (Bio-Rad).
5. Multi-tube vortexer (e.g., VX-2500, VWR International, LLC, Suwanee, GA).
6. Automated pipetting system (e.g., epMotion 5070, Eppendorf, Hauppauge, NY).
7. Class II biosafety cabinet.
8. $CO_2$ incubator.
9. Inverted microscope.
10. Adhesive sealing films.
11. PCR plate sealing film (Bio-Rad).

### 2.3 Host Cells

An adherent ileocecal colorectal adenocarcinoma cell line (HCT-8; ATCC # CCL-244) is used to host the growth of *C. parvum* in vitro. This cell line supports superior development of *C. parvum* in comparison with 10 other cell lines [5, 6]. Other cell lines (e.g., Caco-2 cells) may be used for various purposes. HCT-8 cells are maintained in RPMI 1640 medium supplied with 10% fetal bovine serum (FBS) under 5% $CO_2$ atmosphere and subcultured by 1:4 to 1:8 dilution of cells with the exchange of fresh medium every 2–3 days.

### 2.4 Parasite

Several strains of *C. parvum* may be acquired from commercial sources or research laboratories, e.g., Bunch Grass Farm (Deary, ID) and Waterborne, Inc. (New Orleans, LA). Local or acquired strains can be propagated in calves and purified from calf feces following standard protocols. Partially purified *C. parvum* oocysts, self-propagated or acquired, are stored at 4 °C in phosphate-buffered saline (PBS) containing penicillin (100 units/mL) and streptomycin (100 μg/mL). It is recommended to subtype the parasites, as self-propagated or acquired strains might change due to unexpected contamination during the course of propagation. For example, the original strain propagated at Brunch Grass Farm was the Iowa-1 strain (subtype IIaA15G2R1 based on the *gp60* gene), but it has been replaced by a local isolate with subtype IIaA17G2R1. Oocysts are further purified by a Percoll-based gradient centrifugation protocol (see details in Subheading 3.4). Oocysts used in the in vitro drug efficacy assay should be less than 3 months old with an in vitro excystation rate at >80% (see details in Subheading 3.5).

### 2.5 RNA Isolation Reagents

For HTS or most drug testing assays, several lysis buffers are commercially available to prepare cell lysates as described in Subheading 2.1 above. The cell lysates can be diluted and directly used for one-step qRT-PCR or for synthesizing cDNA. This protocol uses Bio-Rad iScript RT-qPCR sample preparation reagent. When other lysis buffers are used, the procedures need to be optimized. A larger linear dynamic range is needed in cases where samples contain very low parasite loads during invasion and early-stage development. In such cases, total RNA can be isolated using traditional resin column-based isolation kits, such as Qiagen's RNeasy Mini Kit (Germantown, MD).

### 2.6 qRT-PCR Primers

The protocol uses the relative levels of *C. parvum* 18S rRNA (Cp18S; reference sequence in GenBank: NC_006986.1), normalized to those of host cell 18S rRNA (Hs18S; reference sequence in GenBank: NR_003286), as an indicator of relative parasite loads [2]. We prefer using 18S rRNA because 18S and 28S rRNA make up ~80% of total RNA sample, and rRNA levels are less likely to fluctuate under conditions that might affect the expression of housekeeping genes. The primers listed in Table 1 are selected from several pairs of primers in our earlier optimization procedure of the qRT-PCR protocol.

### 2.7 qRT-PCR Kit

Several commercial sources of one-step RT-qPCR kits are available. We previously used the iScript One-Step RT-PCR kit with SYBR (Bio-Rad) and are currently using a more cost-effective qScript One-Step SYBR Green qRT-PCR Kit (Quanta BioSciences, Beverly, MA). Other one-step kits might also be used with optimization.

**Table 1**
**Primers used in determining the relative loads of *C. parvum* in vitro by qRT-PCR**

| Target | Primer name | Primer sequence (5′–3′) | Final conc. | Product length |
|---|---|---|---|---|
| Parasite 18S rRNA (Cp18S) | Cp18S-1011F | TTGTTCCTTACTCCTTCAGCAC | 250 nM | 175 nt |
| | Cp18S-1185R | TCCTTCCTATGTCTGGACCTG | 250 nM | |
| Host cell 18S rRNA (Hs18S) | Hs18S-1F | GGCGCCCCCTCGATGCTCTTA | 700 nM | 189 nt |
| | Hs18S-1R | CCCCCGGCCGTCCCTCTTA | 700 nM | |

## 3 Methods

All procedures are carried out at room temperature unless otherwise specified. The handling of live *C. parvum* should follow biosafety level 2 (BSL2) practices.

### 3.1 Preparation of Cell Culture Medium

1. Add 100 mL of RPMI 1640 medium 10X stock solution (or dissolve powdered RPMI 1640 medium) in ~500 mL ultrapure water into a sterilized 1.0 L conical glass flask containing a magnetic stir bar.
2. Add 2.0 g of sodium bicarbonate, 1.0 g of glucose to the medium, 100 mL of heat-inactivated fetal bovine serum (FBS), and 10 mL of antibiotics mixture containing 10,000 unit/mL of penicillin and 10,000 μg/mL of streptomycin.
3. Gently mix the solution well on a magnetic stir plate.
4. Add ultrapure water to make a final volume of 1000 mL and mix gently.
5. Sterilize the culture medium by filtration with a 0.2 μm filter unit.
6. Store the medium at 4 °C.

### 3.2 Recovery of Cryopreserved Host Cells (HCT-8 Cells)

1. Thaw one vial of cryopreserved cells in a 37 °C water bath with gentle agitation.
2. Remove the vial from the water bath as soon as cells are thawed (usually within ~2 min), and decontaminate the surface of the vial by spraying with 70% ethanol.
3. Move the vial to a biosafety cabinet, and carry out all the remaining procedures under strict aseptic conditions.
4. Transfer thawed cells to a centrifuge tube containing 9.0 mL culture medium, and centrifuge for 5 min at 125× *g*.
5. Resuspend cell pellet with a few mL of cell culture medium with gentle pipetting, and dispense cell suspension into a new culture flask.
6. Incubate cells at 37 °C in a cell culture incubator supplied with 5% $CO_2$ until the cell monolayer grows to confluence. Exchange medium every 2 days.

### 3.3 Subculture of HCT-8 Cells

HCT-8 cells can be used in routine cultivation of *C. parvum* in various experiments including drug efficacy assay because this cell line supports superior development of *C. parvum* in comparison with ten other cell lines [5, 6]. However, other cell lines may be used for various purposes.

All cell lines need to be routinely monitored for potential contaminations, particularly for mycoplasma contamination by fluorescent staining (e.g., DAPI or Hoechst 33258), ELISA, PCR, immunostaining, autoradiography, or microbiological assays. We routinely use DAPI staining or commercially available PCR detection primers for more sensitive and/or specific detection of mycoplasma contamination.

1. Remove old culture medium from the flask containing cultured HCT-8 cells with ~90–100% confluence.
2. Briefly rinse the cell monolayer once with 10 mL sterile PBS (pH 7.4).
3. Add 2.0 mL of Detachin™ cell detachment solution (Genlantis) to a flask, and rock the flask back and forth to cover the entire cell monolayer with Detachin™.
4. Incubate the flask at 37 °C for 5–10 min, and observe cells under an inverted microscope until the cell layer is dispersed.
5. To facilitate dispersal, gently tap the flask three times on a sterilized absorbance pad.
6. Add 8.0 mL of culture medium to neutralize the Detachin™. There is no need to wash the Detachin™ solution away or add any enzyme inhibitors in cells.
7. Inoculate 10–20% of the detached cells into a new 75 $cm^2$ cell culture flask containing 10 mL fresh culture medium (i.e., 1:5–1:10 ratio).
8. Incubate cells until the cell monolayer reaches ~100% confluence (typically in 2 days) with medium exchange every 2 days.
9. Continue subculture of cells, or prepare frozen cells in complete cell culture medium containing 5% DMSO using a standard cryopreservation protocol.

### 3.4 Purification of *C. parvum* Oocysts

The purchased or self-propagated oocysts of *C. parvum* for storage are usually partially purified by sucrose-based gradient centrifugation procedure. The partially purified oocysts can be further purified by a Percoll gradient centrifugation and surface-cleaned with 10% bleach [7, 8]. Surface-cleaned oocysts may be stored at 4 °C for several weeks. All solutions are sterilized either by autoclave or filtration with 0.2 μm filter units.

1. Resuspend a desired number of oocysts in PBS in a 1.5 mL microcentrifuge tube.
2. Centrifuge the oocyst suspension at 2000× *g* for 10 min, remove supernatant, and vortex to loosen the oocyst pellet.
3. Prepare 1.0 mL of 57% isotonic Percoll solution (Percoll/$H_2O$/1.5 M NaCl = **5.7/3.3/1**).
4. Resuspend oocysts in 1.0 mL of 57% Percoll solution in a 1.5 mL microcentrifuge tube, and mix well by pipetting.

5. Centrifuge the suspension at 2000× *g* for 30 min at 4 °C. Dead bacteria and other debris will form a sticky pellet at the bottom, while oocysts remain in the supernatant.
6. Transfer supernatant with oocysts to a 15 mL tube, and dilute the suspension with >4X volume of $dH_2O$.
7. Wash oocysts with $dH_2O$ for three times by centrifugation at 2000× *g* for 10 min at 4 °C.
8. Resuspend oocysts in 0.9 mL of ultrapure water.
9. Add 0.1 mL commercial bleach (~5% hypochlorite) into the suspension, mix well, and incubate the suspension on ice for 5 min. Do not prepare bleach in PBS, which may reduce the viability of oocysts.
10. Centrifuge the bleached oocyst suspension at 10,000× *g* for 2 min at 4 °C and remove the supernatant.
11. Wash oocysts twice with sterilized ultrapure water by centrifugation at 10,000× *g* for 2 min at 4 °C.
12. Resuspend oocysts in ~1.2 mL of PBS in a new microcentrifuge tube.
13. Continue washing oocysts three to five times with PBS by centrifugation.
14. Examine the purified oocysts under a microscope for purity.
15. Count the oocysts in a hemocytometer.
16. Store the cleaned oocysts in PBS with 20 units/mL of penicillin and 20 μg/mL of streptomycin at 4 °C.

### *3.5 Evaluation of Oocyst Viability by In Vitro Excystation*

Before in vitro infection assay, the viability of oocysts may be examined by an in vitro excystation assay in a medium containing taurodeoxycholic acid (TDC). Trypsin (0.25%) may be included, but optional. Some *C. parvum* strains may also undergo excystation in PBS or culture medium when incubated at 37 °C, but at lower excystation rates.

1. Prepare 1.0 mL of excystation medium in a 1.5 mL microcentrifuge tube by dissolving 0.5% taurodeoxycholic acid (TDC) in PBS. Trypsin (0.25%) may be included, but optional.
2. Add ~$10^4$–$10^5$ oocysts into the excystation medium and mix.
3. Incubate the suspension at 37 °C for 45 min to 1 h with occasional shake of the microcentrifuge tube.
4. Add 0.1 mL formalin to fix the parasite.
5. Centrifuge the suspension (5000× *g* for 5 min), remove supernatant, and resuspend the pellet in 0.1 mL PBS.
6. Prepare three microscopic slides with ~10 μL suspension/slide.

7. Count free sporozoites and intact oocysts under a microscope in three or more randomly selected microscopic fields until a total number of >300 sporozoites or oocysts are counted.
8. Calculate the in vitro oocyst excystation rate (OER) by the equation:

$$\text{OER }(\%) = \frac{N\text{ of free sporozoites}}{(N\text{ of free sporozoites}) + ((N\text{ of intact oocysts}) \times 4)} \times 100(\%) \quad (1)$$

9. It is preferred to use oocysts that are less than 3 months old and have an OER higher than 80%.

### 3.6 Standard 44 h Infection Assay for Evaluating Drug Efficacy In Vitro

The 44 h infection time point is preferred for general in vitro drug assay because fewer parasite merozoites or gametes are released into the medium at this time point. It also allows the subsequent preparation of cell lysates and qRT-PCR experiments to start in the morning of the day [2]. The time points may be adjusted according to the purpose of the assay.

1. Grow sufficient subcultured host cells in 75 $cm^2$ cell culture flasks (i.e., ~$3 \times 10^6$ host cells for one 96-well culture plate). One 75 $cm^2$ flasks may provide $10^7$ cells when cell monolayer reaches ~100% confluence. However, overgrowth of the host cells should be avoided.
2. Detach cells, prepare cell suspensions in culture medium, count host cells with a hemocytometer, and resuspend cells in complete culture medium at a concentration of $1.25 \times 10^5$ cells/mL.
3. Seed host cells into 96-well culture plates by adding 200 μL host cell suspension into each well by an 8-channel pipettor (i.e., $2.5 \times 10^4$ host cells/well). Cell suspensions can be placed in a multichannel pipetting reagent reservoir. Mix cell suspension well before each pipetting step.
4. Incubate microplates containing host cells at 37 °C under 5% $CO_2$ atmosphere overnight until the cell monolayers reach ~90% confluence.
5. Prepare cleaned *C. parvum* oocysts ($2 \times 10^5$ oocysts/mL) in complete culture medium at a concentration containing 0.15% TDC. *Note*: TDC was previously not included in the protocol. However, we have observed that some strains of *C. parvum* do not excyst in the absence of bile salt. During the parasite invasion, the addition of 0.15% TDC produces satisfactory excystation without affecting the host cell viability for at least 3 h.

6. Remove culture medium from the 96-well plates containing host cell monolayers.
7. Add 100 μL of oocyst suspension per well (i.e., $2 \times 10^4$ oocysts; 1:2 oocyst-to-host cell ratio) using an 8-channel pipettor. Mix oocyst suspension well before each pipetting step. *Note*: The oocyst-to-host cell ratio may be reassessed in individual laboratories, particularly when different *C. parvum* strains are used. In this test, host cells are inoculated with different oocyst-to-host cell ratios, and the parasite loads over time (e.g., 3, 11, 22, and 44 hpi) are quantified by qRT-PCR. An optimal ratio would produce sufficient parasite loads for detection and maintain parasite growth over the time, while avoiding over-inoculation of oocysts, which is indicated by the decline in parasite loads during the later phase of growth.
8. Incubate the cell culture plates containing *C. parvum* oocysts and host cell monolayers for 3 h at 37 °C to allow excystation of sporozoites from oocysts and invasion of sporozoites into host cells.
9. At 3 h post-infection time (hpi), remove supernatants containing empty oocyst walls and parasites that fail to undergo excystation or invasion.
10. Add 200 μL of culture medium containing drugs at desired concentrations. Appropriate negative control (diluent) and positive control (e.g., paromomycin or nitazoxanide at ~$IC_{70}$ concentrations) should be included in each plates. Each experimental condition should include two to three biological replicates. However, five to six biological replicates are recommended for the negative control because it sets up the baseline for all other experimental groups.
11. Incubate the plates at 37 °C under 5% $CO_2$ atmosphere for an additional 41 h (i.e., 44 h total infection time).
12. Proceed to the preparation of cell lysates as described below.

### 3.7 Preparation of Cell Lysates Containing Total RNA from Host Cells and the Parasite

The volumes described here are based on 96-well plates, which can be proportionally reduced or increased based on the volumes of culture vessels. In addition to this protocol, total RNA may also be isolated using a regular column resin-based RNA isolation kit (e.g., Qiagen's RNeasy RNA Isolation Kit) when a better sensitivity and higher linear dynamic range are needed.

1. Remove plates containing *C. parvum*-infected HCT-8 cells from incubator.
2. Centrifuge the plates for 10 min at 1000 × *g* in a plate centrifuge.
3. Remove medium using an eight-channel multichannel pipette, followed by two gentle rinses with PBS (200 μL/well).

4. Remove PBS, place plates on ice, and add 200 μL of ice-cold iScript RT-qPCR sample preparation reagent into each well.
5. Seal the plates with adhesive sealing films, and place them in an ice bucket containing adequate amount of ice. Multiple plates can be racked in a bucket.
6. Secure the bucket with sealed plates in a multi-tube vortexer, and vortex for 20 min at a speed set at 7 using the CWR VX-2500 vortexer. *Note*: The plates should be firmly clipped between the ice and top cover of the vortexer; otherwise, the loosen plates might not be well vortexed under high-speed shaking.
7. Remove the plates from the vortexer, and incubate them at 75 ° C for 15 min.
8. Centrifuge the plates at 2000 × *g* for 15 min to precipitate cell debris.
9. Supernatants can be used immediately. The plates can be placed on ice for short-term storage or stored at −80 °C until use.

### 3.8 RNA Quality and Quantity

The RNA concentration and quality can be determined using the NanoDrop 1000 spectrophotometer, in which 1.0 μL of RNA sample is sufficient to produce accurate and reproducible results. The lysis buffer is used as a baseline control. RNA concentrations can be estimated according to the absorbance at 260 nm. RNA purity can be estimated by the ratio of absorbance at 260 nm and 280 nm (260/280), in which a ratio of ~2.0 is generally acceptable. The ratio of absorbance at 260 nm and 230 nm (260/230) is a secondary measure of contamination by organic compounds, which is typically in the range of 1.8–2.2 [9].

### 3.9 qRT-PCR Assay

The current protocol has been optimized in our laboratory for reactions in 384-well plates. Different laboratories may consider optimizing the protocol under their own conditions, such as when different cell lines, parasite strains, reagents, and/or kits are used. These include optimal primer concentrations and the dilution factors of the cell lysates for Cp18S and Hs18S rRNA. Because cell lysates are directly used in qRT-PCR, they need to be appropriately diluted to reduce the inhibitory effect of the lysis buffer while maintaining sufficient sensitivity of the assay.

The reaction master mixes for both Cp18S and Hs18S genes are designed for 10 μL reaction volume in 384-well plates, which can be scaled up for reactions performed in 96-well plates. The pipetting of reaction cocktails and RNA templates into 384-well plates may be assisted using an automated pipetting machine (e.g., epMotion 5070 automated pipetting system). qRT-PCR reactions should include two technical replicates for each biological sample.

**Table 2**
**Reaction cocktail for detecting Cp18S transcript in 10 μL reaction volume**

| Component | Volume/reaction | Final conc. |
|---|---|---|
| 2X One-Step SYBR Green Master mix | 5.0 μL | 1X |
| Cp18S-1011F primer (10 μM) | 0.25 μL | 0.25 μM |
| Cp18S-1185R primer (10 μM) | 0.25 μL | 0.25 μM |
| qScript One-Step RTase | 0.2 μL | 1X |
| Nuclease-free water | 1.3 μL | – |
| Total cocktail volume | 7.0 μL | – |

**Table 3**
**Reaction cocktail for Hs18S gene in 10 μL reaction volume**

| Component | Volume/reaction | Final conc. |
|---|---|---|
| 2X One-Step SYBR Green master mix | 5.0 μL | 1X |
| Hs18S-1F primer 10 (μM) | 0.7 μL | 0.7 μM |
| Hs18S-1R primer 10 (μM) | 0.7 μL | 0.7 μM |
| qScript One-Step RT | 0.2 μL | 1X |
| Nuclease-free water | 0.4 μL | – |
| Total cocktail volume | 7.0 μL | – |

1. Thaw all components on ice.
2. Prepare the reaction cocktails for Cp18S and Hs18S as described in Tables 2 and 3. Scale up the volume of the cocktails based on the total numbers of reactions.
3. Mix the cocktails gently and centrifuge them to collect the contents to the bottom of the tube.
4. Load 7.0 μL of the reaction master mix into each well of 384-well plates for real-time PCR system using an automated pipetting system or manually.
5. Prepare RNA templates in 96-well plates or microcentrifuge tubes by diluting the supernatants of individual cell lysates 100X for detecting Cp18S rRNA (~3.0 ng total RNA) and 2000X for Hs18S rRNA (~0.3 ng total RNA).
6. Load 3.0 μL of RNA template to each reaction using an automated pipetting system.
7. Seal plates with PCR plate sealing films (Bio-Rad), and transfer them to a real-time PCR system (e.g., Bio-Rad CFX384 Touch Real-Time PCR Detection System).

**Table 4**
**qRT-PCR thermal cycling protocol**

| Reaction step | Temperature | Duration |
|---|---|---|
| cDNA Synthesis | 50 °C | 10 min |
| RTase inactivation | 95 °C | 5 min |
| Two-step thermal cycles (40X) | 95 °C<br>58 °C | 10 s<br>30 s (data collection) |
| Melting curve | 65 °C–95 °C with increment by 0.5 °C | 0.5 s/step (data collection) |

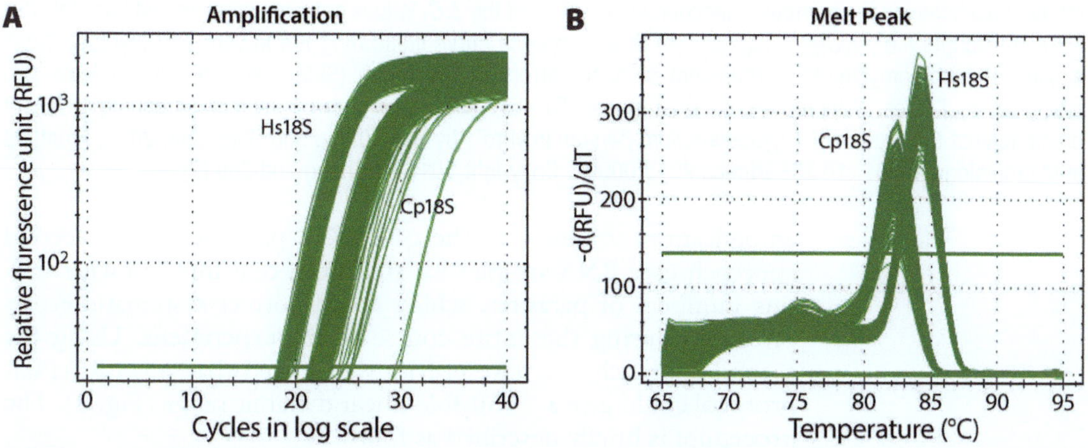


**Fig. 2** Demonstration of specific detection of Cp18S and Hs18S rRNA transcripts by amplification curve (**a**) and melting curve (**b**) analyses of representative plates. Adapted from "Quantitative RT-PCR assay for high-throughput screening (HTS) of drugs against the growth of *Cryptosporidium parvum* in vitro," by Haili Zhang and Guan Zhu, 2015, Frontiers in Microbiology, 6:991. 10.3389/fmicb.2015.00991. Copyright 2015 by Zhang and Zhu [2]

8. Perform real-time qRT-PCR reactions according to the thermal cycling protocol described in Table 4.
9. After qRT-PCR reactions are completed, examine the amplification and melting curves to assess the quality and specificity of the reactions. A specific amplification of an RNA template should produce a sigmoidal amplification curve between relative fluorescence unit (RFU) and the logarithm of thermal cycles and a single peak in the melting curve between the change of RFU (−dRFU/dT) and the temperature (Fig. 2).

### *3.10 Generation of Standard Curves*

A standard curve is useful in determining the linear dynamic range and amplification efficiency in the qRT-PCR assay. There are two approaches to generate standard curves. The first approach uses serially diluted RNA templates, which is indicative of the

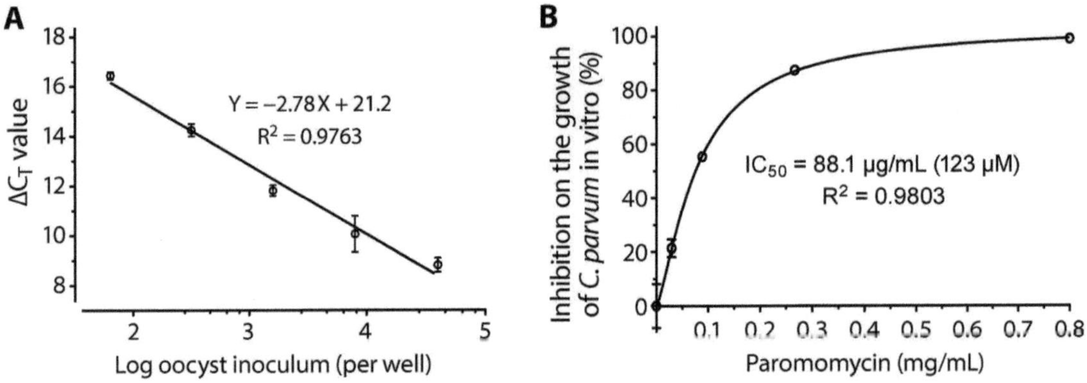


**Fig. 3** Evaluation of the qRT-PCR-based high-throughput screening assay. (**a**) Standard curves showing the relationship between the number of inoculated oocysts and the $\Delta C_T$ values between parasite and host cell 18S rRNA transcripts (i.e., $\Delta C_{T[Cp18S\text{-}Hs18S]}$). (**b**) Dose-response curve illustrating the in vitro anti-cryptosporidial activity of paromomycin. Bars represent standard errors of the mean (SEM), derived from at least two biological replicates. Adapted from "Quantitative RT-PCR assay for high-throughput screening (HTS) of drugs against the growth of *Cryptosporidium parvum* in vitro", by Haili Zhang and Guan Zhu, 2015, Frontiers in Microbiology, 6:991. 10.3389/fmicb.2015.00991. Copyright 2015 by Zhang and Zhu [2]

amplification efficiency in the qRT-PCR procedure. The second approach uses RNA samples isolated from cells infected with various numbers of parasites, which could correct systematic errors produced during the entire course of the experiment. Using the second approach, we have previously showed that this qRT-PCR protocol could give a >150-fold linear dynamic range (Fig. 3). The procedure is briefly described as follows.

1. Seed and grow host cell monolayers in a 96-well cell culture plate until they reached ~90% confluence as described in Subheading 3.6.
2. Prepare five-fold serial dilutions of *C. parvum* oocysts in 300 μL culture medium for 3 replicates of 100 μL suspensions, each containing 40,000, 8000, 1600, 320, and 64 oocysts per 100 μL.
3. Remove culture medium from the cell culture plate, and add 100 μL of oocyst suspensions into individual wells.
4. Place the plate into a 5% $CO_2$ incubator at 37 °C for 3 h to allow the parasite oocysts to undergo excystation and invade host cells.
5. Replace the culture medium at 3 hpi with 200 μL/well of fresh medium to remove free parasites.
6. Incubate the infected cell monolayers at 37 °C with 5% $CO_2$ for additional 41 h.
7. Prepare cell lysates and perform qRT-PCR as described in Subheadings 3.7 and 3.9.

8. Plot the means of $\Delta C_T$ values from technical and biological replicates against the logarithm of the numbers of inoculated oocysts (Fig. 3a).
9. Conduct linear regression to obtain the standard curve, slope value, and $R^2$ value to determine detection efficacy and the coefficient of determination (Fig. 3a).

### 3.11 Calculation of Parasite Loads or Drug Efficacy

The relative parasite loads or inhibition of the parasite growth is calculated based on the cycle threshold ($C_T$) values of Cp18S and Hs18S transcripts [2, 4]. *Note*: Hs18S levels are used to correct RNA isolation efficiency errors between wells and to assess the cytotoxicity of specified compounds to host cells. If a compound is highly cytotoxic, as indicated by a significantly increased $C_T$ value (vs. $C_T$ value of untreated negative control), the calculation of drug efficacy is invalid.

1. Compute the means of $C_{T[\text{Cp18S}]}$ and $C_{T[\text{Hs18S}]}$ values from technical replicates for each biological replicate.
2. Normalize the level of Cp18S with Hs18S in each sample by computing the $\Delta C_T$ between $C_{T[Cp18S]}$ and $C_{T[\text{Hs18S}]}$:

$$\Delta C_T = \Delta C_{T[\text{Cp18S}]} - \Delta C_{T[\text{Hs18S}]} \tag{2}$$

3. Calculate the mean of $\Delta C_T$ values derived from all replicates of the negative control group in each plate or for each set of experimental condition:

$$\begin{aligned} \overline{\Delta C}_{T[\text{ctl}]} &= \frac{1}{n} \sum_{i=1}^{n} \Delta C_{T[\text{ctl}]i} \\ &= \frac{\left(\Delta C_{T[\text{ctl}]1} + \Delta C_{T[\text{ctl}]2} + \ldots + \Delta C_{T[\text{ctl}]n}\right)}{n} \end{aligned} \tag{3}$$

4. Determine the relative level between individual sample and the mean of negative control by computing the $\Delta\Delta C_T$ values:

$$\Delta\Delta C_{T[\text{sample}]} = \Delta C_{T[\text{sample}]} - \overline{\Delta C}_{T[\text{ctl}]} \tag{4}$$

   Individual $\Delta\Delta C_T$ values of all biological replicates from the negative control group should also be calculated for subsequent determination of the standard error of the mean (SEM) or standard deviation of the mean (SD).

5. Calculate the normalized relative parasite loads or percent inhibition by one of the following two methods.

   *Method 1* considers the efficiency of detection derived from the slope value of the standard curve (i.e., parameter $A$ in Eq. 5). The deduction of the equations below has been previously described in more detail [2, 4]:

$$A = 1/\text{Slope} \tag{5}$$

$$\text{Relative parasite load } (\%) = 10^{A\ \Delta\ \Delta\ C_T} \times 100 \tag{6}$$

$$\text{Inhibition } (\%) = \left(1 - 10^{A\ \Delta\ \Delta\ C_T}\right) \times 100 \tag{7}$$

*Method 2* assumes perfect or near-perfect efficiency of detection using the empirical formulae:

$$\text{Relative parasite load } (\%) = 2^{-\Delta\ \Delta\ C_T} \times 100 \tag{8}$$

$$\text{Inhibition } (\%) = \left(1 - 2^{-\Delta\ \Delta\ C_T}\right) \times 100 \tag{9}$$

For comparison between the two methods using data in Fig. 3a (slope = −2.78), when $\Delta\Delta C_T = 1.0$, the percent inhibitions calculated using Eqs. 7 and 9 will be 56.3% and 50.0%, respectively. Therefore, under the same experimental conditions here, the empirical formulae would slightly underestimate the drug efficacy, thus providing more conservative estimation of drug efficacy.

### 3.12 Validation of the Assay Using Paromomycin

Before applying the assay to evaluate the anti-cryptosporidial efficacy in vitro, it is necessary to validate the assay by testing the dose response of a known anti-cryptosporidial compound (e.g., paromomycin or nitazoxanide).

1. Prepare twofold serial dilutions of paromomycin sulfate from 50 to 800 μg/mL (= 70–1121 μM) in cell culture medium. This compound is soluble in aqueous solutions. However, it is recommended to condition all dilutions with 0.5% DMSO (including untreated negative control) for uniformity, since most of the compounds being evaluated are dissolved in DMSO at 0.5% final concentration.
2. Prepare host cell monolayers in 96-well plates, and infect host cells with *C. parvum* oocysts for 3 h at 37 °C as described in Subheading 3.6.
3. At 3 hpi, remove the medium; add fresh medium containing paromomycin sulfate at various concentrations, together with the diluent control, into the plate (200 μL/well); and incubate the plate for 41 h at 37 °C with 5% $CO_2$.
4. At 44 hpi, prepare cell lysates and perform qRT-PCR as described in Subheadings 3.7 and 3.9.
5. Calculate the percentage inhibitions or relative parasite loads as described in Subheading 3.11.
6. Plot dose-response curve on percent inhibition or relative parasite loads against the drug concentrations. Curves can be analyzed by a nonlinear regression using a four-parameter logistic model [10, 11]. The 50% effective concentration ($EC_{50}$ value) for paromomycin sulfate would be around 90 μg/mL (~126 μM) (Fig. 3b) [2].

### 3.13 Validation of Assay for High-Throughput Screening (HTS)

When the assay is applied to HTS screening, it is necessary to evaluate the intra-plate, inter-plate, and inter-day uniformity of the assay as described in the chapter "HTS Assay Validation" at http://www.ncbi.nlm.nih.gov/books/NBK53196/) [2, 12]. This method is recommended by the National Center for Advancing Translational Sciences (NCATS), National Institutes of Health (NIH) (see "Pre-Clinical Research Toolbox" at http://www.ncats.nih.gov/expertise/preclinical).

Briefly, inter-day tests are performed by three independent experiments. Each experiment uses three 96-well plates to assess inter-plate variations, and each plate contains three experimental groups, including uninfected HCT-8 cells (blank "Min" background signal), cells infected with parasites but containing only 0.5% DMSO diluent (untreated, "Max" signal), and cells that are infected with parasites and treated with 120 μg/mL (168 μM) of paromomycin sulfate (drug-treated "Mid" signal). These groups (32 wells each) can be arranged in an interleaved-signal format according to the Assay Guidance Manual.

At 44 hpi as described in Subheading 3.6, qRT-PCR detections are performed with two technical replicates for each biological sample as described in Subheadings 3.7 and 3.9. Based on the qRT-PCR data, mean (AVG), standard deviation (SD), and coefficient of variation (CV) were calculated for each group in each plate. Signal window (SW) and $Z'$ factor for each plate were calculated based on the Assay Guidance Manual as follows [2, 12]:

$$\mathrm{SW} = \frac{(\mathrm{AVG}_{\max} - 3\mathrm{SD}_{\max}/\sqrt{n}) - (\mathrm{AVG}_{\min} + 3\mathrm{SD}_{\min}/\sqrt{n})}{\mathrm{SD}_{\max}/\sqrt{n}} \quad (10)$$

$$Z' = \frac{(\mathrm{AVG}_{\max} - 3\mathrm{SD}_{\max}/\sqrt{n}) - (\mathrm{AVG}_{\min} + 3\mathrm{SD}_{\min}/\sqrt{n})}{\mathrm{AVG}_{\max} - \mathrm{AVG}_{\min}} \quad (11)$$

The Assay Guidance Manual recommends the following acceptable thresholds: SW $\geq 2$, $Z' \geq 0.4$, and CV $\leq 20\%$. However, for the CV values, we recommend using 5% or 10% as the threshold for the $C_T$ values in detecting Cp18S and Hs18S transcripts and the $\Delta C_T$ values between Cp18S and Hs18S.

## Acknowledgments

The work conducted in the authors' laboratory was supported in part by the National Natural Science Foundation of China (Award Number 32250710141 to G.Z.). Previous work was also supported in part by the National Institute of Allergy and Infectious Disease, National Institutes of Health (award number R21AI099850 to G.Z.).

**Competing Interest Statement** The authors declare no conflicts of interest relevant to the content of this chapter.

## References


1. Bessoff K, Sateriale A, Lee KK, Huston CD (2013) Drug repurposing screen reveals FDA-approved inhibitors of human HMG-CoA reductase and isoprenoid synthesis that block *Cryptosporidium parvum* growth. Antimicrob Agents Chemother 57(4): 1804–1814. https://doi.org/10.1128/AAC.02460-12
2. Zhang H, Zhu G (2015) Quantitative RT-PCR assay for high-throughput screening (HTS) of drugs against the growth of *Cryptosporidium parvum* in vitro. Front Microbiol 6:991. https://doi.org/10.3389/fmicb.2015.00991
3. Vinayak S, Pawlowic MC, Sateriale A, Brooks CF, Studstill CJ, Bar-Peled Y, Cipriano MJ, Striepen B (2015) Genetic modification of the diarrhoeal pathogen *Cryptosporidium parvum.* Nature 523(7561):477–480. https://doi.org/10.1038/nature14651
4. Cai X, Woods KM, Upton SJ, Zhu G (2005) Application of quantitative real-time reverse transcription-PCR in assessing drug efficacy against the intracellular pathogen *Cryptosporidium parvum* in vitro. Antimicrob Agents Chemother 49(11):4437–4442. https://doi.org/10.1128/AAC.49.11.4437-4442.2005
5. Upton SJ, Tilley M, Brillhart DB (1994) Comparative development of *Cryptosporidium parvum* (Apicomplexa) in 11 continuous host cell lines. FEMS Microbiol Lett 118(3):233–236
6. Upton SJ, Tilley M, Brillhart DB (1995) Effects of select medium supplements on in vitro development of *Cryptosporidium parvum* in HCT-8 cells. J Clin Microbiol 33(2):371–375
7. Arrowood MJ, Sterling CR (1987) Isolation of *Cryptosporidium* oocysts and sporozoites using discontinuous sucrose and isopycnic Percoll gradients. J Parasitol 73(2):314–319
8. Suresh P, Rehg JE (1996) Comparative evaluation of several techniques for purification of *Cryptosporidium parvum* oocysts from rat feces. J Clin Microbiol 34(1):38–40
9. Glasel JA (1995) Validity of nucleic acid purities monitored by 260nm/280nm absorbance ratios. BioTechniques 18(1):62–63
10. Sebaugh JL (2011) Guidelines for accurate EC50/IC50 estimation. Pharm Stat 10(2): 128–134. https://doi.org/10.1002/pst.426
11. Yang B, Yan Y, Wang D, Zhang Y, Yin J, Zhu G (2023) On-target inhibition of *Cryptosporidium parvum* by nitazoxanide (NTZ) and paclitaxel (PTX) validated using a novel MDR1-transgenic host cell model and algorithms to quantify the effect on the parasite target. PLoS Negl Trop Dis 17(3):e0011217. https://doi.org/10.1371/journal.pntd.0011217
12. Iversen PW, Benoit Beck B, Yun-Fei Chen Y, Dere W, Devanarayan V, Eastwood BJ, Farmen MW, Iturria SJ, Montrose C, Moore RA, Weidner JR, Sittampalam GS (2012) Validation. Eli Lilly & Company and the National Center for Advancing Translational Sciences. http://www.ncbi.nlm.nih.gov/books/NBK53196/

# Chapter 19

# Establishing TurboID Proximity Labeling and mAID Conditional Knockdown System in *Cryptosporidium parvum*

**Rui Xu and L. David Sibley**

## Abstract

Protein interaction networks and subcellular proteomes can provide key insights into the biological functions of protein complexes. Here, we describe a method for proximity-based labeling using TurboID to define protein interaction networks in *Cryptosporidium parvum*. In addition, we have developed an auxin-inducible degron (AID) labeling system for conditional protein degradation in *C. parvum* to facilitate functional studies. This chapter describes the procedures for generating miniTurbo or mAID-TIR1-tagged *C. parvum* strains and validating the TurboID and mAID systems for target genes.

**Key words** CRISPR/Cas9, Proximity labeling, Mass spectrometry, Conditional knockdown, Degron

## 1 Introduction

*Cryptosporidium parvum* can be reliably genetically manipulated using CRISPR/Cas9 technology [1], which has been successfully employed in multiple laboratories over the past 5 years. Additional technologies based on genetic manipulation are also being developed. These advances will enable the clarification of virulence factors and pathogenic pathways, as well as the validation of drug targets.

Enzyme-catalyzed proximity labeling has emerged as a novel approach to investigate the spatial and interaction characteristics of proteins within living cells [2]. Two enzymes are commonly used for proximity labeling: APEX2 [3] and BioID [4, 5]. APEX2 employs an engineered soybean ascorbate peroxidase, and proximal proteins can be tagged in 1 min by adding $H_2O_2$ in cultures [3]. Moreover, it has yet to be implemented in *Cryptosporidium.* BioID utilizes a promiscuous *Escherichia coli* biotin protein ligase (BirA) fused to a target protein of interest. The major disadvantage of BioID is the slow kinetics, which needs 18–24 h for labeling with biotin. Recently, two newer promiscuous ligases, TurboID and

Jan R. Mead and Michael J. Arrowood (eds.), *Cryptosporidium: Methods and Protocols*, Methods in Molecular Biology, vol. 2978, https://doi.org/10.1007/978-1-0716-4824-7_19,

miniTurbo, were developed for proximity labeling. These modified version of BioID can efficiently biotinylate target molecules in 10 min [6], with a static labeling radius of ≤10 nm [7]. TurboID contains 15 mutations relative to the wild-type BirA, and miniTurbo results from a deletion of the N-terminal domain and 13 mutations relative to the wild-type BirA [6]. The miniTurbo demonstrates a 1.5- to two-fold reduction in activity relative to TurboID, but it represents a smaller tag, which is sometimes better tolerated as a fusion [6]. These changes result in an enhanced level of precision for the miniTurbo in comparison to TurboID. Here we provide a method for the generation of a transgenic *C. parvum* where a target protein is linked with miniTurbo tag using CRISPR/Cas9. We then describe the procedures of biotin labeling and purification of biotinylated proteins from host cells and parasites, as well as the identification of captured biotinylated proteins through mass spectrometry.

Conditional knockdown procedures are essential for the functional analysis of essential proteins of interest. Apicomplexans do not have functional RNAi pathways, leaving regulated transcriptional expression or protein degradation as the available options. The tetracycline-controlled Tet on [8] and Tet off [9] systems have been developed for *Toxoplasma*, but not yet implemented in *Cryptosporidium*. For rapid protein depletion, the auxin-inducible degron (AID) system has been used in yeast [10] and mammalian cells [11]. When *Oryza sativa* TIR1 (OsTIR1) is expressed in cells, it can modulate endogenous components to facilitate the formation of an operational SCF ubiquitin ligase complex (Skp1-Cul1-F-box) [11]. Subsequently, proteins fused with mAID can be rapidly degraded within 30 min in the presence of auxin, and the reaction was reversible and tunable [11, 12]. Moreover, the AID system has been previously employed in the study of essential proteins in *Plasmodium* [13] and *Toxoplasma* [14]. Here, we outline a protocol for the generation of a transgenic parasite in which the target gene has been labelled with a mAID tag and expresses the OsTIR1 gene downstream of the target gene. This system enables the functionality of the parasite protein to be evaluated by the measurement of the parasite growth in vitro.

## 2 Materials

### 2.1 Stock Reagents and Working Solutions

1. Excystation buffer: 0.075 g sodium taurocholate hydrate dissolved in Dulbecco's phosphate-buffered saline (DPBS) to a final volume of 5 mL, filter-sterilized, and stored at 4 °C.
2. HCT-8 media: 450 mL RPMI-1640 (ATCC modification) and 50 mL FBS (heat-inactivated), filter-sterilized, and stored at 4 °C.

3. RPMI 1640 media without biotin: 5.2 g RPMI 1640 without biotin and phenol red powder, 1 mL 0.5% phenol red solution, 25 mL 100 g/L D-(+)-glucose solution, 5 mL 100 mM sodium pyruvate solution, 5 mL 1 M HEPES solution, and 10 mL 7.5% sodium bicarbonate solution dissolved into 400 mL distilled $H_2O$; adjust pH to 7.2 with 1 M NaOH, then add distilled $H_2O$ (q.s. to 500 mL), filter-sterilize, and store at 4 °C.
4. HCT-8 media without biotin: 450 mL RPMI 1640 media without biotin and 50 mL dialyzed fetal bovine serum (DFBS), filter-sterilized, and stored at 4 °C.
5. 500 mM biotin (1000× stock): 122.16 mg biotin into 1 mL DMSO, stored at −20 °C.
6. 1% NP40 lysis buffer: 250 μL 1 M Tris–HCl, pH 7.5, 150 μL 5 M NaCl, 500 μL 10% NP40, 500 μL 20× protease inhibitor, 5 μL benzonase, and distilled $H_2O$ (q.s. to 5 mL).
7. 10× phosphate-buffered saline (PBS): 2 g KCl, 80 g NaCl, 14.4 g $Na_2HPO_4$, and 2 g $KH_2PO_4$ dissolved in deionized $H_2O$ (q.s. to 1000 mL) and stored at room temperature (RT).
8. Phosphate-buffered saline + Tween-20 (PBST): 100 mL 10× PBS, 1 mL Tween-20, and deionized $H_2O$ (q.s. to 1000 mL); stored at RT.
9. Reagent A: 4% formaldehyde in PBST.
10. Reagent B: 0.2 M sodium cyanoborohydride in PBST.
11. Reagent CHD: 120 mg cyclohexanedione into 14 mL PBST, adjust pH to 13 with 1 M NaOH.
12. TSS blocking buffer: 120 mL PBS, 1.5 g BSA, 1.5 mL 10% Triton X-100, and PBS (q.s. to 150 mL); filter-sterilize and store at 4 °C.
13. Fecal lysis buffer: 2.5 mL 1 M Tris–HCl, pH 8.0, 100 μL 1 M DL-Dithiothreitol (DTT), 200 μL 0.5 M EDTA, pH 8.0, 10 mL 50% glycerol, 5 mL 10% Triton X-100, and 37.2 mL distilled $H_2O$; store up to 3 months at 4 °C.
14. 1 M KCl: 3.728 g KCl and distilled $H_2O$ (q.s. to 50 mL); filter-sterilize and store at RT.
15. 0.1 M $Na_2CO_3$: 0.53 g $Na_2CO_3$ and distilled $H_2O$ (q.s. to 50 mL); filter-sterilize and store at RT.
16. 2 M urea in 10 mM Tris–HCl, pH 8.0: 6 g urea, 0.5 mL 1 M Tris–HCl, pH 8.0, and distilled $H_2O$ (q.s. to 50 mL); filter-sterilize and store at RT.
17. 500 mM 3-indoleacetic acid (IAA) (1000× stock): 87.59 mg 3-indoleacetic acid added into 1 mL 100% EtOH; store at −20 °C and protect from light.

18. SF Cell Line 4D-Nucleofector X Kit (Lonza Bioscience, Walkersville, MD).
19. Nano-Glo Luciferase Assay System kit (Promega Corp., Madison, WI).
20. BioTek Cytation 3 instrument or equivalent (Agilent Technologies Inc., Santa Clara, CA).
21. QIAamp PowerFecal DNA Kit (Qiagen, Germantown, MD).
22. SYBR Green JumpStart Taq ReadyMix (Sigma-Aldrich, St. Louis, MO).
23. PrimeSTAR GXL Premix (Takara Bio, San Jose, CA).
24. ProLong Diamond Antifade Mountant or equivalent (Thermo Fisher Scientific, Waltham, MA).
25. IRDye 800CW Streptavidin (LI-COR Biosciences, Lincoln, NE).

### 2.2 Cell and Mouse Lines

1. Human ileocecal adenocarcinoma (HCT-8) cells (ATCC, Cat. No. CCL-244).
2. *Ifng*$^{-/-}$ mice (referred to as GKO) (Jackson Laboratories, Cat. No. 002287).
3. *Ifngr1*$^{-/-}$ mice (Jackson Laboratories, Cat. No. 003288).
4. NOD scid gamma mice (referred to as NSG) (Jackson Laboratories, Cat. No. 005557).

### 2.3 Parasites

*Cryptosporidium parvum* isolate AUCP-1 was maintained by repeated passage in male Holstein calves and purified from fecal material as previously described [15]. All calf procedures were approved by the Institutional Animal Care and Use Committee (IACUC) at the University of Illinois at Urbana-Champaign. Purified oocysts were stored at 4 °C in PBS plus 50 mM Tris-HCl and 10 mM EDTA (pH 7.2) for up to 6 months after fecal collection.

## 3 Methods

### 3.1 The miniTurbo/mAID-TIR1 Tagging Using CRISPR Genome Editing

For designing CRISPR/Cas9 plasmids targeting a gene of interest (GOI) for C-terminal tagging with miniTurbo or mAID-TIR1, we are following the protocol as previously described [16]. The diagrams of miniTurbo and mAID-TIR1 tagging strategy are shown below (Figs. 1 and 2).

### 3.2 Creating Stable mTurbo or mAID-TIR1 Transgenic *C. parvum* Strain

To create a stable transgenic *C. parvum* strain, we typically test the transfection efficiency in HCT-8 cells with transient expression of nanoluciferase. After testing, the stable transgenic strain is amplified and selected in GKO or *Ifngr1*$^{-/-}$ mice with paromomycin treatment in drinking water.

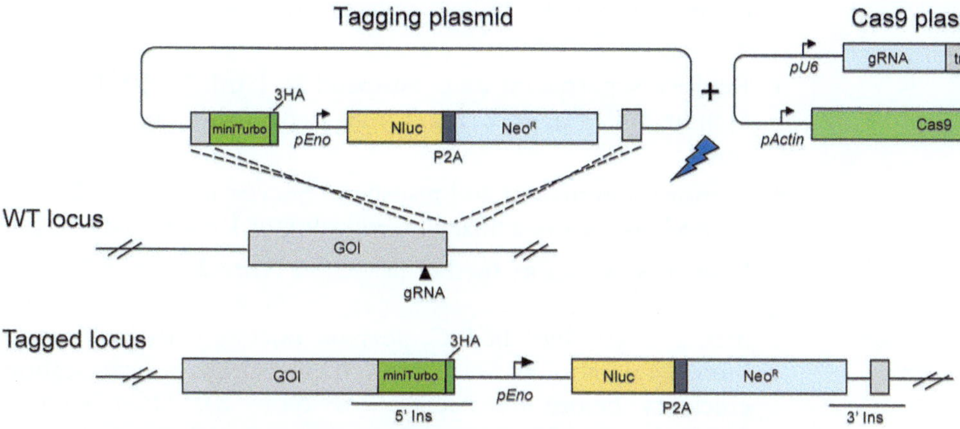


**Fig. 1** Diagram of CRISPR/Cas9 tagging strategy for miniTurbo
To make miniTurbo tagging of gene of interest (GOI), a repair cassette containing the C-terminus of GOI fused to the miniTurbo-3HA tag (*see* **Note 1**), a Nluc-P2A-neo cassette, and the 3'UTR of GOI is cloned into a plasmid. And the sgRNA for the GOI is cloned into a pACT1:Cas9-GFP, U6:sgTK (Addgene no. 122852) plasmid. Tagging plasmid and Cas9 plasmid were transfected into *Cp* sporozoites, followed by drug selection in immunocompromised mice. The proper genomic insertion was identified by diagnostic PCR

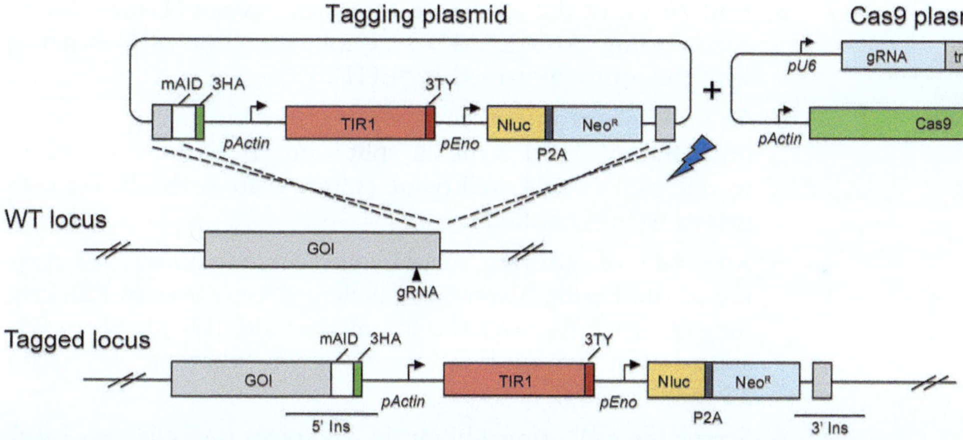


**Fig. 2** Diagram of the CRISPR/Cas9 tagging strategy for mAID conditional knockdown. To make mAID-TIR1 tagging of GOI, a repair cassette containing the C-terminus of GOI fused to the mAID-3HA tag, a TIR1-3TY cassette (*see* **Note 1**), a Nluc-P2A-neo cassette, and the 3'UTR of GOI are cloned into a plasmid. The sgRNA of GOI is cloned into a pACT1:Cas9-GFP, U6:sgTK (Addgene 122,852) plasmid. Tagging plasmid and Cas9 plasmid were transfected into *Cp* sporozoites, followed by drug selection in immunocompromised mice. The proper genomic insertion was identified by diagnostic PCR

#### 3.2.1 Bleaching C. parvum Oocysts

1. Prepare 3 mL of 40% bleach solution (containing 8.25% sodium hypochlorite) diluted in DPBS, then add unbleached *C. parvum* oocysts into this tube.
2. Incubate for 10 min on ice, then spin at 800 × *g*, 4 °C for 3 min.
3. Remove bleach solution, and resuspend in 1 mL DPBS containing 1% bovine serum albumin (BSA). Transfer resuspended

oocysts to a new 15 mL polystyrene conical tube. Spin at 800 × *g* for 3 min.

4. Remove supernatant and resuspend in 1 mL DPBS-1% BSA. Spin at 800 × *g*, 4 °C for 3 min. Repeat this step 1–2 more times.
5. Remove supernatant and resuspend oocysts in enough DPBS-1% BSA to achieve a final concentration of 1 × $10^8$ oocysts/ml.
6. Store at 4 °C before further usage (*see* **Note 2**).

*3.2.2 Test Transfection Efficiency with Transient Expression of Nanoluciferase*

1. Mix 1 × $10^7$ bleached *C. parvum* oocysts with excystation buffer, 1:1 ratio, at 37 °C for 1 h, and check the excystation efficiency before transfection (*see* **Note 3**). Then, spin at 1250 × *g* for 3 min at RT, and resuspend sporozoites in 15 μL Complete SF buffer (Complete SF buffer: SF Buffer with Supplement #1 at a 4.5:1 ratio).
2. Prepare transfection sample: Transfer 15 μL of resuspended sporozoites into a 0.7 mL tube, add 10 μg of vector, and then add Complete SF buffer if needed to bring the final volume to 20 μL.
3. Add 20 μL of the sample to the small cuvette (Lonza Bioscience), using AMAXA 4D-Nucleofector (Lonza Bioscience) with the protocol named as "EH100."
4. Cell monolayer infection: Transfer the transfected sporozoites into 1000 μL HCT-8 media, split it into two, and add 500 μL to one well of a 24-well tissue culture plate with HCT-8 cells grown to 90% confluency.
5. After 24 h of culturing, measure luminescence value for targeting plasmid using Nano-Glo Luciferase Assay System Kit (Promega Corp.): Remove the old media, add 100 μL Nano-Glo assay buffer (without substrate), and incubate at 37 °C for 15 min.
6. Scrape the well using a blunt tip, and pipet the cells into 1 well of a clear-bottom, white 96-well plate (Corning).
7. Dilute the Nano-Glo substrate 1:25 in Nano-Glo buffer. Make enough reagent for 100 μL per well.
8. Add 100 μL Nano-Glo reagent (with substrate) to each well. Incubate at RT in the dark for 3 min.
9. Read luminescence values on a BioTek Cytation 3 instrument or an equivalent.

*3.2.3 Creation of Stable Transgenic* C. parvum

1. Mix 2.5 × $10^7$ bleached *C. parvum* oocysts with excystation buffer at a 1:1 ratio and incubate at 37 °C for 1 h, and check the excystation efficiency before transfection (*see* **Note 3**). Then, 1250 × *g* at RT for 3 min, resuspend sporozoites in 150 μL

Complete SF buffer (Complete SF buffer: SF Buffer with Supplement #1 at a 4.5:1 ratio).

2. Prepare transfection sample: Transfer 150 μL of resuspended sporozoites into a 1.7 mL tube, add 60 μg of Cas9 vector (e.g., pACT:Cas9-GFP, U6:sg4070) and 100 μg of target plasmid (e.g., p4070-miniTurbo-3HA-Nluc-P2A-neo), and then add some Complete SF buffer if needed to bring the final volume to 200 μL.
3. Split the liquid into two large cuvettes (Lonza Bioscience), with 100 μL sample into each cuvette, using the AMAXA 4D-Nucleofector (Lonza Bioscience) with the protocol named "EH100."
4. Transfer the samples to cold DPBS (1:1 dilution, 200 μL sample to 200 μL cold DPBS), and put it on ice before gavage.
5. Prepare 1 mL syringes with needle tips for gavage; normally, use three mice for two large cuvettes of transfected sporozoites. Gavage each GKO or *Ifngr1*$^{-/-}$ mouse with 200 μL of 8% sodium bicarbonate solution and wait 5 min. Then, gavage each mouse with 100 μL of transfected sporozoites.
6. After allowing the sporozoites to infect overnight, change the cage water with 16 g/L paromomycin drinking water.

### 3.3 Feces Collection and Oocyst Burden Measurement

Mice typically begin to shed transgenic parasites between the 7th and 14th day post-infection [16]. Oocyst burden can be monitored from collected fecal samples using qPCR or nanoluciferase. Collect feces pellet at 3rd, 9th and 12th day post-infection for monitoring oocyst burden. From each mouse, collect two pellets (one for nanoluciferase and one for qPCR) remembering to change gloves for each mouse. Pellets are stored at 4 °C for nanoluciferase assay and at −80 °C for qPCR. Once mice begin to shed transgenic parasites, start daily fecal collection. Store feces at 4 °C until purification.

#### 3.3.1 Nanoluciferase Test with Mouse Feces

1. Remove fecal pellets from storage at 4 °C. Crush each pellet with a pipet tip or plastic pestle.
2. Add ten 3-mm glass beads (Thermo Fisher Scientific) and 1 mL fecal lysis buffer to each tube. Incubate at 4 °C for 30 min.
3. Vortex each tube until homogenized. Spin at 16,000 × *g* for 1 min at 4 °C.
4. Pipet supernatant into 2 wells of a white 96-well plate, 100 μL per well.
5. Dilute Nano-Glo substrate 1:25 in Nano-Glo buffer. Make enough reagent for 100 μL per well.

6. Add 100 μL Nano-Glo reagent to each well and incubate at RT for 3 min.
7. Read luminescence on Cytation 3. Average the two technical duplicates to get a single value for each fecal pellet.

#### 3.3.2 qPCR Test [17]

1. Remove fecal pellets from storage at −80 °C. Extract fecal DNA using the QIAamp PowerFecal DNA Kit.
2. Purified DNA was eluted in 100 μL of Buffer AE and then diluted 1:10 with nuclease-free water. A standard 20 μL reaction consists of: 2 μL of the diluted DNA + 0.8 μL (2 μM) Forward Primer (*Cp* GAPDH forward primer 5'-CGGATGGC CATACCTGTGAG) + 0.8 μL (2 μM) Reverse Primer (*Cp* GAPDH reverse primer 5'-GAAAACACGGGGGCAATGAG ) + 10 μL SYBR Green JumpStart Taq Ready Mix +6.4 μL nuclease-free water.
3. Reactions were performed on a QuantStudio 3 Real-Time PCR System (Thermo Fisher Scientific) with the following amplification conditions: Priming at 95 °C for 2 min, followed by 40 cycles of denaturing at 95 °C for 10 s, annealing at 60 °C for 20 s, and extension at 72 °C for 30 s, followed by a melt curve analysis to detect nonspecific amplification.
4. Genomic DNA equivalents in each sample were determined using the QuantStudio Design & Analysis software, based on the standard curve samples present on each plate.

#### 3.3.3 Diagnostic PCR

1. Primer design for diagnostic PCR: To confirm proper template integration after CRISPR genetic modification, PCR is used to amplify genomic regions of transgenic or wild-type parasites. The 5′ Ins and 3′ Ins pair of primes are specific for amplification of the 5′ and 3′ insertion sites of GOI. One primer anneals to the genome of GOI, and the other anneals to the sequence of insert (Figs. 1 and 2). A short 588-bp sequence from an unrelated gene (TK, cgd5_4450) is amplified as a control.
2. A standard 20 μL diagnostic PCR reaction consists of: 1 μL fecal DNA + 1 μL (10 μM) Forward primer +1 μL (10 μM) Reverse primer +10 μL PrimeSTAR GXL premix +7 μL nuclease-free water.
3. Thermal cycling: 98 °C for 30 s; 35 cycles of 98 °C for 10 s, 60 °C for 15 s, and 68 °C for 1 min; and a final extension of 72 °C for 2 min. Resolve 5 μL of PCR reactions on a 1% agarose gel (TBE) with a 100 bp or 1 kb DNA ladder to check for positive amplification.

### 3.4 Transgenic Parasite Purification

To amplify transgenic parasites, three to four NSG mice are orally gavaged with a suspension of bleached fecal slurry containing $2 \times 10^4$ oocysts/mouse, and 16 g/L paromomycin drinking water is used for selection immediately after infection. Fecal materials containing transgenic *Cp* is collected for several days (e.g., from 12 dpi to 40 dpi). Transgenic oocysts are purified from fecal material as previously described [18].

### 3.5 Generation of Protease-Resistant Streptavidin Beads

Generation of protease-resistant streptavidin beads through a two-step chemical modification (*see* **Note 4**) (Adapted from methods previously described [19]):

1. Resuspend the desired volume of streptavidin beads at 4 mg beads/ml. Split equal volumes of beads between protein LoBind tubes.
2. Put the tube on the magnet, wait 5 min, and then discard the liquid. Wash the beads with 1 mL of PBST.
3. Put the tube on the magnet, wait 5 min, and then discard the liquid. Resuspend the beads in 1.5 mL of Reagent CHD.
4. Rotate at RT for 4 h.
5. Put the tube on the magnet, wait 5 min, and then discard the liquid. Wash the beads with 1 mL PBST.
6. Put the tube on the magnet, wait 5 min, and then discard the liquid. Resuspend the beads in 750 μL of Reagent A. Then, add 750 μL of Reagent B.
7. Rotate at RT for 2 h.
8. Put the tube on the magnet, wait 5 min, and then discard the liquid properly. Wash the beads with 1 mL Tris–HCl 0.1 M, pH 7.5.
9. Put the tube on the magnet, wait 5 min, and then discard the liquid properly. Wash the beads with PBST.
10. Put the tube on the magnet, wait 5 min, and then discard the liquid properly. Resuspend the beads at 10 mg beads/ml in PBST.
11. Keep the beads in the fridge. The beads are stable at 4 °C for 3 months.

### 3.6 Proximity Labeling

To check the biotinylated proteins, we recommend performing immunofluorescence staining to confirm that the endogenous localization of the miniTurbo-conjugated protein hasn't changed and that the biotinylated proteins can be detected by streptavidin. Then, pull-down and mass spectrometry experiments can be performed.

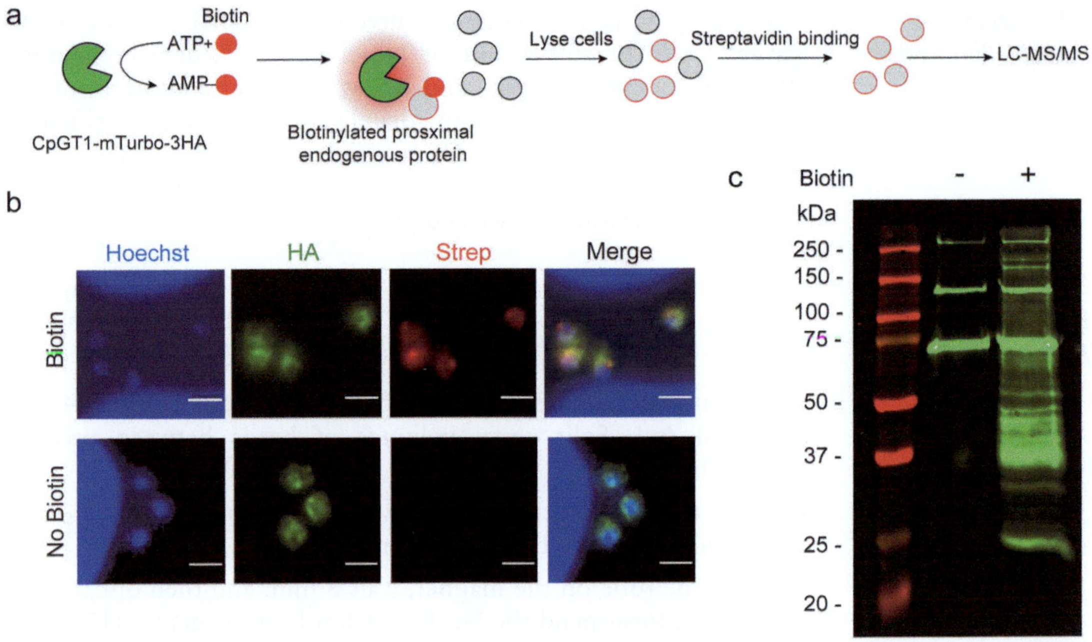


**Fig. 3** Strategy and examples of TurboID in *C. parvum*. (**a**) Schematic representation of proximity-dependent biotinylation in *Cp*. HCT-8 cells were infected with CpGT1-mTurbo-3HA parasites, labeled with biotin, lysed and affinity purified using streptavidin beads, and captured proteins identified by LC-MS/MS. (**b**) Immunofluorescence of biotin-labeled parasites grown in HCT-8 cells. Cells were infected with CpGT1-mTurbo-3HA parasites. After 19 h, cells were treated with 500 μM biotin or vehicle (DMSO) for 1 h and then fixed and stained with rat anti-HA (green), streptavidin-Alex Fluor 568 (red), and Hoechst (blue). Scale bars = 2 μm. (**c**) Western blot analysis of biotinylated proteins in *Cp* cell lysates. Cells were infected with CpGT1-mTurbo-3HA parasites. After 19 h, cells were treated with 500 μM biotin or vehicle (DMSO) for 1 h, lysed in 1% NP-40 lysis buffer, captured on streptavidin beads and detected by Western blot with IRDye 800CW-labeled streptavidin (green). Figure modified and reproduced from the reference [21] under CC BY 4.0 (https:/creativecommons.org/licenses/by/4.0/)

*3.6.1 Immunofluorescence Microscopy (Fig. 3b)*

1. Put 12-mm-diameter glass coverslips in 24-well tissue culture plates.
2. Plate HCT-8 cells at $4 \times 10^5$ cells per well using HCT-8 media without biotin, and grow for ~24 h until confluent (*see* **Note 5**).
3. Infect each well with $1 \times 10^5$ bleached transgenic oocysts (e.g., CpGT1-miniTurbo-3HA).
4. After 19 h, replace the media with HCT-8 media without biotin containing 500 μM biotin or vehicle (0.1% DMSO) for 1 h (*see* **Note 6**).
5. Fix cells with 4% formaldehyde for 10 min.
6. Wash cells twice with PBS.
7. Permeabilize and block cells with TSS buffer for 20 min.

8. Dilute primary antibodies (rat anti-HA (Roche) used at 1:500; mouse 1B5 (mAb) used at 1:500; and rabbit PanCp (pAb) used at 1:2000) in blocking buffer and stain for 60 min at RT.
9. Washed three times with PBS.
10. Dilute secondary antibodies (goat anti-rat Alexa Fluor 488 at 1:1000; streptavidin conjugate Alexa Fluor 568 at 1:1000; goat anti-rabbit Alexa Fluor 647 at 1:1000) in blocking buffer and stain for 60 min at RT.
11. Washed three times with PBS.
12. Stain with Hoechst 33342 (5 μg/mL) for 15 min.
13. Wash once with PBS.
14. Mount samples using ProLong Diamond Antifade Mountant or an equivalent (Thermo Fisher Scientific).
15. Examine samples using a Zeiss Axioskop Mot Plus fluorescence microscope equipped with a 100×, 1.4 N.A. Zeiss Plan Apochromat oil objective lens or an equivalent microscope.
16. Image processing via ImageJ software (https://fiji.sc/).

#### 3.6.2 *C.* parvum *Infection and Biotin Labeling for Pull-Down*

1. Plate HCT-8 cells at $2 \times 10^6$ cells per well using HCT-8 media without biotin in a 6-well plate and grow for ~24 h until confluent.
2. Infect each well with $2 \times 10^6$ bleached transgenic oocysts (e.g., CpGT1-miniTurbo-3HA transgenic oocysts).
3. After 19 h, replace the media with HCT-8 media without biotin containing 500 μM biotin or vehicle (0.1% DMSO) for 1 h (*see* **Note 6**).
4. Stop labeling by placing cells on ice.
5. Wash five times with ice-cold DPBS.

#### 3.6.3 *Pull-Down*

1. Add 1% NP40 lysis buffer into the well, 300 μL per well.
2. Lyse cells for 30 min on ice, pipetting the buffer/lysate every 5 min to help the cells dissociate.
3. Transfer the lysates into a 1.7 mL protein LoBind tube and centrifugation at 14,000 × *g* for 10 min at 4 °C.
4. Transfer the supernatant into a new 1.7 mL protein LoBind tube.
5. Dilute the supernatant two times with PBS to reduce the detergent concentration.
6. Transfer protease-resistant streptavidin beads into a 1.7 mL protein LoBind tube, 50 μL/sample.
7. Wash the beads three times with 1 mL 1% NP40 lysis buffer.

8. Incubated with the diluted lysate overnight at 4 °C with gentle rotation.
9. Wash the beads (1 mL each, 5 min, 4 °C, on a rotor): twice with 0.5% NP40 lysis buffer; once with 1 M KCl; once with 0.1 M $Na_2CO_3$; once with 2 M urea in 10 mM Tris–HCl, pH 8.0; and twice with 0.5% NP40 lysis buffer at 4 °C.
10. Change tubes, then wash four times with PBS.
11. Transfer the beads into a new protein LoBind tube.
12. Wash once more with PBS.
13. Take 10% of the beads for WB.
14. Store the rest of the beads at −80 °C for mass spectrometry.

#### 3.6.4 Western Blot Analysis (Fig. 3c)

1. Resuspend the beads in 20 μL PBS and 5 μL 5× SDS sample buffer containing 100 mM dithiothreitol (DTT), and boil for 10 min.
2. Resolve by SDS-PAGE and transfer to nitrocellulose membranes.
3. Block the membranes with 3% fat-free milk in PBS.
4. Probe with IRDye 800CW streptavidin (LI-COR Biosciences) diluted 1:2000 in PBST overnight at 4 °C with gentle rotation.
5. Wash three times with PBST.
6. Scan the membranes on Li-Cor Odyssey imaging system (LI-COR Biosciences).

### 3.7 Mass Spectrometry Analysis of Biotinylated Proteins

Mass spectrometry was performed by Dr. Michael Naldrett at the Proteomics & Metabolomics Facility, Center for Biotechnology, University of Nebraska, Lincoln (https://biotech.unl.edu/).

#### 3.7.1 Sample Preparation and LC-MS/MS

1. Reduce beads in 50 mM ammonium bicarbonate with 5 mM DTT for 1 h at 37 °C and then alkylate with 20 mM iodoacetamide for 30 min.
2. Quench with DTT and digest with trypsin overnight at 37 °C.
3. Dry the samples.
4. Redissolve in 5% acetonitrile and 0.5% formic acid, and then analyze by nanoscale liquid chromatography (nanoLC)-MS/MS using a 2 h gradient on a 0.075 mm × 250 mm Waters CSH C18 column into a Thermo Eclipse mass spectrometer.

#### 3.7.2 MS Analysis

1. Search data against the following proteomes using Mascot (v2.7.0, Matrix Science): *Cryptosporidium parvum* Iowa II (UniProt number UP000006726; 3805 sequences), *Homo sapiens* (UniProt number UP000005640; 20, 594 sequences), and common contaminants (cRAP 20150130; 124 sequences).

2. Use Scaffold (v4.8.9, Proteome Software Inc.) to validate MS/MS-based peptide and protein identifications. The validated data sets were filtered using a peptide threshold of 95%, a protein threshold of 99%, and a minimum of 2 peptides.
3. The fold change in peptide counts was determined using the ratio of total unique peptide counts in biotin-labeled samples relative to no-biotin samples, averaged over three independent experiments.
4. Adjust the average total unique peptide count to 1 for proteins with no peptides detected in no biotin samples for fold enrichment calculation.
5. Perform statistical analyses in Scaffold software using unpaired Student's $t$ tests ($P < 0.05$), two-tailed (*see* **Note 7**).

### 3.8 Validation of Putative Interactor Proteins

This step is suggested as a follow-up to TurboID experiments, but some experiments may only require a survey of "interactors" or "neighbors" without validating specific interactors.

#### 3.8.1 Localization of Putative Interactor Proteins

1. Interactors identified in the protein interactome may not have been previously localized. To verify the possible interactions at the right location, verifying their localization would be the first step.
2. The candidate protein coding genes can be endogenously tagged by HA or Ty epitope tags using a CRISPR tagging strategy, as described above for the miniTurbo tagging. The only difference is that the miniTurbo-3HA tag is replaced with HA or Ty tags in the PCR template for generation of the tagging cassette.
3. Once the HA- or Ty-tagged strain is generated and amplified from in vivo, it can be grown on HCT-8 cells on coverslips in a 24-well plate for indirect immunofluorescence microscopy.
4. Interactors that co-localized with the bait proteins are subjected to bioinformatic analysis in *Crypto*DB (https://cryptodb.org/cryptodb/app), such as cellular compartment and molecular function.
5. The stability and function of the protein complex can be determined by knockout or mutant of individual substituents using CRISPR genome editing [17, 20].

### 3.9 Depletion of AID-Tagged Proteins

To confirm conditional knockdown, immunofluorescence staining can be performed to verify the endogenous expression of the mAID-3HA-conjugated target and the expression of TIR1 protein. Once target degradation is confirmed, the growth assay could be used to quantify the biological functions of the parasite protein.

*3.9.1 Quantification of AID-Tagged Protein Levels by Immunofluorescence Microscopy*

1. Put 12-mm-diameter glass coverslips in 24-well tissue culture plates.
2. Plate HCT-8 cells at 4 × $10^5$ cells per well using HCT-8 media and grow for ~24 h until confluent.
3. Infect cells with 1 × $10^5$ bleached mAID-tagged transgenic oocysts (e.g., CpGP-mAID-TIR1 oocysts) per well.
4. The incubation time needed to achieve complete protein knockdown will vary based on protein localization, stability, and abundance. To test the degradation efficiency of mAID-tagged proteins by IAA, we set four groups, with two wells for each group (Fig. 4). One group of media was changed to HCT-8 media with vehicle (0.1% EtOH). The other three groups of media are changed to HCT-8 media with 500 μM IAA for 1 h, 6 h, and 24 h treatment.
5. After 24 h, fix cells with 4% formaldehyde for 10 min.

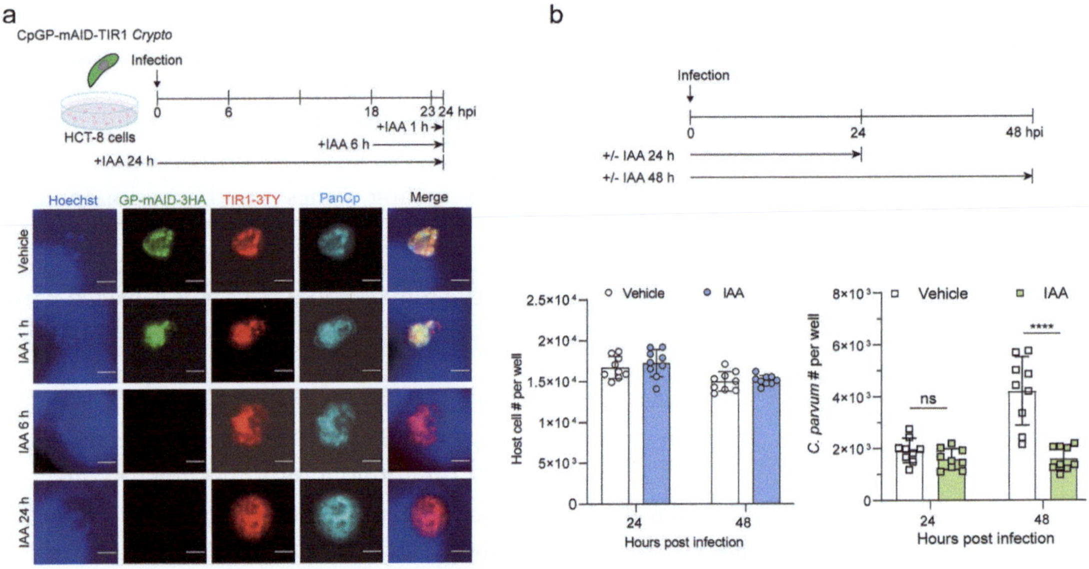


**Fig. 4** CpGP-mAID protein knockdown determined by IFA and growth assay. (**a**) Immunofluorescence localization of CpGP-mAID-3HA expression. HCT-8 cells were infected with CpGP-mAID-TIR1 parasites and treated with 500 μM IAA or the vehicle (EtOH) for 24 h. Cells then were fixed and stained with rabbit anti-HA (green), mouse anti-TY (red), rat PanCp (cyan), and Hoechst (blue). Scale bars = 2 μm. (**b**) Growth of CpGP-mAID-TIR1 parasites following treatment with IAA. HCT-8 cells were infected with CpGP-mAID-TIR1 parasites, treated with 500 μM IAA or the vehicle (EtOH) for 24 h or 48 h. Cells were fixed and labeled with rabbit PanCp and Hoechst, and the number of host cell and *Cp* in each well was determined using a Cytation 3 imager. Each bar represents the mean ± SD for nine replicates in total from three experiments. Statistical analysis was performed by two-way ANOVA corrected for multiple comparisons by Sidak's method. ns, not significant; ****, $P < 0.001$. (Figure modified and reproduced from reference [21] under CC BY 4.0 (https:/creativecommons.org/licenses/by/4.0/))

6. Washed twice with PBS.
7. Permeabilized and block with TSS buffer for 20 min.
8. Dilute primary antibodies (rabbit anti-HA used at 1:1000; mouse anti-TY (mAb) used at 1:500; and rat PanCp (pAb) used at 1:2000) in blocking buffer and stain for 60 min at RT.
9. Wash three times with PBS.
10. Dilute secondary antibodies (goat anti-rabbit Alexa Fluor 488 diluted 1:1000; goat anti-mouse Alexa Fluor 568 diluted 1:1000; goat anti-rat Alexa Fluor 647 diluted 1:1000) in blocking buffer and stain for 60 min at RT.
11. Wash three times with PBS.
12. Stain with Hoechst 33342 (5 μg/mL) for 15 min.
13. Wash once with PBS.
14. Mount samples using ProLong Diamond Antifade Mountant or an equivalent (Thermo Fisher Scientific).
15. Examine samples using a Zeiss Axioskop Mot Plus fluorescence microscope equipped with a 100×, 1.4 N.A. Zeiss Plan Apochromat oil objective lens or an equivalent microscope.
16. Image processing via ImageJ software (https://fiji.sc/).

*3.9.2 Growth Assay Quantification mAID-Tagged Proteins In Vitro (Fig. 4)*

1. Plate HCT-8 cells at $1 \times 10^5$ cells per well in black-sided, optically clear-bottomed 96-well plates and grow for ~24 h until confluent.
2. Infect cells with $5 \times 10^4$ bleached mAID-tagged transgenic oocysts (e.g., CpGP-mAID-TIR1 oocysts) per well.
3. Change the media and treat with 500 μM IAA or vehicle (0.1% EtOH) control for 24 h or 48 h, with three replicates per treatment.
4. After 24 or 48 h, fix cells with 4% formaldehyde for 10 min.
5. Wash twice with PBS.
6. Permeabilize and block with TSS buffer for 20 min.
7. Stain with primary antibody (rabbit PanCp diluted 1:2000) for 60 min at RT.
8. Wash three times with PBS.
9. Stain with secondary antibody (goat anti-rabbit Alexa Fluor 488 diluted 1:1000) for 60 min at RT.
10. Wash three times with PBS.
11. Stain with Hoechst 33342 (5 μg/mL) for 20 min.
12. Wash once with PBS.

13. Image plates with a 20× objective on a Cytation 3 cell imager (36 images per well in a 6 × 6 grid).
14. Use BioTek Gen5 software version 3.08 (Agilent) to quantify the total number of parasites (puncta in the GFP channel) and host cells (nuclei in the DAPI channel) per well.

## 4 Notes

1. The sequence of miniTurbo and TIR1 was codon optimized for *C. parvum*. We recommended doing codon optimization for heterologous sequences.
2. To get a high transfection efficiency, we suggest using fresh bleached oocysts. Otherwise, the bleached oocysts can be stored at 4 °C for 2 weeks before in vivo or in vitro infection.
3. After 1 h incubation, the excystation efficiency should be approximately 90–95%. If the efficiency is lower than that, we suggest prolonging the incubation for an additional 15 min.
4. Previous studies show that protease-resistant streptavidin beads are resistant to proteolysis by trypsin with over 100-fold reduction in streptavidin contamination and better coverage of proteins interacting with various biotinylated molecules [19]. This approach is likely important where biological input is low, as will organisms like *Cryptosporidium* where obtaining large numbers of recombinant oocysts is a challenge.
5. Before the formal experiment, we recommend passing the HCT-8 cells for one to two generations with HCT-8 media without biotin to eliminate any trace of biotin in the cells or media.
6. The miniTurbo tag is approximately twofold less active than Turbo tag. We evaluated miniTurbo labeling in *C. parvum* and found an optimal labeling time of ~30 min with 500 μM biotin. If biotinylated proteins are not detected by IFA or Western blot, it may be helpful to extend the labeling time or change the miniTurbo tag to Turbo tag.
7. For optimal data analysis, it is important to have two or more biological replicates for each miniTurbo-tagged protein. In addition, we have successfully used the statistical features provided within Scaffold software to define potential interactors between replicate data sets. When performing comparisons of replicate experiments, we analyze data using Student's *t* test with or without correction for multiple comparisons.

## Acknowledgments

We are grateful to William H. Witola, Daniel Goldberg, and the members of the Sibley laboratories for helpful advice. This work was supported by a grant from the National Institutes of Health (AI145496).

**Competing Interest Statement** The authors declare no conflicts of interest relevant to the content of this chapter.

### References

1. Vinayak S, Pawlowic MC, Sateriale A, Brooks CF, Studstill CJ, Bar-Peled Y, Cipriano MJ, Striepen B (2015) Genetic modification of the diarrhoeal pathogen *Cryptosporidium parvum*. Nature 523(7561):477–480. https://doi.org/10.1038/nature14651
2. Kim DI, Roux KJ (2016) Filling the void: proximity-based labeling of proteins in living cells. Trends Cell Biol 26(11):804–817. https://doi.org/10.1016/j.tcb.2016.09.004
3. Lam SS, Martell JD, Kamer KJ, Deerinck TJ, Ellisman MH, Mootha VK, Ting AY (2015) Directed evolution of APEX2 for electron microscopy and proximity labeling. Nat Methods 12(1):51–54. https://doi.org/10.1038/nmeth.3179
4. Choi-Rhee E, Schulman H, Cronan JE (2004) Promiscuous protein biotinylation by *Escherichia coli* biotin protein ligase. Protein Sci 13(11):3043–3050. https://doi.org/10.1110/ps.04911804
5. Roux KJ, Kim DI, Raida M, Burke B (2012) A promiscuous biotin ligase fusion protein identifies proximal and interacting proteins in mammalian cells. J Cell Biol 196(6):801–810. https://doi.org/10.1083/jcb.201112098
6. Branon TC, Bosch JA, Sanchez AD, Udeshi ND, Svinkina T, Carr SA, Feldman JL, Perrimon N, Ting AY (2018) Efficient proximity labeling in living cells and organisms with TurboID. Nat Biotechnol 36(9):880–887. https://doi.org/10.1038/nbt.4201
7. Qin W, Cho KF, Cavanagh PE, Ting AY (2021) Deciphering molecular interactions by proximity labeling. Nat Methods 18(2):133–143. https://doi.org/10.1038/s41592-020-01010-5
8. Meissner M, Brecht S, Bujard H, Soldati D (2001) Modulation of myosin A expression by a newly established tetracycline repressor-based inducible system in *toxoplasma gondii*. Nucleic Acids Res 29(22):E115. https://doi.org/10.1093/nar/29.22.e115
9. Meissner M, Schluter D, Soldati D (2002) Role of *toxoplasma gondii* myosin a in powering parasite gliding and host cell invasion. Science 298(5594):837–840. https://doi.org/10.1126/science.1074553
10. Yesbolatova A, Saito Y, Kitamoto N, Makino-Itou H, Ajima R, Nakano R, Nakaoka H, Fukui K, Gamo K, Tominari Y, Takeuchi H, Saga Y, Hayashi KI, Kanemaki MT (2020) The auxin-inducible degron 2 technology provides sharp degradation control in yeast, mammalian cells, and mice. Nat Commun 11(1):5701. https://doi.org/10.1038/s41467-020-19532-z
11. Nishimura K, Fukagawa T, Takisawa H, Kakimoto T, Kanemaki M (2009) An auxin-based degron system for the rapid depletion of proteins in nonplant cells. Nat Methods 6(12):917–922. https://doi.org/10.1038/nmeth.1401
12. Natsume T, Kiyomitsu T, Saga Y, Kanemaki MT (2016) Rapid protein depletion in human cells by auxin-inducible degron tagging with short homology donors. Cell Rep 15(1):210–218. https://doi.org/10.1016/j.celrep.2016.03.001
13. Philip N, Waters AP (2015) Conditional degradation of *plasmodium* calcineurin reveals functions in parasite colonization of both host and vector. Cell Host Microbe 18(1):122–131. https://doi.org/10.1016/j.chom.2015.05.018
14. Brown KM, Long SJ, Sibley LD (2017) Plasma membrane association by N-acylation governs PKG function in *Toxoplasma gondii*. mBio 8(3). https://doi.org/10.1128/mBio.00375-17

15. Kuhlenschmidt TB, Rutaganira FU, Long S, Tang K, Shokat KM, Kuhlenschmidt MS, Sibley LD (2016) Inhibition of calcium-dependent protein kinase 1 (CDPK1) *In vitro* by pyrazolopyrimidine derivatives does not correlate with sensitivity of *Cryptosporidium parvum* growth in cell culture. Antimicrob Agents Chemother 60(1):570–579. https://doi.org/10.1128/AAC.01915-15
16. Sateriale A, Pawlowic M, Vinayak S, Brooks C, Striepen B (2020) Genetic manipulation of *Cryptosporidium parvum* with CRISPR/Cas9. Methods Mol Biol 2052:219–228. https://doi.org/10.1007/978-1-4939-9748-0_13
17. Wilke G, Funkhouser-Jones LJ, Wang Y, Ravindran S, Wang Q, Beatty WL, Baldridge MT, VanDussen KL, Shen B, Kuhlenschmidt MS, Kuhlenschmidt TB, Witola WH, Stappenbeck TS, Sibley LD (2019) A stem-cell-derived platform enables complete *Cryptosporidium* development *in vitro* and genetic tractability. Cell Host Microbe 16(1):123–134. https://doi.org/10.1016/j.chom.2019.05.007
18. Zhang X, Kim CY, Worthen T, Witola WH (2018) Morpholino-mediated *in vivo* silencing of *Cryptosporidium parvum* lactate dehydrogenase decreases oocyst shedding and infectivity. Int J Parasitol 48(8):649–656. https://doi.org/10.1016/j.ijpara.2018.01.005
19. Rafiee MR, Sigismondo G, Kalxdorf M, Forster L, Brugger B, Bethune J, Krijgsveld J (2020) Protease-resistant streptavidin for interaction proteomics. Mol Syst Biol 16(5):e9370. https://doi.org/10.15252/msb.20199370
20. Xu R, Feng Y, Xiao L, Sibley LD (2021) Insulinase-like protease 1 contributes to macrogamont formation in *Cryptosporidium parvum*. mBio 12(2):e03405–e03420. https://doi.org/10.1128/mBio.03405-20
21. Xu R, Beatty WL, Greigert V, Witola WH, Sibley LD (2024) Multiple pathways for glucose phosphate transport and utilization support growth of *Cryptosporidium parvum*. Nat Commun 15(1):380. https://doi.org/10.1038/s41467-024-44696-3

# Chapter 20

# In Vitro Culture of *Cryptosporidium* spp. Using Stem Cell-Derived Intestinal Epithelial Monolayers

**Valentin Greigert, Georgia Wilke, Yi Wang, Soumya Ravindran, Matthew Freese, Thaddeus Stappenbeck, William Witola, and L. David Sibley**

## Abstract

*Cryptosporidium parvum* has a complex life cycle consisting of asexual and sexual phases that culminate in oocyst formation in vivo. The most widely used cell culture platforms to study *C. parvum* only support a few days of growth and do not allow the parasite to proceed past the sexual stages to complete oocyst formation. Additionally, these cell culture platforms are mostly adenocarcinoma cell lines, which do not adequately model the parasite's natural environment in the small intestine. We present a method for generating mouse primary intestinal epithelial cell monolayers that support long-term *Cryptosporidium parvum* growth, as well as human primary intestinal epithelial cell monolayers that facilitate long-term growth of *C. parvum* and *C. hominis.* Stem cells are grown as spheroids and plated onto transwells, allowing for separate apical and basolateral compartments. In the apical chamber, the cell growth medium was removed to create an "air–liquid interface" that enhanced host cell differentiation and supported *Cryptosporidium* growth including all stages of the life cycle. The use of primary intestinal cells to grow *Cryptosporidium* in vitro will be a valuable tool for studying host–parasite interactions using a convenient in vitro model that more closely resembles the natural niche in the intestine.

**Key words** *Cryptosporidium*, Transwell, Long-term growth, Primary cell, Air, liquid interface

## 1 Introduction

A significant barrier to *Cryptosporidium* research is the lack of a cell culture system that enables long-term growth in vitro. Historically, human adenocarcinoma cells have been used to study the asexual and early sexual development of the parasite [1]. These cell lines support the growth of the parasite over 2–3 days, but after the parasite enters the sexual part of its life cycle, the numbers rapidly diminish because oocyst formation does not occur [1–3]. This limitation greatly affects the type of studies researchers can pursue when investigating *Cryptosporidium* biology.

Jan R. Mead and Michael J. Arrowood (eds.), *Cryptosporidium: Methods and Protocols*, Methods in Molecular Biology, vol. 2978, https://doi.org/10.1007/978-1-0716-4824-7_20,

Recently, there have been several reports of cell culture platforms capable of supporting complete life cycle development of *C. parvum*, a species of *Cryptosporidium* frequently associated with disease in humans and animals [4–6]. One platform utilizes a hollow fiber system to mimic the intestine by allowing for separate control of the apical and basolateral compartments [7]; this system enables the creation of an environment with redox and nutrient conditions optimal for parasite growth. However, it requires the use of expensive equipment, and the hollow fiber cannot be "opened" during the experiment, making direct observation studies difficult. Also, the expense of the hollow fiber system means that experiments are not easily scalable, and perturbations to the system are often done serially rather than in parallel, which is a major limitation for many studies. Another novel platform is a 3D intestinal model that uses silk protein as a scaffold for cell growth [8]; this model also has a lumen, creating two separate compartments. This platform involves the use of specialized equipment to create the 3D scaffolds, making it more challenging for other researchers to adopt. The two new platforms use adenocarcinoma cells (HCT-8 and Caco-2 cells), which are likely very different from the cell types *C. parvum* infects in vivo, in terms of differentiation status, metabolism, and proliferation.

To accurately reproduce in vivo host–parasite interactions in cell culture, it is desirable to utilize primary cells. Advancements in stem cell biology have made it possible to propagate intestinal stem cells as spheroids in medium containing specific growth factors [9, 10]. The required cell culture medium has become relatively easy to make due to the creation of a supportive cell line, L-WRN, that secretes Wnt, R-spondin, and noggin, creating conditioned medium that is used to support spheroid growth [9]. Cells derived from spheroids can be used to create primary intestinal epithelial cell monolayers grown on transwell membranes that recapitulate aspects of the gut, including the development of specific cell lineages such as goblet and Paneth cells [11, 12]. Murine ileal monolayers derived from an early version of this method support robust *C. parvum* growth, which enabled the creation of novel antibodies to intracellular parasite stages [13]. These monolayers, however, typically only last 2–5 days, which make them unsuitable for studying long-term *C. parvum* growth. A more recent report utilized organoids grown in Matrigel and demonstrated that *C. parvum* will undergo its entire life cycle within a closed system [14]. However, the difficulty in having to micro-inject individual organoids, combined with the relative inaccessibility of the system, is likely to limit this platform for larger-scale studies.

More recently, a protocol for creating primary intestinal monolayers that exhibit dramatic differentiation and long-term viability has been published [15]. This method relies on the inclusion of a feeder cell layer to support stem cell growth and the introduction of an air–liquid interface (ALI) at the surface of the cell monolayer. ALI has been used to enhance the differentiation of airway tract epithelial cells for many years [16–18] and has also recently been applied to intestinal epithelial cell culture, resulting in increased expression of markers associated with cell differentiation, as well as significant changes in cell metabolism [19, 20].

Epithelial monolayers grown with an ALI and feeder cell layer can survive for weeks [15], making them attractive candidates for a cell culture platform for *C. parvum*. We have found that mouse primary intestinal monolayers grown under ALI conditions on transwells in L-WRN-conditioned medium support long-term *C. parvum* growth [21]. Parasite growth can last for weeks in this system without affecting host cell viability. Recent studies have taken advantage of this system to generate genetic crosses in vitro and perform the first tetrad experiments in any apicomplexan parasite [22]. Additionally, we recently developed a similar system using human intestinal stem cells that showed strong cellular differentiation and a phenotype resembling the physiological human intestinal epithelium [23]. We have demonstrated that this system allows for the in vitro completion of the *C. parvum* life cycle, as well as that of *C. hominis*, a parasite that is restricted to the human host. This newly developed system for *C. hominis* opens up new approaches for studying the pathophysiology of human cryptosporidiosis.

Collectively, these cell culture systems will support further research into the biology of *Cryptosporidium*, including the development of in vitro genetic approaches and testing mechanism of action for potential growth inhibitors.

## 2 Materials

### *2.1 Creating Irradiated 3T3 (i3T3) Cell Stocks for Feeder Cell Layer (Murine System)*

1. Biosafety cabinet and 37 °C, 5% $CO_2$ incubator.
2. Mouse fibroblast cells (NIH/3T3, ATCC CRL-1658).
3. 37 °C water bath.
4. 15 and 50 mL polypropylene conical tubes.
5. 3 T3 growth medium: For 500 mL, combine 445 mL Dulbecco's Modified Eagle's Medium (DMEM, high glucose) with 50 mL heat-inactivated fetal bovine serum (FBS) and 5 mL 100× penicillin/streptomycin.
6. Centrifuge capable of holding 15 and 50 mL conical tubes that can spin at 200 × *g*.
7. Tissue culture supplies such as 25, 75, and 150 $cm^2$ cell culture flasks (T-25, T-75, T-150) and disposable plastic pipettes.
8. Dulbecco's Phosphate-Buffered Saline (DPBS)–EDTA: For 500 mL, add 500 μL of 0.5 M EDTA (pH 8.0) solution to 500 mL tissue culture-grade DPBS.

9. Trypsin–EDTA: For 100 mL, add 10 mL 10× trypsin solution to 90 mL DPBS–EDTA. Aliquot and store at −20 °C (stable for months); aliquots can be thawed and stored at 4 °C for 7–10 days.
10. Machine capable of irradiating 15 mL conical at 3000 rads (e.g., Xstrahl Small Animal Radiation Research Platform or SARRP).
11. Hemocytometer and cover glass.
12. 3 T3 freezing medium: For 50 mL, combine 10 mL FBS, 2.5 mL DMSO, and 37.5 mL 3T3 growth medium.
13. Cryovials, 1.5 mL volume.
14. −80 °C freezer.
15. $LN_2$ storage tank.

### 2.2 Creating Irradiated Human Intestinal Myofibroblast (iHIMF) Cell Stocks for Feeder Cell Layer (Human System)

1. Biosafety cabinet and 37 °C, 5% $CO_2$ incubator.
2. Human intestinal myofibroblasts (Lonza; reference discontinued. It is being maintained and shared with the community by Valentin Greigert, University of Reims, Champagne Ardenne, France).
3. 37 °C water bath.
4. 15 and 50 mL polypropylene conical tubes.
5. HIMF growth medium: Combine 500 mL of SmBM-2™ (Lonza, CC-3181) and SmGM-2 SingleQuots™ (Lonza, CC-4149). Once SingleQuots are added, the medium should be used within a month.
6. Centrifuge capable of holding 15 and 50 mL conical tubes that can spin at 200 × *g*.
7. Tissue culture supplies such as 25, 75, and 150 $cm^2$ cell culture flasks (T-25, T-75, T-150) and disposable plastic pipettes.
8. Dulbecco's Phosphate-Buffered Saline (DPBS)–EDTA: For 500 mL, add 500 μL of 0.5 M EDTA (pH 8.0) solution to 500 mL tissue culture-grade DPBS.
9. Trypsin–EDTA: For 100 mL, add 10 mL of 10× trypsin solution to 90 mL DPBS–EDTA. Aliquot and store at −20 °C (stable for months); aliquots can be thawed and stored at 4 °C for 7–10 days.
10. Machine capable of irradiating 15 mL conical at 5000 rads (e.g., the Xstrahl Small Animal Radiation Research Platform or SARRP).
11. Hemocytometer and cover glass.
12. 3T3 freezing medium: For 50 mL, combine 10 mL FBS, 2.5 mL DMSO, and 37.5 mL 3T3 growth medium.

13. Cryovials, 1.5 mL volume.
14. −80 °C freezer.
15. $LN_2$ storage tank.

#### 2.3 Plating i3T3 Cells or iHIMF on Transwells

1. Biosafety cabinet.
2. Matrigel.
3. Cold DPBS.
4. Transwell with 0.4 μm pore polyester membrane insert.
5. Tissue culture supplies such as 15 mL conical centrifuge tubes and 1.7 mL microcentrifuge tubes.
6. Vacuum flask and sterile disposable glass Pasteur pipettes.
7. 37 °C, 5% $CO_2$ incubator.

#### 2.4 Preparing Conditioned Medium

Conditioned medium was prepared as previously described [10] with slight modifications presented here.

1. Biosafety cabinet.
2. L-WRN cells producing Wnt, R-Spondin, and Noggin (ATCC CRL-3276).
3. L-cell media: To 500 mL of DMEM high-glucose media, add 50 mL heat-inactivated FBS and 5 mL 100× penicillin/streptomycin.
4. Washing medium: Add 50 mL FBS, 5 mL 100× penicillin/streptomycin, and 5 mL 100× L-glutamine to DMEM/F12 with HEPES.
5. Primary medium: Add 6.25 mL 100× L-glutamine, 6.25 mL 100× penicillin/streptomycin, and 125 mL heat-inactivated FBS to 500 mL Advanced DMEM/F12.
6. DPBS–EDTA, trypsin–EDTA.
7. Selective agents Hygromycin B Gold and the aminoglycoside antibiotic G418.
8. FBS for conditioned medium.
9. T150 flasks.
10. EasyFill Cell Factory (Nalge Nunc Int., Rochester, NY).
11. Autoclaved 2 L glass media bottles.
12. Magnetic stir bars.
13. Corning 1 L storage bottles.
14. 50 mL conical centrifuge tubes.
15. Autoclavable bottle-top dispenser (5–50 mL).
16. Inlet tube for autoclavable bottle-top dispenser.

17. Centrifuge capable of holding 50 mL conical centrifuge tubes that can spin at 200 and 1800 × *g*.
18. 37 °C, 5% $CO_2$ tissue culture incubator.

### 2.5 Passaging Mouse Spheroids

1. Spheroid line established as previously described [10]; for our purposes, we used a line isolated from the ileum of an 8–10-week-old C57BL/6J mouse. Lines can be frozen 1–2 passages after the original derivation and thawed as needed; use single line for 20–25 passages.
2. 37 °C water bath.
3. Tissue culture supplies such as 12-well plates, disposable pipettes, 15 and 50 mL conical centrifuge tubes, and 1.7 mL microcentrifuge tubes.
4. Washing medium, as described above.
5. Matrigel.
6. Biosafety cabinet.
7. Cell scrapers.
8. Centrifuge capable of holding 15 conical tubes that can spin at 200 × *g*.
9. Centrifuge capable of holding 1.7 mL microcentrifuge tubes that can spin at 200 × *g*.
10. DPBS–EDTA and trypsin–EDTA.
11. 50% conditioned medium.
12. Primary medium, as described above.
13. Rho-associated protein kinase (ROCK) inhibitor Y-27632.
14. 37 °C, 5% $CO_2$ incubator.

### 2.6 Passaging Human Spheroids

1. Spheroid line established as previously described [10]; for our purposes, cells were isolated from biopsies obtained from de-identified healthy donors through the BioBank Core of the Digestive Diseases Research Core Center at Washington University School of Medicine in St. Louis. The lines can be frozen 1–2 passages after the original derivation and thawed as needed; use single line for 20–25 passages.
2. 37 °C water bath.
3. Tissue culture supplies such as 12-well plates, disposable pipettes, 15 and 50 mL conical centrifuge tubes, and 1.7 mL microcentrifuge tubes.
4. Washing medium, as described above.
5. Matrigel.
6. Biosafety cabinet.
7. Cell scrapers.

8. Centrifuge capable of holding 15 conical tubes that can spin at 200 × *g*.
9. Centrifuge capable of holding 1.7 mL microcentrifuge tubes that can spin at 200 × *g*.
10. DPBS–EDTA.
11. TrypLE™ Express (Thermo Fisher Scientific Inc., Waltham, MA, 12604013).
12. 50% conditioned medium.
13. Primary medium, as described above.
14. Rho-associated protein kinase (ROCK) inhibitor Y-27632.
15. TGF-β inhibitor (ALK5 inhibitor SB431542).
16. 37 °C, 5% $CO_2$ incubator.

### 2.7 Creating Mouse Primary Cell Monolayers for Cryptosporidium parvum Infection

1. Biosafety cabinet.
2. Cell scrapers.
3. Tissue culture supplies such as 12-well plates, disposable pipettes, and 15 and 50 mL conical centrifuge tubes.
4. Centrifuge capable of holding 15 mL conical tubes that can spin at 200 × *g*.
5. DPBS–EDTA and trypsin–EDTA.
6. 37 °C water bath.
7. Washing medium, as described above.
8. 50% conditioned medium prepared as previously described [10]; requires L-WRN cell line (ATCC CRL-3276).
9. Primary medium: Add 6.25 mL 100× L-glutamine, 6.25 mL 100× penicillin/streptomycin, and 125 mL heat-inactivated FBS to 500 mL Advanced DMEM/F12.
10. Rho-associated protein kinase (ROCK) inhibitor Y-27632.
11. Sterile cell strainers, 40 μm.
12. 37 °C, 5% $CO_2$ incubator.

### 2.8 Creating Human Primary Cell Monolayers for Cryptosporidium parvum and Cryptosporidium hominis Infection

1. Biosafety cabinet.
2. Cell scrapers.
3. Tissue culture supplies such as 12-well plates, disposable pipettes, and 15 and 50 mL conical centrifuge tubes.
4. Centrifuge capable of holding 15 mL conical tubes that can spin at 200 × *g*.
5. DPBS–EDTA.
6. TrypLE™ Express (Thermo Fisher Scientific, 12604013).
7. 37 °C water bath.

8. Washing medium, as described above.
9. 50% conditioned medium prepared as previously described [10]; requires L-WRN cell line (ATCC CRL-3276).
10. Primary medium: Add 6.25 mL 100× L-glutamine, 6.25 mL 100× penicillin/streptomycin, and 125 mL heat-inactivated FBS to 500 mL Advanced DMEM/F12.
11. Rho-associated protein kinase (ROCK) inhibitor Y-27632.
12. TGF-β inhibitor (ALK5 inhibitor SB431542).
13. Sterile cell strainers, 40 μm.
14. 37 °C, 5% $CO_2$ incubator.

### 2.9 Preparing and Maintaining Cryptosporidium Oocysts for Infection

1. Biosafety cabinet.
2. Purified oocysts, stored in 50 mM Tris-10 mM EDTA, pH 7.2.
3. Cold DPBS.
4. 15 mL conical centrifuge tubes.
5. Commercial laundry bleach containing 8.25% sodium hypochlorite.
6. Centrifuge capable of holding 15 conical tubes that can spin at 800 and 1250 × *g*.
7. Centrifuge buckets with sealed biohazard caps.
8. Cold DPBS +1% BSA.
9. Excystation buffer: To make 5 mL, dissolve 0.75 g synthetic sodium taurocholate into 5 mL DPBS (1.5% w/v sodium taurocholate solution). May need vortexing to dissolve completely. Filter-sterilize with 0.2 μm filter. Can be stored at 4 °C for several weeks.
10. 37 °C water bath.
11. Glass slides and square coverslips (22 × 22 mm).
12. Swin-Lok or Swinnex filter holders, 25 mm.
13. Track-etch polycarbonate hydrophilic membranes, 25 mm circle, 1.0 μm pore size.
14. 3 or 5 mL plastic syringe.
15. Cell growth medium.

### 2.10 Measuring Host and Parasite Replication in Primary Cell Monolayers by qPCR

1. QIAamp DNA Mini Kit (Qiagen Inc., Germantown, MD).
2. Blunt pipette tips (20–200 μL pipette tips with ends snipped off).
3. 1.7 mL microcentrifuge tubes.
4. 56 °C water bath.
5. 70 °C heat block.
6. Ethanol, 200 proof.

7. Centrifuge capable of spinning microcentrifuge tubes and spin columns at 7000 × g.
8. Sterile $H_2O$.
9. 2× SYBR Green mix or similar qPCR reaction mix.
10. 96-well reaction plate compatible with qPCR machine.
11. Adhesive film for sealing reaction plate.
12. Centrifuge capable of spinning 96-well plates.
13. qPCR machine with 96-well block.

#### 2.11 Monitoring Life Cycle Progression in Primary Cell Monolayers by Microscopy

1. DPBS.
2. Fixative such as 4% formaldehyde.
3. Permeabilization agent such as Triton X-100 or saponin.
4. Normal goat serum (NGS).
5. Fetal bovine serum (FBS).
6. Hoechst 33258 dye or other DNA stain.
7. Scalpel and forceps.
8. Mounting reagent with fluorescence anti-quenching agent.
9. Glass slides and square coverslips (22 × 22 mm).

## 3 Methods

#### 3.1 Creating Irradiated 3T3 (i3T3) Cell Stocks for Feeder Cell Layer (Mouse System)

1. Thaw a frozen vial of 3T3 cells (*see* **Note 1**) in 37 °C water bath. Remove vial immediately when cells have thawed.
2. Add thawed cell mixture to 15 mL conical containing 10 mL 3T3 growth medium.
3. Spin at 200 × *g* for 3 min. Remove supernatant.
4. Resuspend pellet in 5 mL growth medium and pipette 8–10 times to evenly distribute cells.
5. Transfer to T-25 flask and culture in 37 °C, 5% $CO_2$ incubator.
6. When cells are ~90% confluent, remove the medium from the flask, and wash with 10 mL DPBS–EDTA.
7. Add trypsin–EDTA to flask in a volume sufficient to cover cell monolayer (1 mL for T-25, 3 mL for T-75, 6 mL for T-150). Swirl flask to ensure complete cell coverage.
8. Keep flask at room temperature for 2–3 min. Swirl flask to see if cells have detached.
9. Once cells have detached, add sufficient volume of growth medium to cover flask, and resuspend the cells (4 mL for T-25, 6 mL for T-75, 7 mL for T-150). Serum in the medium will stop the action of the trypsin.

10. Pipette the medium over the surface of the flask three times to ensure all cells are in suspension. Pipette up and down eight to ten times to create a single-cell suspension.
11. Transfer cell mixture to 15 mL conical and spin at 200 × g for 3 min.
12. Remove supernatant and resuspend pellet in the desired amount of medium.
13. Pipette mixture up and down five to ten times to ensure even distribution of cells, and transfer the desired amount of cells to new flasks. Swirl flasks to ensure cells are covering flask surface.
14. The passage ratio determines the length of time it takes the flask to reach 90–100% confluency. Passage ratio of 1:2.5 will reach confluency in 2 days, 1:5 in 3 days, and 1:10 in 4 days.
15. Expand the cells into the number of flasks desired for the irradiation process.
16. When final flasks become confluent, trypsinize cells as described in **Steps 1–6**. The irradiation protocol can be scaled up or down depending on the number of irradiated cells required. For example, three T-T150s, once trypsinized, can be pooled into one 50 mL conical and spun down at 200 × *g* for 3 min.
17. Remove supernatant and resuspend pellet in 10 mL growth medium, and transfer cell mixture to 15 mL conical.
18. Irradiate cells at 3000 rads. We use the Xstrahl SARRP for this purpose.
19. Pipette cells to ensure even distribution, and take 10 μL to determine cell density using a hemocytometer. Cells should not be clumpy and should appear uniformly round.
20. Spin down cells at 200 × *g* for 3 min.
21. Resuspend pellet in sufficient volume of freezing medium to achieve 2 × 106 cells/mL.
22. Aliquot cells for freezing, 1.0 mL per cryovial. Three T-150 s typically yield 20–25 vials at the above cell density.
23. Move cryovials to −80 °C freezer for short-term storage (days to weeks) and then to LN2 tank for long-term storage (months).

#### 3.2 Creating Irradiated Human Intestinal Myofibroblasts (iHIMF) Cell Stocks for Feeder Cell Layer (Human System)

1. Thaw a frozen vial of HIMF (*see* **Note 2**) in 37 °C water bath. Remove vial immediately when cells have thawed.
2. Add thawed cell mixture to 15 mL conical containing 10 mL HIMF growth medium.
3. Spin at 200 × *g* for 3 min. Remove supernatant.

4. Resuspend pellet in 5 mL growth medium and pipette 8–10 times to evenly distribute cells.
5. Transfer to T-25 flask and culture in 37 °C, 5% $CO_2$ incubator.
6. When cells are ~90% confluent, remove the medium from the flask, and wash with 10 mL DPBS–EDTA.
7. Add trypsin–EDTA to flask in a volume sufficient to cover cell monolayer (1 mL for T-25, 3 mL for T-75, 6 mL for T-150). Swirl flask to ensure complete cell coverage.
8. Keep flask at room temperature for 2–3 min. Swirl flask to see if cells have detached.
9. Once cells have detached, add sufficient volume of growth medium to cover flask and resuspend the cells (4 mL for T-25, 6 mL for T-75, 7 mL for T-150). Serum in the medium will stop the action of the trypsin.
10. Pipette the medium over the surface of the flask three times to ensure all cells are in suspension. Pipette up and down eight to ten times to create a single-cell suspension.
11. Transfer cell mixture to 15 mL conical and spin at 200 × g for 3 min.
12. Remove supernatant and resuspend pellet in the desired amount of medium.
13. Pipette mixture up and down five to ten times to ensure even distribution of cells, and transfer the desired amount of cell to new flasks. Swirl flasks to ensure cells are covering flask surface.
14. The passage ratio determines the length of time it takes the flask to reach 90–100% confluency. Passage ratio of 1:2.5 will reach confluency in 2 days, 1:5 in 3 days, and 1:10 in 4 days.
15. Expand the cells into the number of flasks desired for the irradiation process.
16. When final flasks become confluent, trypsinize cells as described in **Steps 1–6**. The irradiation protocol can be scaled up or down depending on the number of irradiated cells required. For example, three T-T150s, once trypsinized, can be pooled into one 50 mL conical and spun down at 200 × *g* for 3 min.
17. Remove supernatant and resuspend pellet in 10 mL growth medium and transfer cell mixture to 15 mL conical.
18. Irradiate cells at 5000 rads. We use the Xstrahl SARRP for this purpose.
19. Pipette cells to ensure even distribution and take 10 μL to determine cell density using a hemocytometer. Cells should not be clumpy and should appear uniformly round.
20. Spin down cells at 200 × *g* for 3 min.

21. Resuspend pellet in sufficient volume of freezing medium to achieve $2 \times 10^6$ cells/mL.
22. Aliquot cells for freezing, 1.0 mL per cryovial. Three T-150 s typically yield 20–25 vials at the above cell density.
23. Move cryovials to −80 °C freezer for short-term storage (days to weeks) and then to LN2 tank for long-term storage (months).

### 3.3 Plating i3T3 Cells (Mouse System) or iHIMF (Human System) on Transwells

This step occurs the day before the stem cell spheroids are split and plated onto the transwells. All cell manipulation is performed inside a biosafety cabinet under sterile conditions.

1. Make 10% Matrigel solution: Dilute Matrigel (*see* **Note 3**) in cold DPBS (keep DPBS and Matrigel on ice), then pipette to ensure even distribution. Prepare sufficient volume for 100 μL of solution per transwell.
2. Add 100 μL 10% Matrigel solution to each transwell. Move plate with transwells to 37 °C incubator for 20–30 min to allow the Matrigel to polymerize while finishing the following steps.
3. Thaw vial(s) of frozen cells (either i3T3 cells or iHIMF) in 37 °C water bath. Once thawed, immediately add to 15 mL conical containing 10 mL 3T3 growth medium.
4. Spin down cells at 200 × *g* for 3 min.
5. Remove supernatant and resuspend cells in 1–3 mL of growth medium.
6. Count cells using a hemocytometer. Each transwell will be plated with $8 \times 10^4$ cells in 100 μL volume; resuspend cells accordingly.
7. After transwells have been in 37 °C incubator for at least 20 min, remove and aspirate Matrigel using vacuum flask and glass pipette. Alternatively, Matrigel can be removed with a pipette tip.
8. Add cell suspension to each transwell.
9. Add 400 μL growth medium to bottom chamber of each transwell.
10. Place transwells in 37 °C, 5% $CO_2$ incubator.

### 3.4 Preparing Conditioned Medium

The protocol for this step has been published in detail previously [10]. All cell manipulation is performed inside a biosafety cabinet under sterile conditions.

1. Thaw a vial of L-WRN cells in a 37 °C water bath.
2. Transfer to 25 mL of warm L-cell media and transfer into a T150 flask.

3. Allow the cells to grow in a 37 °C, 5% $CO_2$ incubator for 3 days without media change.
4. On day 3, aspirate media and wash with DPBS–EDTA.
5. Aspirate DPBS, add 1 mL trypsin–EDTA, and incubate for 5 min in the 37 °C incubator.
6. Tap the flask to ensure that cells have dislodged. Check under an inverted tissue culture microscope to confirm.
7. Add 12 mL of L-cell media to wash the cells for complete cell retrieval.
8. Add the above media containing cells to 140 mL of pre-warmed L-cell media containing 750 μl of 100 mg/mL Hygromycin B Gold and 1.5 mL of 50 mg/mL G418.
9. Swirl contents and distribute 25 mL each into six T150 flasks.
10. Allow the cells to grow for 3 days in a 37 °C, 5% $CO_2$ incubator.
11. Once cells reach confluency, follow the same procedure as above for trypsinizing cells from the flasks and transferring to the EasyFill Cell Factory.
12. Allow the cells to grow for 3 days in a 37 °C, 5% $CO_2$ incubator.
13. On day 3, wash the cell factory with 1000 mL of washing media before adding 1050 mL of primary culture media.
14. Incubate cell factory in a 37 °C, 5% $CO_2$ incubator for 24 h.
15. Collect the conditioned medium every 24 h from the cell factory into 50 mL conical tubes.
16. Replace with fresh primary culture media.
17. Spin down the condition medium in 50 mL tubes in a centrifuge at 1800 × *g*.
18. Transfer supernatant equally to each of the four glass media bottles avoiding the cell pellet.
19. Dilute 100% conditioned media 1:1 ratio with primary culture medium generating "50% conditioned medium", that is used for cultivation below.
20. Keep on a stir plate for 30 min to ensure uniform mixing.
21. Dispense 40 mL of 50% conditioned medium into 50 mL conical tubes. Store at −20 °C.

### 3.5 Passaging Mouse Spheroids

This is a very brief summary of the protocol described previously [10]. Growing the mouse and human spheroids (and mouse and human primary cell monolayers described in Subheadings 3.7 and 3.8) requires the use of conditioned medium produced by the L-WRN cell line. The production of this medium is described above and as previously published [10]; it requires the L-WRN cell line (ATCC CRL-3276). The final step of preparing conditioned medium is to add the conditioned medium to an equal volume of primary medium (recipe included in the *Materials* section); therefore, all conditioned medium used in these protocols is "50% conditioned medium."

1. Thaw vial of spheroids isolated and frozen as previously described [10]. Place vial in 37 °C water bath until contents are thawed (1–2 min).
2. Transfer cells to 15 mL conical containing 5 mL pre-warmed washing medium.
3. Centrifuge at 200 × *g* for 5 min.
4. Aspirate supernatant, being careful not to disturb pellet.
5. Add 1 mL washing medium and resuspend cells by gentle pipetting. Transfer to 1.5 mL microcentrifuge tube.
6. Centrifuge at 200 × *g* for 5 min.
7. Remove supernatant with a pipette, being careful not to disturb pellet.
8. Place tube on ice and chill for 5 min.
9. Add Matrigel to cells for desired volume. For example, if plating a 12-well plate, the volume of Matrigel/cell suspension for each well is 30 μL. Pipette cells gently, avoiding bubbles. Hold tube by rim while pipetting to avoid warming Matrigel.
10. Place tissue culture plate (e.g., a 12-well plate) on ice. This can be done inside or outside the biosafety cabinet.
11. Remove lid from plate and place on sterile surface.
12. Pipette the appropriate volume of Matrigel/cell suspension into the middle of each well. Carefully spread out Matrigel using a pipette tip; do not let Matrigel touch the sides of the well.
13. Cover plate with lid, remove from ice, flip upside down (this is done to prevent the cells from touching the plate bottom), and place into 37 °C incubator for 10 min. Matrigel will solidify.
14. Remove plate from incubator, flip right side-up, and add appropriate volume of 50% conditioned medium with 10 μM ROCK inhibitor (e.g., 800 μL/well for 12-well plate).
15. Place cells in 37 °C, 5% $CO_2$ incubator.
16. Two days after plating, aspirate medium and add fresh 50% conditioned medium.
17. Three days after plating, scrape Matrigel/cell suspension into medium using a cell scraper or pipette tip. Combine wells into 15 mL conical (6 wells of a 12-well plate can be combined into 1 conical).

18. Spin cells at 200 × *g* for 3 min.
19. Aspirate supernatant and resuspend cells in 5 mL DPBS–EDTA.
20. Spin cells at 200 × *g* for 5 min (longer spin time is necessary to ensure adequate exposure to DPBS–EDTA).
21. Aspirate supernatant and add 300 μL trypsin–EDTA to cells. Pipette two to three times to ensure even mix.
22. Add cells to 37 °C water bath, and incubate for 2–3 min.
23. Remove cells from water bath, pipette gently 10–12 times. Add 5 mL washing medium.
24. Spin cells at 200 × *g* for 3 min.
25. Aspirate supernatant and resuspend cells in 1 mL 50% conditioned medium with 10 μM ROCK inhibitor. Move cell suspension to 1.7 mL microcentrifuge tube.
26. Spin cells at 200 × *g* for 3 min.
27. Remove supernatant with a pipette tip. Place tube on ice. Let chill for 5 min.
28. Add the desired volume of Matrigel to cell pellet (volume is dependent on the number of wells being plated). Spheroid lines have different optimal passage ratios; for the murine ileal lines used to grow *C. parvum*, a 1:6 passage ratio is often used. Passage ratios will need to be adjusted in response to spheroid density; murine ileal spheroids should not be maintained at very high density, or else they will not grow well. Pipette Matrigel gently to mix, avoiding bubbles.
29. Plate spheroids as described in **Steps 1–15**.
30. Passage spheroids every 3 days.

### 3.6 Passaging Human Spheroids

This is an adaptation of the passaging mouse spheroid protocol (*see* Subheading 3.5) and as previously described [23]. The main modifications are the use of SB431542 in addition to ROCK inhibitor to complete the 50% conditioned medium, as well as the lower frequency of passaging due to slower cell growth.

1. Thaw vial of spheroids isolated and frozen as previously described [10]. Place vial in 37 °C water bath until contents are thawed (1–2 min).
2. Transfer cells to 15 mL conical containing 5 mL pre-warmed washing medium.
3. Centrifuge at 200 × *g* for 5 min.
4. Aspirate supernatant, being careful not to disturb pellet.
5. Add 1 mL washing medium and resuspend cells by gentle pipetting. Transfer to 1.5 mL microcentrifuge tube.

6. Centrifuge at 200 × *g* for 5 min.
7. Remove supernatant with a pipette, being careful not to disturb pellet.
8. Place tube on ice and chill for 5 min.
9. Add Matrigel to cells for the desired volume. For example, if plating a 12-well plate, the volume of Matrigel/cell suspension for each well is 30 μL. Pipette cells gently, avoiding bubbles. Hold tube by rim while pipetting to avoid warming Matrigel.
10. Place tissue culture plate (e.g., a 12-well plate) on ice. This can be done inside or outside the biosafety cabinet.
11. Remove lid from plate and place on sterile surface.
12. Pipette appropriate volume of Matrigel/cell suspension into the middle of each well. Carefully spread out Matrigel using a pipette tip; do not let Matrigel touch the sides of the well.
13. Cover plate with lid, remove from ice, flip upside down (this is done to prevent the cells from touching the plate bottom), and place into 37 °C incubator for 10 min. Matrigel will solidify.
14. Remove plate from incubator, flip right side-up, and add appropriate volume of 50% conditioned medium with 10 μM ROCK inhibitor and 10 μM SB431542 (e.g., 800 μL/well for 12-well plate).
15. Place cells in 37 °C, 5% $CO_2$ incubator.
16. Three and five days after plating, aspirate medium and add fresh 50% conditioned medium.
17. Seven days after plating, scrape Matrigel/cell suspension into medium using a cell scraper or pipette tip. Combine wells into 15 mL conical (6 wells of a 12-well plate can be combined into 1 conical).
18. Spin cells at 200 × *g* for 3 min.
19. Aspirate supernatant and resuspend cells in 5 mL DPBS–EDTA.
20. Spin cells at 200 × *g* for 5 min (a longer spin time is necessary to ensure adequate exposure to DPBS–EDTA).
21. Aspirate supernatant and add 300 μL TrypLE to cells. Pipette two to three times to ensure even mix.
22. Add cells to 37 °C water bath, incubate for 2–3 min.
23. Remove cells from water bath, pipette gently 10–12 times. Add 5 mL washing medium.
24. Spin cells at 200 × *g* for 3 min.
25. Aspirate supernatant and resuspend cells in 1 mL 50% conditioned medium with 10 μM ROCK inhibitor and

10 μM SB431542. Move cell suspension to 1.7 mL microcentrifuge tube.

26. Spin cells at 200 × *g* for 3 min.
27. Remove supernatant with a pipette tip. Place tube on ice. Let chill for 5 min.
28. Add the desired volume of Matrigel to cell pellet (volume is dependent on the number of wells being plated). Spheroid lines have different optimal passage ratios; for the human ileal lines used to grow *C. parvum* and *C. hominis*, a 1:6 passage ratio is often used. Passage ratios will need to be adjusted in response to spheroid density; murine ileal spheroids should not be maintained at a very high density, or else they will not grow well. Pipette Matrigel gently to mix, avoiding bubbles.
29. Plate spheroids as described in **Steps 10–15**.
30. Passage spheroids every 7 days.

### 3.7 Creating Mouse Primary Cell Monolayers for C. parvum Infection

1. Scrape spheroids and Matrigel into medium using a pipette tip or cell scraper. Combine wells with 5 mL plastic pipette, and move cell suspension into 15 mL conical tube. Six wells of a 12-well plate can be combined in one 15 mL tube.
2. Spin spheroids down at 200 × *g* for 3 min.
3. Remove medium and resuspend spheroids in 5 mL DPBS–EDTA. At this time, warm an aliquot of trypsin–EDTA in 37 °C water bath.
4. Spin spheroids down at 200 × *g* for 5 min (a longer spin time is necessary to ensure adequate exposure to DPBS–EDTA).
5. Remove DPBS–EDTA and add 300 μL warm trypsin–EDTA to pellet. Pipette up and down to ensure cells are evenly exposed to trypsin.
6. Move cells to 37 °C water bath, and incubate for 2–3 min.
7. Remove cells from bath and pipette 50 times to break up spheroids.
8. Add 5 mL washing medium to cells.
9. Place sterile cell strainer (40 μm) on top of 50 mL conical tube. Pre-wet strainer with 2 mL washing medium. Pipette cell mixture on top of strainer. Two 15 mL conical tubes worth of cells (equivalent to one 12-well plate of spheroids) can be combined onto one cell strainer.
10. Tap strainer against conical to force liquid through; move filtrate from 50 mL conical to 15 mL conical.
11. Spin at 200 × *g* for 3 min.
12. Remove washing medium and resuspend cells in 1–3 mL of 50% conditioned medium with 10 μM ROCK inhibitor

(Y-27632; *see* **Note 4**). Count cells using a hemocytometer. Cells might be clumpy. Each transwell will be plated with $5 \times 10^4$ cells in 200 μL medium. Resuspend cells accordingly (*see* **Note 5**).

13. Remove transwells plated with i3T3 cells from 37 °C incubator; confirm that the i3T3 monolayer is uniform by examining it under an inverted light microscope equipped with phase-contrast optics (4× objective, 10× ocular).
14. Remove the 3T3 growth medium from the top chamber of the compartment using a pipette tip (do not use vacuum flask because aspiration will disturb the monolayer).
15. Add stem cells to transwells.
16. Remove the 3T3 growth medium from the bottom chamber of the transwell compartment, and add 400 μL conditioned medium with 10 μM ROCK inhibitor.
17. Move transwells to 37 °C, 5% $CO_2$ incubator.
18. Change the medium in both the top and bottom chambers of transwells every 2 days. Add 200 and 400 μL 50% conditioned medium with 10 μM ROCK inhibitor to the top and bottom chambers, respectively. Remove medium from the top chamber with a pipette tip only; aspiration can disturb cell layer. Cells can tolerate a 2-day gap between feedings; medium can be changed on a Monday–Wednesday–Friday schedule with no ill effects.
19. After 7 days, remove medium from the top chamber with a pipette tip and do not replace. This is "day 0 of top media removal" and the creation of the ALI.
20. Continue to change medium in the bottom chamber three times weekly (*see* **Note 6**).
21. On day 3 post-top media removal, remove liquid/mucus from the top chamber of transwells (*see* **Note** 7). Resuspend the desired infectious dose of *C. parvum* oocysts or sporozoites in 30 μL 50% conditioned medium per transwell. Add oocysts or sporozoites to transwell.
22. After incubation period (1–2 h for sporozoites, 3–4 h for oocysts), remove liquid and wash with DPBS if required for assay.

### 3.8 Creating Human Primary Cell Monolayers for C. parvum and C. hominis Infection

1. Scrape spheroids and Matrigel into medium using a pipette tip or cell scraper. Combine wells with 5 mL plastic pipette and move cell suspension into 15 mL conical tube. Four wells of a 12-well plate can be combined in one 15 mL tube.
2. Spin spheroids down at 200 × *g* for 3 min.

3. Remove medium and resuspend spheroids in 5 mL DPBS–EDTA. At this time, warm TrypLE in 37 °C water bath.
4. Spin spheroids down at 200 × *g* for 5 min (a longer spin time is necessary to ensure adequate exposure to DPBS–EDTA).
5. Remove DPBS–EDTA and add 300 μL warm TrypLE to pellet. Pipette up and down to ensure cells are evenly exposed to trypsin.
6. Move cells to 37 °C water bath, and incubate for 2–3 min.
7. Remove cells from bath and pipette 50 times to break up spheroids.
8. Add 5 mL washing medium to cells.
9. Place sterile cell strainer (40 μm) on top of 50 mL conical tube. Pre-wet strainer with 1 mL washing medium. Pipette cell mixture on the top of strainer. No more than one 15 mL conical tubes worth of cells (equivalent to 4 wells of a 12-well plate of spheroids) should be combined onto one cell strainer.
10. Tap strainer against conical to force liquid through; if the strainer seems clogged, pipetting up and down a few times can help. Move filtrate from 50 mL conical to 15 mL conical.
11. Spin at 200 × *g* for 3 min.
12. Remove washing medium, and resuspend cells in 1–3 mL of 50% conditioned medium with 10 μM ROCK inhibitor (Y-27632; *see* **Note 4**) and 10 μM SB431542 (*see* **Note 8**). Count cells using a hemocytometer. Cells might be clumpy. Each transwell will be plated with $5 \times 10^4$ cells in 200 μL medium. Resuspend cells accordingly (*see* **Note 5**).
13. Remove transwells plated with i3T3 cells from 37 °C incubator; confirm that the i3T3 monolayer is uniform by examining with an inverted light microscope equipped with phase-contrast optics (4× objective, 10× ocular).
14. Remove 3T3 growth medium from the top chamber of the compartment using a pipette tip (do not use vacuum flask because aspiration will disturb the monolayer).
15. Add stem cells to transwells.
16. Remove 3T3 growth medium from the bottom chamber of the transwell compartment, and add 400 μL conditioned medium with 10 μM ROCK inhibitor.
17. Move transwells to 37 °C, 5% $CO_2$ incubator.
18. Change medium in the top and bottom chambers of transwells every 2 days. Add 200 and 400 μL 50% conditioned medium with 10 μM ROCK inhibitor to the top and bottom chambers, respectively. Remove medium from the top chamber using a pipette tip only; aspiration can disturb cell layer. Cells can

tolerate a 2-day gap between feedings; medium can be changed on a Monday–Wednesday–Friday schedule with no ill effects.

19. After 7 days, remove medium from the top chamber with a pipette tip and do not replace. This is "day 0 of top media removal" and the creation of the ALI.
20. Continue to change medium in the bottom chamber three times weekly (*see* **Note 6**).
21. On day 3 post-top media removal, remove liquid/mucus from the top chamber of transwells (*see* **Note 7**). Resuspend the desired infectious dose of *C. parvum* oocysts or sporozoites in 30 μL 50% conditioned medium per transwell. Add oocysts or sporozoites to transwell.
22. After the incubation period (1–2 h for sporozoites, 3–4 h for oocysts), remove liquid and wash with DPBS if required for assay.

### 3.9 Preparing and Maintaining Cryptosporidium Oocysts for Infection

We have used the AUCP-1 isolate of *C. parvum* and the TU502 isolate of *C. hominis* for growth in the ALI system (*see* **Note 9**), and other lines would need to be tested for efficiency. It is a requirement of our biosafety protocol to use plastic caps on the centrifuge buckets when spinning the oocysts. Only open the caps inside a biosafety cabinet. Autoclave the buckets and caps after the protocol in case of accidental contamination. All pipette tips and other materials contaminated with oocysts should be disposed of in a container that can be sealed before removing from the biosafety cabinet and placing in biohazard waste. An absorbent mat pad can be placed in the biosafety cabinet during this procedure in case of spills.

1. Store purified oocysts at 4 °C in 50 mM Tris-10 mM EDTA, pH 7.2.
2. Before infection, dilute $1 \times 10^8$ oocysts into 1 mL cold DPBS in a 15 mL conical. Place on ice.
3. Add 1 mL of 40% commercial bleach (diluted in cold DPBS) to oocysts. Pipette multiple times to ensure even mix.
4. Bleach oocysts on ice for 10 min.
5. Spin at 800 × *g* for 3 min. Remove bleach, and resuspend pellet in cold DPBS +1% BSA. Move the oocyst mixture to a new 15 mL conical.
6. Spin at 800 × *g* for 3 min. Remove supernatant and resuspend pellet in cold DPBS +1% BSA. Repeat the centrifugation and washing steps three times to remove all traces of bleach.
7. After the final centrifugation step, resuspend oocysts in cold DPBS +1% BSA at a concentration of $1 \times 10^8$/mL. Bleached

oocysts can be stored at 4 °C for up to 2 weeks. After 2 weeks, the excystation rate will decline.

8. Oocysts can be diluted into cell culture medium at a desired concentration for infection.
9. If sporozoites are needed for infection, add the required amount of oocysts to the same volume of excystation buffer in a 15 mL conical. For example, if 500 μL of oocysts ($0.5 \times 10^8$ oocysts total) is required, add the oocysts to 500 μL of excystation buffer. The final concentration of sodium taurocholate in the solution will be 0.75% w/v.
10. Incubate oocysts in 37 °C water bath for 45–60 min. Check excystation rate by pipetting 10 μL onto glass slide, cover sporozoites with glass coverslip, and look under an inverted light microscope equipped with phase-contrast optics (20× objective, 10× ocular). Sporozoites are best seen at ≥400× combined magnification. Excystation rate should be high (80–100%) by 60 min. Lower excystation rates indicate that the viability of the oocysts is declining and the subsequent growth in vitro will not be optimal.
11. After the incubation period is over, remove plunger from 3 or 5 mL plastic syringe and place upright in biosafety cabinet. Attach syringe to plastic filter holder containing filter membrane with 1 μm pore size. Place filter with syringe on top of the 15 mL conical.
12. Pipette 1 mL of cell growth medium into syringe. Pipette excysted oocysts into syringe. Insert plunger into syringe and push contents through filter. Remove filter with syringe and place aside. Filter can be carefully disassembled and dropped into an autoclave-safe container filled with $H_2O$ and autoclaved. Autoclave cycle must not include any drying time, because this can warp the filters. Membrane can be discarded after autoclaving.
13. Spin-filter sporozoites at 1250 × *g* for 3 min.
14. Remove supernatant and resuspend sporozoites in the desired volume of cell growth medium. For infecting ALI monolayers, resuspend sporozoites in 50% conditioned medium (without ROCK inhibitor), with 30 μL per transwell.
15. To confirm that oocysts have been completely removed from the inoculum, check filtered sporozoites under an inverted tissue culture microscope (20× objective, 10× ocular) by pipetting onto glass slide and covering with glass coverslip. Infected monolayers can also be fixed on the same day of infection and stained with an antibody that recognizes *C. parvum* oocysts.

### *3.10 Measuring Host and Parasite Replication in Primary Cell Monolayers by qPCR*

This method relies on measuring parasite replication by quantifying *C. parvum* genomic DNA equivalents. Host viability can be assessed by measuring host genomic DNA equivalents from the same sample. The following is a modified protocol from the Qiagen QIAamp DNA Mini Kit. Buffers ATL, AL, AW1, AW2, and AE are provided in the kit.

1. Add 100 μL of Buffer ATL to transwell. Thoroughly scrape cells into buffer with blunt pipette tip; be careful not to puncture transwell membrane. Lysed cells will be sticky and clumpy.
2. Move the entire contents of the transwell (will likely be more than 100 μL due to the presence of mucus/liquid in the top chamber) to a 1.7 mL microcentrifuge tube. Add 20 μL proteinase K. Vortex to mix.
3. Incubate in 56 °C water bath for 3–24 h.
4. Briefly centrifuge tube to remove drops from inside the lid.
5. Add 200 μL Buffer AL to sample, and vortex briefly to mix. A white precipitate is likely to form. Briefly centrifuge tube to remove drops from inside the lid.
6. Incubate in 70 °C heat block for 10 min.
7. Add 200 μL ethanol to sample, and vortex briefly to mix. Briefly centrifuge tube to remove drops from inside the lid.
8. Move sample from microcentrifuge tube to spin column. When pipetting sample, be careful not to wet the rim of the spin column. Spin at 7000 × *g* for 1 min.
9. Place the spin column in a new collection tube, and discard the tube containing filtrate.
10. Add 500 μL Buffer AW1 to spin column (be sure not to wet the rim). Centrifuge at 7000 × *g* for 1 min.
11. Place the spin column in a new collection tube, and discard the tube containing filtrate.
12. Add Buffer AW2 to spin column (be sure not to wet the rim). Centrifuge at maximum speed for 3 min.
13. Place the spin column in a clean 1.7 mL microcentrifuge tube, and discard the tube containing filtrate. Add 100 μL Buffer AE to the spin column, incubate at room temperature for 1 min, and then centrifuge at 7000 × *g* for 1 min.
14. Dilute DNA sample tenfold with sterile $H_2O$.
15. To make primer mix for qPCR reaction mix: Make 200 μM master stocks of forward and reverse primers in sterile $H_2O$. Combine equal volume of forward and reverse master stocks to make 100 μM master mix. Dilute 100 μM master mix in sterile $H_2O$ to make 5 μM working mix.

16. To make qPCR reaction mix: 9.0 μL 2× SYBR Green mix (or equivalent), 1.8 μL 5 μM primer mix, 5.2 μL $H_2O$ (16 μL total volume). Scale up volume for the number of wells needed. Pipette mix into the required number of wells of a 96-well qPCR reaction plate.
17. Add 2 μL of diluted DNA sample to well. Seal plate with optical film. Centrifuge briefly to ensure DNA and reaction mix are combined.
18. Primers to measure *Cryptosporidium* GAPDH: Forward—GAA GATGCGCTGGGAACAAC ; reverse— CGGATGGCCA TACCTGTGAG.
19. Primers to measure mouse GAPDH: Forward— GCCAT GAGTGGACCCTTCTT ; reverse— GAAAACACGGGGG CAATGAG.
20. Primers to meqsure human GAPDH: Forward— TGAG TACGTCGTGGAGTCCA ; reverse—CTGAGTCAGCTTC CCCTCC.
21. qPCR thermocycler protocol (*see* Table 1).
22. Place reaction plate in qPCR machine and run thermocycler protocol.
23. In order to correlate Ct values with actual genomic DNA numbers, it is necessary to make a standard for *Cryptosporidium* and mouse cells. For *Cryptosporidium*, DNA from a known number of oocysts can be extracted using the Qiagen QIAamp DNA Mini Kit and diluted to make a standard. For example, DNA from $1 \times 10^7$ oocysts can be extracted. After extraction, the number of gDNA equivalents will be $4 \times 10^7$ per 100 μL (since each oocyst contains 4 sporozoites and the DNA is eluted in a 100 μL volume) or $4 \times 10^5$/μL. Create a dilution series (1:2 to 1:16, for a total of 5 values including the undiluted sample to make up the curve). For mouse cells, a

**Table 1**
**qRT-PCR thermal cycling protocol**

| Reaction step | | Temperature | Duration |
|---|---|---|---|
| Priming | | 95 °C | 2 min |
| Three-step thermal cycles (40×) | Denaturing | 95 °C | 10 s |
| | Annealing | 60 °C | 20 s |
| | Extension | 72 °C | 30 s (read fluorescence) |
| Melt curve | | 95 °C | 15 s |
| | | 60–95 °C | 15 s |
| Hold | | 4 °C | |

known number of cells can be lysed after the cells are quantified in Subheading 2.2, **step 12** and a dilution series set up in the same way as for *Cryptosporidium*.

24. Because the DNA sample was diluted 10× before running the qPCR reaction, the number of gDNA equivalents determined to be present in 2 μL of sample needs to be multiplied by 10 to be accurate.

### 3.11 Monitoring Life Cycle Progression in Primary Cell Monolayers by Fluorescence Microscopy

1. Move transwells to a 24-well plate. Add 500 μL PBS to the bottom chamber of transwell.
2. Fix transwells in the desired fixative. For example, 4% formaldehyde for 10 min.
3. Wash transwells 2× with PBS. Use pipette to remove liquid from transwell because aspiration may disturb cell layer. If there is a significant mucus layer present (likely if the transwells are at day 7 or later after top media removal), the mucus layer can be removed through pipetting.
4. After fixing, transwells can be stored at 4 °C for up to 10 days before staining (helpful in experiments that span over multiple days).
5. Permeabilize and block cells with desired agents. For example, 0.1% Triton X-100 for 3–5 min or 0.05% saponin for 10 min. If using Triton X-100, dilute in PBS containing 1% BSA. This buffer can be used throughout the staining protocol as a blocking buffer and as a vehicle for primary and secondary antibodies. If using saponin, block with 0.05% saponin in PBS containing 5% normal goat (NGS) and 5% fetal bovine serum (FBS). For antibody dilution and wash steps, drop the saponin concentration down to 0.01% in PBS containing 1% NGS.
6. After blocking, dilute primary antibody in buffer and stain for 60 min at room temperature or overnight at 4 °C.
7. Wash 3× with buffer.
8. Dilute secondary antibody (e.g., goat anti-mouse IgG conjugated to Alexa Fluor dyes) in buffer, and stain for 60 min at room temperature. Protect from light.
9. Wash 3× with buffer.
10. Stain with 1 μg/mL Hoechst 33258 (1:1000 dilution of stock solution) for 10–30 min. Protect from light.
11. Wash 1× with PBS.
12. Dot 30–50 μL mounting medium fluorescence anti-quenching agent onto glass slide. Cut membrane from transwell insert with scalpel. Using forceps, place membrane onto mounting medium, cell side up.

13. Place square coverslip over membrane and push firmly down to remove air pockets. If there is an air pocket that will not go away, check membrane for plastic "curlicues" that can be created during the cutting process—these will stop the coverslip from lying flush with the membrane. Two membranes can fit under one coverslip, but this increases the chance that the coverslip will not lie flush with the membranes, and air pockets can develop overnight as the mounting medium dries. It is important to thoroughly press down on the coverslip to remove all bubbles. Excess mounting medium will be pushed out and can be wiped off.
14. Protect slides from light and let dry overnight at room temperature.
15. Examine using a light microscope equipped with epi-fluorescence illumination and appropriate filters. An example of fluorescent staining of *C. parvum* grown in mouse ALI monolayers is shown in Fig. 1c.

## 4 Notes

1. Stocks of 3T3 cells should be kept at low passage numbers (<10). Check the appearance of 3T3 cells; unhealthy cells will grow slowly and have vacuolated cytoplasm.
2. Original cells from Lonza are not available commercially anymore, but are being maintained in laboratories, such as ESCAPE, UR7510 (University of Reims, Champagne Ardenne, France). Initial provider suggested to keep passage number under 15, even though higher passage numbers have been reported [24].
3. Matrigel typically arrives from the manufacturer in glass bottles containing 10 mL and should be stored at −20 °C. To prepare Matrigel aliquots, thaw Matrigel overnight on ice in a polystyrene foam container in a 4 °C fridge. Remove cap and pipette Matrigel up and down three times using 10 mL pipette (only draw up 9 mL to avoid creating bubbles). Aliquot Matrigel in 1 mL aliquots in 1.7 mL microcentrifuge tubes and store at −20 °C. Move aliquots to 4 °C to thaw. Do not keep aliquots in "warm spots" in fridge such as the door because they may solidify. Aliquots can be kept at 4 °C for several weeks, but it is best to use them quickly. Frozen aliquots can be thawed rapidly (30–60 min) by placing in an ice slurry or placing on top of crushed ice. Do not thaw aliquots in 37 °C water bath or room temperature because the Matrigel will solidify. Solidified Matrigel can be refrozen at −20 °C and will thaw into a liquid, though it is not clear if this negatively affects the product.

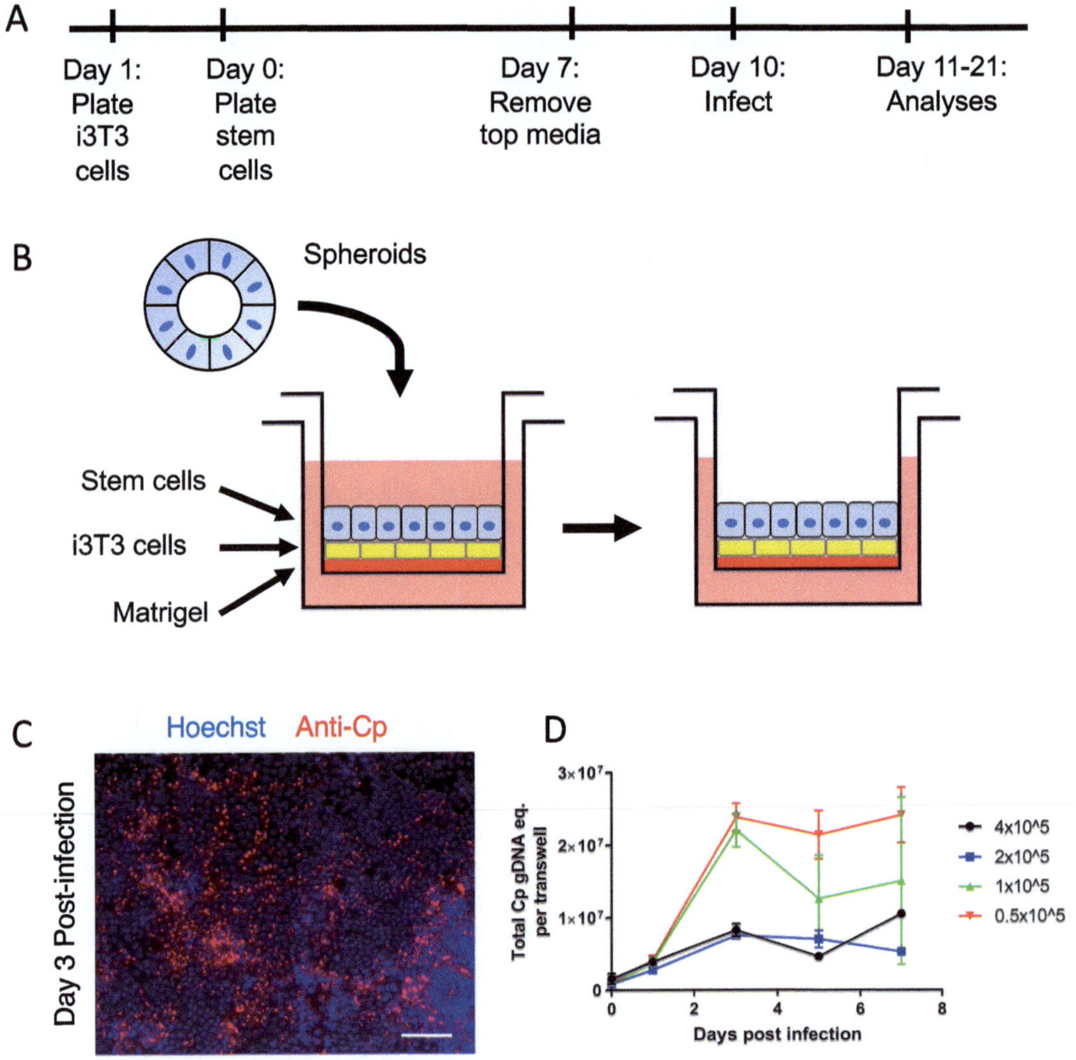


**Fig. 1** Timeline and schematic for the production of ALI monolayers and infection with *C. parvum* (**a**) Timeline for production of irradiated 3T3 feeder cells, plating of transwells to form spheroids, and construction of ALI cultures. Optimal production of monolayers is observed with submerged culture in 50% conditioned medium for 7 days, followed by removal of the top medium to allow the ALI to form. Optimal infection is achieved by infecting at 3 days post-top medium removal (**b**) Schematic of the construction of ALI monolayers from feeder cells and spheroids (i.e., stem cells). Irradiated 3T3 cells (feeder cells) are plated on transwell filters containing a layer of Matrigel. The feeder cells are overlaid with a layer of stem cells isolated from parallel cultures of spheroids. The top and bottom chambers both contain 50% conditioned medium. After 7 days, the medium is removed from the top chamber to allow development of the ALI culture (**c**) Immunofluorescent staining of *C. parvum*-infected ALI monolayers. ALI monolayers were infected on day 3 post-top media removal with $2 \times 10^5$ oocysts. Monolayers were incubated with oocysts for 4 hr. and then washed three times. On 3 days post-infection, monolayers were fixed and stained with pan-Cp (rabbit polyclonal raised against *C. parvum* oocysts and sporozoites) and Hoechst. Scale bar = 50 μm (**d**) Lower seeding density improves *C. parvum* growth. Stem cells were seeded at different densities. Seven days post-seeding, top medium was removed to create ALI cultures. Three days post-top media removal, monolayers were infected with $2 \times 10^5$ oocysts. Monolayers were incubated with oocysts for 4 h and then washed three times. DNA samples were collected every other day. The graph depicts means ± S.D. for qPCR measurements of *C. parvum* in DNA samples, where each data point represents the mean ± S.D. of two transwells.

When handling thawed aliquots of Matrigel, do not grip the tube with hand, because this will cause it to solidify. Keep the aliquots on ice as much as possible. Always dilute Matrigel into cold DPBS; if the DPBS is warm, the Matrigel will solidify upon contact and will not create a homogenous solution.

4. Y-27632 arrives from the manufacturer as a powder that can be stored at room temperature. Add sterile $H_2O$ to the powder to make 10 mM stock solution. Aliquot 50–150 μL amounts and store at −20 °C. Aliquots can be thawed and refrozen. Do not store aliquots at 4 °C.
5. The amount of spheroids required for different numbers of transwells can vary slightly from run-to-run. Fifteen to 20 wells (12-well plate format) of healthy spheroids should give enough cells to plate 48 transwells at the $5 \times 10^4$ seeding density. If it takes significantly more wells, the spheroids are likely too sparse. The $5 \times 10^4$ seeding density was determined to be optimal for *C. parvum* growth (*see* Fig. 1d); higher seeding densities lead to decreased cell growth. If *C. parvum* is not growing well, try a dilution series of different seeding densities.
6. It is important to observe the color of the medium when feeding the cells, since it is an indicator of viability. Four to six days after plating the stem cells, the medium should become bright yellow in between feedings. After the ALI is created (7 days after plating the stem cells), the medium will continue to turn yellow between feedings for 3–5 days, but the color change will gradually become less dramatic. Eventually the medium will remain pink in between feedings. If the medium is remaining pink before the creation of the ALI, it is likely the cells are unhealthy and the monolayer is too sparse. Additionally, mucus will form in the top chamber after the start of the ALI; copious liquid in the top chamber in addition to pink medium indicates a dead monolayer.
7. The day of infection strongly affects the outcome of infection. Infecting too early after the creation of the ALI (i.e., day 0) leads to poor growth and cell death. Infecting too late (day 7 and later) leads to poor growth, possibly due to the presence of the mucus layer, which may act as a physical barrier to the oocysts and sporozoites. Optimal growth is seen when cultures are infected at day 3 after medium removal.
8. SB431542 arrives from the manufacturer as a powder that can be stored at room temperature. Add DMSO to the powder to make 10 mM stock solution. Aliquot 50–150 μL amounts and store at −20 °C. Aliquots can be thawed and refrozen. Do not store aliquots at 4 °C.

9. In our experience, it is preferable that the *C. parvum* oocysts are purified from cow feces using Sheather's sugar flotation and discontinuous sucrose density gradient centrifugation; oocysts purified using cesium chloride gradients are less viable and exhibit lower excystation and infection rates in the ALI system.

## Acknowledgments

Supported by grants from The Open Philanthropy—Good Ventures (GV67360389) and The Open Philanthropy—Silicon Valley Community Foundation (2023-327770). G. Wilke was partially supported by an institutional training grant to Washington University (AI007172). We are grateful to Mark Kuhlenschmidt, Theresa Kuhlenschmidt, Lisa Funkhouser-Jones, and Kelli vanDussen for their helpful comments and assistance during the initial phase of this project.

**Competing Interest Statement** The authors declare no conflicts of interest relevant to the content of this chapter.

### References

1. Arrowood MJ (2002) In vitro cultivation of *Cryptosporidium* species. Clin Microbiol Rev 15:390–400
2. Flanigan TP, Aji T, Marshall R et al (1991) Asexual development of *Cryptosporidium parvum* within a differentiated human enterocyte cell line. Infect Immun 59:234–239
3. Gut J, Petersen C, Nelson R et al (1991) *Cryptosporidium parvum*: in vitro cultivation in Madin-Darby canine kidney cells. J Protozool 38:72S–73S
4. Sow SO, Muhsen K, Nasrin D et al (2016) The burden of *Cryptosporidium* diarrheal disease among children <24 months of age in moderate/high mortality regions of Sub-Saharan Africa and South Asia, utilizing data from the Global Enteric Multicenter Study (GEMS). PLoS Negl Trop Dis 10:e0004729
5. Chalmers RM, Smith R, Elwin K et al (2011) Epidemiology of anthroponotic and zoonotic human cryptosporidiosis in England and Wales, 2004-2006. Epidemiol Infect 139: 700–712
6. Cacciò SM, Chalmers RM (2016) Human cryptosporidiosis in Europe. Clin Microbiol Infect 22:471–480
7. Morada M, Lee S, Gunther-Cummins L et al (2016) Continuous culture of *Cryptosporidium parvum* using hollow fiber technology. Int J Parasitol 46:21–29
8. DeCicco RePass MA, Chen Y, Lin Y et al (2017) Novel bioengineered three-dimensional human intestinal model for long-term infection of *Cryptosporidium parvum*. Infect Immun 85:e00731–e00716
9. Miyoshi H, Ajima R, Luo CT et al (2012) Wnt5a potentiates TGF-β signaling to promote colonic crypt regeneration after tissue injury. Science 338:108–113
10. Miyoshi H, Stappenbeck TS (2013) In vitro expansion and genetic modification of gastrointestinal stem cells in spheroid culture. Nat Protoc 8:2471–2482
11. Moon C, VanDussen KL, Miyoshi H et al (2014) Development of a primary mouse intestinal epithelial cell monolayer culture system to evaluate factors that modulate IgA transcytosis. Mucosal Immunol 7:818–828
12. VanDussen KL, Marinshaw JM, Shaikh N et al (2015) Development of an enhanced human gastrointestinal epithelial culture system to facilitate patient-based assays. Gut 64:911–920
13. Wilke G, Ravindran S, Funkhouser-Jones L et al (2018) Monoclonal antibodies to intracellular stages of *Cryptosporidium parvum* define

life cycle progression in vitro. mSphere 3: e00124–e00118

14. Heo I, Dutta D, Schaefer DA et al (2018) Modelling *Cryptosporidium* infection in human small intestinal and lung organoids. Nat Microbiol 3:814–823
15. Wang X, Yamamoto Y, Wilson LH et al (2015) Cloning and variation of ground state intestinal stem cells. Nature 522:173–178
16. Whitcutt MJ, Adler KB, Wu R (1988) A biphasic chamber system for maintaining polarity of differentiation of cultured respiratory tract epithelial cells. In Vitro Cell Dev Biol 24:420–428
17. Pezzulo AA, Starner TD, Scheetz TE et al (2011) The air-liquid interface and use of primary cell cultures are important to recapitulate the transcriptional profile of in vivo airway epithelia. Am J Physiol Lung Cell Mol Physiol 300:L25–L31
18. O'Boyle N, Sutherland E, Berry CC et al (2017) Temporal dynamics of ovine airway epithelial cell differentiation at an air-liquid interface. PLoS One 12:e0181583
19. Nossol C, Diesing A-K, Walk N et al (2011) Air-liquid interface cultures enhance the oxygen supply and trigger the structural and functional differentiation of intestinal porcine epithelial cells (IPEC). Histochem Cell Biol 136:103–115
20. Klasvogt S, Zuschratter W, Schmidt A et al (2017) Air-liquid interface enhances oxidative phosphorylation in intestinal epithelial cell line IPEC-J2. Cell Death Discov 3:17001
21. Wilke G, Funkhouser-Jones LJ, Wang Y et al (2019) A stem-cell-derived platform enables complete *cryptosporidium* development in vitro and genetic tractability. Cell Host Microbe 26:123–134.e8
22. Kimball A, Funkhouser-Jones L, Huang W et al (2024) Mendelian segregation and high recombination rates facilitate genetic analyses in *Cryptosporidium parvum*. PLoS Genet 20: e1011162
23. Greigert V, Saraav I, Son J et al (2024) *Cryptosporidium* infection of human small intestinal epithelial cells induces type III interferon and impairs infectivity of rotavirus. Gut Microbes 16:2297897
24. Kurahara LH, Hiraishi K, Hu Y et al (2017) Activation of myofibroblast TRPA1 by steroids and pirfenidone ameliorates fibrosis in experimental Crohn's disease. Cell Mol Gastroenterol Hepatol 5:299–318

# Chapter 21

# Two- and Three-Dimensional Human Intestinal Tissue Models for *Cryptosporidium* Infection

**Daviel Cardenas, Seema Bhalchandra, Hymlaire Lamisere, and Honorine D. Ward**

## Abstract

Conventional cell cultures utilizing transformed or immortalized cell lines or primary human epithelial cells have played a fundamental role in furthering our understanding of *Cryptosporidium* infection. However, they remain inadequate with respect to their inability to emulate in vivo conditions and support long-term growth and completion of the parasite's life cycle. Previously, we developed a three-dimensional (3D) silk scaffold-based model using transformed human intestinal epithelial cells (IECs). This model supported *C. parvum* infection for up to 2 weeks and resulted in completion of the life cycle of the parasite. However, transformed IECs are not representative of primary human IECs.

Human intestinal organoids (hIOs) or human intestinal enteroids (hIEs) are 3D cultures derived from $Lgr5^{+}$ stem cells isolated from the crypts of human intestinal biopsies or surgical intestinal tissues. These multicellular cultures can be induced to differentiate into enterocytes, enteroendocrine cells, goblet cells, Paneth cells, and tuft cells. hIEs better represent human intestinal structure and function than immortalized, IEC lines. Recently, significant progress has been made in the development of technologies to culture hIEs in vitro. When grown in a 3D matrix, hIEs provide a spatial organization resembling the native human intestinal epithelium. Additionally, they can be dissociated and grown as monolayers in tissue culture plates, permeable supports, or silk scaffolds that enable mechanistic studies of pathogen infections. They can also be co-cultured with other human cells such as macrophages, myofibroblasts, and immune cells. The hIEs grown in these novel culture systems recapitulate the physiology, the 3D architecture, and functional diversity of native intestinal epithelium and provide a powerful and promising new tool to study *Cryptosporidium*-host cell interactions and screen for interventions ex vivo. In this chapter, we describe a 3D silk scaffold-based model using transformed IECs co-cultured with human intestinal myofibroblasts as well as 2D and 3D hIE-derived models of *C. parvum* infection.

**Key words** Two dimensional (2D), Three dimensional (3D), Intestinal, Tissue, Organoid, Enteroid, Stem cell, Monolayer, Drug screen, Permeable support, Transwell, Silk scaffold

---

Daviel Cardenas, Seema Bhalchandra and Hymlaire Lamisere contributed equally.

Jan R. Mead and Michael J. Arrowood (eds.), *Cryptosporidium: Methods and Protocols*, Methods in Molecular Biology, vol. 2978, https://doi.org/10.1007/978-1-0716-4824-7_21,

## 1 Introduction

*Cryptosporidium*, an intestinal apicomplexan parasite, is a major cause of diarrhea and death in young children and immunocompromised hosts in low- and middle-income countries and causes waterborne outbreaks worldwide [1–3]. Pathogenic mechanisms underlying *Cryptosporidium* infection and host responses are poorly understood due in part to a lack of appropriate human model systems that accurately recapitulate the complex multicellular intestinal epithelium. Existing models of *C. parvum* infection in vitro employ transformed cancer-derived IECs or primary human intestinal epithelial cells on 2D [4–6] or 3D culture platforms [7, 8]. However, most of these culture models can only support short-term (<5 days) infection and do not allow continuous propagation of the parasite or completion of its life cycle [9]. Previously, we developed a 3D silk scaffold-based model for *C. parvum* infection using Caco-2 and HT29-MTX transformed IECs, which supported infection for up to 2 weeks and permitted completion of the parasite life cycle [7]. However, these IEC lines are not representative of native intestinal tissue in many ways since they lack the cellular diversity encountered in the intestinal epithelium and display variations in protein and gene expression [10, 11]. Therefore, an alternative source of primary human epithelial cells is needed to model physiological 3D human intestinal tissue ex vivo.

In seminal work, Sato et al. showed that intestinal crypts containing leucine-rich repeat-containing G-protein coupled receptor 5 positive ($Lgr5^{+}$) stem cells can be cultured indefinitely as multicellular intestinal epithelial cultures that form self-organizing 3D units called intestinal organoids [12] or later enteroids [13]. Organoids/enteroids accurately recapitulate the cell complexity and functionality that are observed in vivo and can differentiate into enterocytes, goblet cells, enteroendocrine cells, and Paneth cells [12, 14, 15]. M cells, which are specialized cells located in the Peyers' patches of the intestinal epithelium and function as important antigen-presenting cells, can also be differentiated from hIEs by supplementing the medium with receptor activator of NFκB ligand (RANKL) [16]. In addition, differentiated hIEs polarize to form brush border microvilli apically, display intact tight junctions, and exhibit basolateral expression of $NA^{+}/K^{+}$ ATPase, NKCC1, β-catenin, and E-cadherin [17–19].

hIOs or hIEs provide a novel way to study host–microbe interactions [17, 19, 20] or innate intestinal immune responses [21] in a number of different formats. Their 3D architecture preserves the physiological nature of intestinal epithelia, enabling them to be utilized as model systems for evaluating the dynamic processes occurring at the host–microbe interface. hIEs form 3D units naturally in culture but can also be mechanically dissociated and seeded as monolayers onto tissue culture plates, transwell inserts, or bioengineered silk scaffolds as 2D and 3D models respectively grown with or without other cell types such as human intestinal myofibroblasts (H-InMyoFib) [22, 23]. These culture platforms have distinct apical/basolateral cell surfaces allowing easy access to the apical surface for infection. Since cell-seeding densities can be standardized, these systems permit reproducible and rapid large-scale analyses of the interaction between pathogens and host cells. Recently, human intestinal and lung organoids have been shown to support the growth of *C. parvum* [24–28]. Murine intestinal organoids have also been used to model and study *C. parvum* infection [29] and utilized in co-culture with intraepithelial lymphocytes to study innate immune responses [30]. Thus, by incorporating hIEs (with or without co-culture with other human or murine cells) into 2D and 3D model systems, we can better study host–*C. parvum* interactions during infection and discover and test new interventions against this pathogen.

Here, we describe methods for growing transformed human IECs and hIEs and incorporating them into 2D and 3D models to study host–*C. parvum* interactions. In particular, we focus on the use of 2D models of differentiated and undifferentiated hIEs to study the effect of *C. parvum* infection on intestinal epithelial integrity [31]. The hIEs grown in these novel culture systems recapitulate the physiology, 3D architecture, and functional diversity of native intestinal epithelium and provide a powerful and promising new tool to study *Cryptosporidium*–host cell interactions and screen for interventions in vitro or ex vivo.

## 2 Materials

All materials are sterile, and all procedures are performed in a biosafety cabinet under biosafety level 2 conditions, unless otherwise stated.

### 2.1 Reagents, Buffers, Working Solutions, and Kits

1. 4′,6-Diamidino-2-phenylindole (DAPI): prepared as a 300 nM solution by diluting a 1000× stock in phosphate-buffered saline (PBS; 137 mM NaCl, 2.7 mM KCl, 10 mM $Na_2HPO_4$, 1.8 mM $KH_2PO_4$, pH 7.4).
2. Antibiotic–antimycotic: 10,000 U/mL penicillin, 10,000 μg/mL streptomycin, and 25 μg/mL amphotericin B.
3. Antibodies:
   (a) Primary antibodies
      (i) Markers of IEC differentiation are detected using monoclonal antibodies (mAbs) against chromogranin A, lysozyme, mucin 2 (MUC-2), and sucrose isomaltose (all from Santa Cruz Biotechnology, Dallas, TX). A mAb to LgrGR5 (Invitrogen, Thermo Fisher Scientific, Waltham, MA) is used as a marker of stem cells.

(ii) Markers of tight or adherens junction proteins can be detected using mAbs against Zonula Occludens-1 (ZO-1) (Invitrogen) or E-cadherin respectively (BD Biosciences, Woburn, MA).

(iii) Crypt-a-Glo Antibody (Waterborne, Inc., New Orleans, LA).

(iv) Sporo-Glo Antibody, (Waterborne, Inc., New Orleans, LA).

(b) Secondary antibodies

(i) Goat anti-mouse IgG or IgM or goat anti-rabbit IgG, as appropriate, conjugated to fluorescent dyes Alexa Fluor 488, 594, or 647 (Invitrogen).

4. Blocking/permeabilization solution: 0.5% Triton X-100, 5% normal goat serum in PBS.
5. Bovine serum albumin (BSA): 5% solution in PBS to block non-specific binding and a 1% solution in PBS to dilute antibodies.
6. CellTiter 96 MTT Assay kit (Promega Corp., Madison, WI).
7. Collagen type I: 10% 1X DMEM (Gibco), 10% 10X DMEM (Sigma-Aldrich, St. Louis, MO), and 80% type I rat tail collagen 2.01 mg/mL as described previously [23].
8. Collagen type IV: prepared as a 1 mg/mL solution in 0.6% acetic acid, which is then diluted 1:30 in water.
9. Commercial bleach: contains between 5% and 6% (by weight) sodium hypochlorite.
10. Dimethyl sulfoxide (DMSO).
11. Ethylenediaminetetraacetic acid (EDTA): 0.5 mM in water.
12. Fetal bovine serum (FBS).
13. Fluoro-Gel II (with DAPI), slide mounting medium with fluorescence quenching inhibitor (Electron Microscopy Sciences, Inc., Hatfield, PA).
14. Glutaraldehyde: 2.5% in water.
15. Human transferrin: 10 μg/mL used in complete DMEM (Table 1).
16. Hoechst 33258 (Invitrogen) 2 μg/mL in PBS.
17. Matrigel, Growth Factor Reduced (GFR), phenol-free (Corning Inc., Corning, NY) (*see* **Note 2**).
18. Methylene blue: used at 0.2% in PBS.
19. Nano-Glo Luciferase Assay System (Promega).
20. Nitazoxanide: make a stock solution of 65 mM in DMSO and dilute with differentiation medium (Table 1) to 20 μM or as needed.

**Table 1**
**Composition of media for growth and differentiation of hIEs**

| Reagent | Final concentration | Origin |
|---|---|---|
| *Complete medium* | | |
| Advanced DMEM/F12 | N/A | Invitrogen |
| GlutaMAX-1 | 2 mM | Invitrogen |
| HEPES | 10 mM | Invitrogen |
| Penicillin/streptomycin | 100 U/mL | Invitrogen |
| *Growth medium* | | |
| L-WRN-conditioned medium | 50% | ATCC |
| Noggin-conditioned medium | 10% | A gift from Dr. Gijs van den Brink (University of Amsterdam) |
| B27 | 1X | Invitrogen |
| N2 | 1X | Invitrogen |
| *N*-acetylcysteine | 1 mM | Sigma-Aldrich |
| Mouse recombinant EGF | 50 ng/mL | Invitrogen |
| [Leu15]-Gastrin I | 10 nM | Sigma-Aldrich |
| Nicotinamide | 10 mM | Sigma-Aldrich |
| A-83-01 | 500 nM | Tocris, Biotechne Corporation, Minneapolis, MN |
| SB202190 | 10 μM | Sigma-Aldrich |
| Y-27632 | 10 μM | Sigma-Aldrich |
| *Differentiation Medium* | | |
| L-WRN-conditioned medium | 50% | ATCC |
| Noggin-conditioned medium | 5% | A gift from Dr. Gijs van den Brink (University of Amsterdam) |
| B27 | 1X | Invitrogen |
| N2 | 1X | Invitrogen |
| *N*-acetylcysteine | 1 mM | Sigma-Aldrich |
| Mouse recombinant EGF | 50 ng/mL | Invitrogen |
| [Leu15]-Gastrin I | 10 nM | Sigma-Aldrich |
| Nicotinamide | 10 mM | Sigma-Aldrich |
| A-83-01 | 500 nM | Tocris |

Modified and reproduced from reference [32] with permission from Springer Nature Methods in Molecular Biology

21. No-Fade mounting medium (Waterborne).
22. Normal goat serum (NGS): used at 5% in PBS to block non-specific binding and at 1% in PBS for diluting antibodies.
23. Optimal Cutting Temperature (OCT) embedding medium.
24. Pancreatin: used at a concentration of 0.5 mg/mL in growth medium (Table 1) to dissociate hIEs.
25. Paraformaldehyde (PFA): make a 4% solution in PBS while heating to 60 °C in a glass beaker on a stir plate in a ventilated hood. The pH is raised by adding 1 N NaOH dropwise from a pipet until the solution clears. Once the paraformaldehyde is dissolved, the solution should be cooled and filtered. Recheck the pH and adjust it with small amounts of dilute HCl to approximately pH 6.9. The solution can be aliquoted and frozen or stored at 2–8 °C for up to 1 month.
26. Fluorescein Phalloidin (Invitrogen): prepare a 6.6 μM stock solution in methanol according to the manufacturer's instructions. Dilute 5 μL of the stock solution into 200 μL PBS for each coverslip to be stained.
27. Poly-L-lysine: use at a concentration of 100 μg/mL in PBS.
28. Polydimethylsiloxane (PDMS): mix the base reagent and curing reagent at a 10:1 mass ratio (Dow Corning, Midland MI).
29. RNeasy Plus Mini Kit (Qiagen Inc., Valencia, CA).
30. Recovery Cell Culture Freezing Medium (Thermo Fisher Scientific Inc., Waltham, MA).
31. Sodium hydroxide (NaOH): 1–5 N solution in water is used to adjust pH.
32. Sodium taurocholate: 0.75% sodium taurocholate in PBS is used for excystation of oocysts.
33. Trypsin–EDTA: 0.25% trypsin, 0.5 mM EDTA in water.
34. TURBO DNase Kit (Thermo Fisher Scientific).
35. QIAamp DNA Mini Kit (Qiagen).
36. QuantiTect SYBR Green Master Mix (Qiagen).
37. Y-27632, Rho-associated protein kinase (ROCK) inhibitor (Sigma-Aldrich).

### 2.2 Media

1. Complete Dulbecco's Modified Eagle's Medium (DMEM): DMEM supplemented with 10% FBS, 10 μg/mL human transferrin, 100 U/mL penicillin, 100 μg/mL streptomycin, and 0.25 μg/mL amphotericin B (Gibco).
2. Smooth Muscle Growth Medium-2 (SmGM-2) (Lonza, Walkersville, MD).
3. Growth medium: (*see* Table 1, **Note 1**).

4. Complete medium (*see* Table 1, **Note 1**).
5. Differentiation medium: growth medium without L-WNT3a, R-spondin, nicotinamide, and SB202190; Noggin reduced to 5%.

### 2.3 Cell Lines

1. HT29-MTX (European Collection of Authenticated Cell Cultures, Public Health England, Salisbury, UK).
2. Caco-2 cells (ATCC, Manassas, VA).
3. Primary human intestinal myofibroblasts (H-InMyoFib) (Lonza).
4. L-WNT3a cells (ATCC).
5. L-WRN cells (ATCC).
6. R-spondin cells: obtained from Dr. Calvin J. Kuo under a Materials Transfer Agreement with Stanford University.
7. Noggin cells: a gift from Dr. Gijs van den Brink (University of Amsterdam) [33].

### 2.4 Parasites

1. *C. parvum* (Iowa isolate) oocysts: stored at 4 °C in PBS with 100 U/ml penicillin, 100 μg/mL streptomycin, and 0.25 μg/mL amphotericin B and used within 3 months (Bunch Grass Farms, Deary, ID).
2. Transgenic *C. parvum* expressing nanoluciferase (NLuc) gene obtained as previously described [34].

## 3 Methods

### 3.1 3D Silk Scaffold-Based System Using Transformed Intestinal Epithelial Cell (IEC) Lines and Human Intestinal Myofibroblasts (H-InMyoFib) for C. parvum Infection (Fig. 1) [7]

#### 3.1.1 Growth and Maintenance of IEC Lines

1. Caco-2 and HT29-MTX cells are grown in complete DMEM.
2. H-InMyoFib are grown in SmGM-2.
3. All cell lines are cultured in T-75 or T-175 $cm^2$ tissue culture flasks at 37 °C, 5% $CO_2$ in a humidified atmosphere, and harvested with trypsin–EDTA prior to seeding.

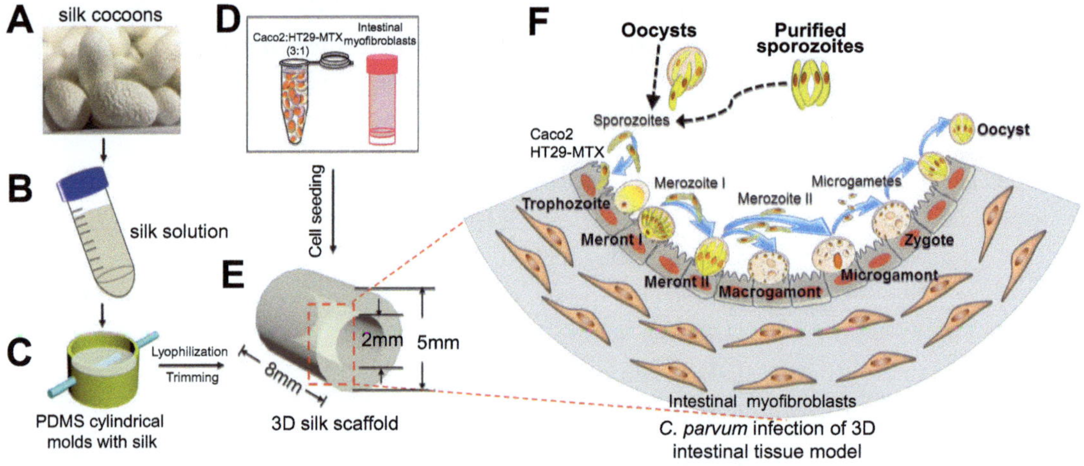


**Fig. 1** Bioengineered 3D human intestinal tissue model for *C. parvum*. (**a** and **b**) Silk cocoons (**a**) are processed to yield a viscous silk solution (**b**). (**c**) The silk solution is poured into cylindrical molds, and a wire is inserted to develop a lumen equivalent. (**d** and **e**) Caco-2 and HT29-MTX or human intestinal enteroid (hIE) cells (**d**) are seeded into the lumen (**e**), while the porous bulk space is seeded with H-InMyoFibs. (**f**) Caco-2 and HT29-MTX or hIE cells in the lumen are infected with *C. parvum* oocysts or purified sporozoites. Intracellular development through asexual and sexual cycles occurred to complete the life cycle, resulting in the formation of oocysts. (Figure modified and reproduced from reference [35] with permission of the publisher and reference [23] from Scientific Reports)

*3.1.2 Freezing Stocks of IEC*

1. Ideally, cell viability should be in excess of 90% in order to achieve a good recovery after freezing.
2. Treat a T75 (or T175) flask of adherent Caco-2 or HT29-MTX cells with 4 mL (or 10 mL, respectively) of trypsin–EDTA, and incubate at 37 °C for 1–2 min or until the cells detach.
3. Add a volume of complete DMEM equivalent to twice the volume of trypsin–EDTA to inactivate the enzyme.
4. Remove a small aliquot of cells (100–200 μL) and perform a cell count using a hemocytometer. Centrifuge the remaining culture at 200 × *g* for 5 min.
5. Resuspend the pelleted cells at a concentration of 2–4 × $10^6$ cells per mL in Recovery Cell Culture Freezing Medium.
6. Pipet 1 mL aliquots of cells into cryogenic vials that have been labeled with the cell line name, passage number, cell concentration, and date.
7. Place cryogenic vials inside a passive freezing container. Fill container with isopropyl alcohol and place at −80 °C overnight. Frozen cryogenic vials should be transferred to the vapor phase of a liquid nitrogen storage container and the locations of vials recorded.

*3.1.3 Reviving Cryopreserved IECs*

1. Remove a frozen stock of IECs from liquid nitrogen storage and immediately transfer to a 37 °C water bath.
2. Thaw cells rapidly (<1 min) by gently swirling the vial in a 37 °C water bath.
3. Transfer vial to a biosafety cabinet. Before opening, wipe the outside of the vial with 70% ethanol.
4. Transfer thawed cells into 10 mL of warm (37 °C) complete medium in a 15 mL conical tube.
5. Centrifuge cell suspension at 200 × *g* for 5 min at RT.
6. Check the clarity of the supernatant and visibility of a complete pellet. Aseptically decant the supernatant without disturbing the cell pellet.
7. Gently resuspend cells in 4 or 10 mL complete DMEM for T75 or T175 flasks, respectively, transfer them into a tissue culture flask, and incubate at 37 °C, 5% $CO_2$.

*3.1.4 Generation of 3D Silk Scaffolds*

1. 3D silk scaffolds are prepared from silk fibroin protein extracted from *Bombyx mori* silkworm cocoons as described previously [23].
2. To prepare silk scaffolds with hollow channels, special cylindrical molds are cast from PDMS (Fig. 1).
3. Insert a Teflon-coated stainless-steel wire (2 mm diameter; McMaster–Carr, Douglasville, GA) through the cross-section of the cylinder to develop a hollow channel in the silk scaffold.
4. Allow the PDMS to cure at 60 °C for 2 h.
5. Pour a 4–5% (w/v) viscous silk solution into the PDMS molds.
6. Freeze molds at −20 °C overnight, then transfer to a lyophilizer for drying.
7. Autoclave the dried silk scaffolds to induce β-sheet conformation.
8. Soak in distilled water overnight, and then trim along the axis of the hollow channel to obtain a cuboid 5 by 5 by 8 mm.
9. The resulting scaffold is 8 mm in length and has a hollow channel space, 2 mm in diameter, and surrounded by a bulk space containing interconnected pores.

*3.1.5 Preparing Silk Scaffolds for Seeding*

1. Transfer scaffolds to a petri dish using forceps, and aspirate excess fluid gently (use a 2 mL pipet tip) until the scaffolds turn white in color.
2. For every scaffold to be seeded, prepare 400 μL of collagen type I solution. To neutralize the pH, add 5 N NaOH dropwise and swirl slowly until the solution turns a peach color. Care must be taken not to overshoot the end point of titration (pH 8.0), in which case the solution will turn bright pink in color. Keep collagen type I solution on ice until ready to use.

3. Introduce 200 μL of collagen type I solution into the lumen of a scaffold from one end. Add another 200 μL of collagen type I solution from the other end. Dispense any remaining collagen type I gel on the outer walls of the scaffold.
4. Using forceps, insert Teflon-coated stainless steel into each scaffold to create a lumen as described previously [23].
5. Place the petri dish containing scaffolds into a 5% $CO_2$ incubator at 37 °C for at least 15 min for the collagen type I to set. Do not wait longer than 1.5 h.

*3.1.6 Seeding Cells onto Silk Scaffolds*

1. Harvest Caco-2, HCT29-MTX, and H-InMyoFib cells by treating with trypsin–EDTA for 2 min at 37 °C as described in Subheading 3.1, "Freezing stocks of IEC."
2. Resuspend $2 \times 10^5$ H-InMyoFib cells per mL of collagen type I solution.
3. Introduce H-InMyoFib cells into the porous bulk space of the scaffold using a micropipette (20–200 μL).
4. Using forceps carefully transfer a Teflon-coated stainless-steel wire into the opening of the collagen type I-coated scaffold, to leave the hollow channels open for seeding with Caco-2/HT29-MTX cells in the next step.
5. After 20–25 min of gelation at 37 °C, carefully remove the Teflon-coated stainless-steel wire from the scaffold to create a hollow channel (lumen).
6. Using a micropipette (20–200 μL), seed the hollow channel of the scaffold with Caco-2/HT29-MTX cells (3:1) at a density of $4 \times 10^6$ cells/mL.
7. Flip the scaffolds upside down and incubate for 1 h at 37 °C, 5% $CO_2$ to facilitate cell adherence.
8. During the incubation, drip a small amount of SmGM-2 medium onto the scaffolds to keep them moistened.
9. Incubate seeded scaffolds in 24-well plates with SmGM-2 culture medium (complete DMEM and SmGM-2 at a 1:1 ratio) for up to 2 weeks (changing the medium every other day) at 37 °C, 5% $CO_2$ prior to infection with *C. parvum*.

*3.1.7 Infection of 3D IEC-Based Silk Scaffolds*

Preparation of *C. parvum* Oocysts for Infection

1. Take $1 \times 10^7$ oocysts per scaffold to be infected. Add an extra 10% of oocysts to account for loss during processing. Centrifuge oocysts at 10,000 × *g* for 2 min at 4 °C in a microcentrifuge and discard the supernatant.
2. Resuspend the required number of oocysts in 1 mL of 10% commercial bleach (sodium hypochlorite) in sterile PBS, and incubate on ice for 10 min.

3. Centrifuge at 10,000 × *g* for 2 min at 4 °C and discard the supernatant.
4. Wash the oocyst pellet in 1 mL of sterile PBS, centrifuge at 10,000 × *g* for 2 min at 4 °C, and discard the supernatant. Repeat this step once to remove all traces of bleach.
5. Resuspend bleach-treated oocysts in 1 mL of 0.75% sodium taurocholate in PBS, and incubate at 15 °C for 15 min to allow excystation.
6. Centrifuge at 10,000 × *g* for 2 min at 4 °C.
7. Wash the intact oocyst and excysted sporozoite pellet in 1 mL of sterile PBS by centrifuging at 10,000 × *g* for 2 min at 4 °C.
8. Resuspend the pellet in culture medium before infection.

Isolation of *C. parvum* Sporozoites for Infection

9. Incubate bleach- and taurocholate-treated oocysts in sterile PBS for 1 h at 37 °C.
10. Filter the suspension through a 3 μM filter (EMD Millipore, Billerica, MA) to remove unexcysted oocysts and oocyst shells.
11. Centrifuge at 16,000 × *g* for 3 min at 4 °C and discard supernatant.
12. Resuspend the pellet in 1 mL of PBS, centrifuge at 16,000 × *g* for 3 min at 4 °C, and discard the supernatant.
13. Resuspend the pellet in 1 ml of warm (37 °C) complete DMEM.
14. Resuspend the final pellet in enough warm differentiation medium to give 40 μL per scaffold to be infected.
15. Dilute an aliquot of oocyst suspension 1:100 in differentiation medium and count using a hemocytometer.
16. If necessary, adjust the volume to give $1 \times 10^7$ oocysts per 40 μL per scaffold.
17. Infect scaffolds following the protocol in Subheading 3.1 "Infection of 3D IEC-based silk scaffolds."
18. Leftover oocyst suspension can be incubated for 1 h at 37 °C to determine the excystation rate (determined by counting numbers of intact and excysted oocysts using a hemocytometer).

#### 3.1.8 *C. parvum* Infection of 3D IEC-Based Scaffold Model with Pre-treated Oocysts

1. Using forceps carefully transfer scaffold(s) to a petri dish. Avoid squeezing the spongy silk scaffold during this process.
2. Gently remove the medium from inside the lumen of the scaffold using a micropipette (20–200 μL), taking care not to touch the walls to avoid dislodging the seeded cells within.
3. Carefully dispense 40 μL of oocyst suspension ($10^7$ total oocysts) into the scaffold lumen without touching the seeded cell layer within. To keep a steady hand while aspirating and

dispensing suspensions into the lumen, it helps to rest the upper end of the pipet shaft on the raised edge of the petri dish.

4. Cover the petri dish and incubate for 2 h at 37 °C, 5% $CO_2$ to allow excystation of the oocysts and attachment of the sporozoites.
5. After 2 h, use forceps to carefully transfer each scaffold to a well of a 24-well plate containing 2 mL of warm differentiation medium.
6. Incubate at 37 °C, 5% $CO_2$ for the duration of infection.
7. Keep an uninfected control scaffold for each time point or as required.

*3.1.9 Isolation of* C. parvum *Sporozoites for Infection*

1. Resuspend oocysts in 1 mL of PBS and allow them to excyst by incubating for 1 h at 37 °C.
2. Monitor excystation as described in Subheading 3.1 "Preparation of *C. parvum* oocysts for infection."
3. Once ~80% of oocysts have excysted, centrifuge the excystation mixture at 500 × *g* for 3 min at 4 °C to pellet oocyst walls (shells) and unexcysted oocysts. Sporozoites will remain in suspension.
4. Filter sporozoites through a 3 μm polycarbonate membrane to remove the remaining unexcysted oocysts and empty shells. Check an aliquot of the filtrate under the microscope to ensure there are no visible oocysts. Filter again if necessary.
5. Centrifuge purified sporozoites at 16,000 × *g* for 5 min at 4 °C.
6. Resuspend in 500 μL of culture medium and count the number of sporozoites using a hemocytometer. Adjust the volume to obtain $1 \times 10^5$ sporozoites/μL.
7. To ensure that no oocysts are present in the filtrate, check by IFA using an oocyst-specific mAb (Crypt-a-Glo) and a sporozoite-reactive rat polyclonal antibody (Sporo-Glo). Spot aliquots of the filtrate onto poly-L-lysine-coated 8-well slides. Once the spots are dry, fix filtrate in ice-cold methanol for 30 min at RT, then wash off the fixative with excess PBS. Incubate with Crypt-a-Glo or Sporo-Glo for 30 min at RT according to the manufacturer's instructions. Place a drop of mounting medium on each well containing the fixed sporozoites prior to sealing with a coverslip. Secure the coverslip with clear nail polish. Proceed with imaging by differential interference contrast (DIC) or fluorescence microscopy.

*3.1.10* C. parvum *Infection of 3D IEC-Based Scaffolds with Purified Sporozoites*

1. Using forceps carefully transfer a scaffold to a petri dish. Avoid squeezing the spongy silk scaffold during this process.
2. Gently remove the medium from inside the lumen using a micropipette (20–200 μL), taking care not to touch the walls to avoid dislodging the seeded cells within.

3. Dispense 40 μL of purified sporozoites (4 × $10^6$) into the lumen of the scaffolds.
4. Incubate the infected scaffolds for 2 h at 37 °C, 5% $CO_2$.
5. After 2 h, use forceps to carefully transfer each scaffold to a well of a 24-well plate containing 2 mL warm (37 °C) differentiation medium.
6. Incubate at 37 °C 5% $CO_2$ for the duration of infection, changing the medium every other day.

#### 3.1.11 Processing Infected 3D IEC-Based Scaffolds for Confocal Microscopy

1. At each time point, remove scaffold from culture plate and transfer to a new 24-well plate.
2. Add 1 mL of freshly prepared 4% PFA and incubate for 45 min at RT.
3. Remove PFA and wash three times with PBS.
4. Cut the scaffold lengthwise into two halves using a pair of small scissors (*see* **Note 3**).
5. Submerge the scaffold halves (lumen face down) in 1–2 mL of 0.5% Triton X-100, 5% BSA in PBS for 10–15 min to permeabilize the cells within the scaffold.
6. Wash three times with PBS.
7. Block non-specific binding with 5% BSA for 30 min to 2 h at RT.
8. Wash three times with PBS.
9. Incubate scaffold with primary anti-*C. parvum* antibody of choice or Sporo-Glo in 1% BSA (lumen face down) in a closed, moist chamber overnight at 4 °C.
10. Wash three times with PBS.
11. If using primary anti-*C. parvum* antibody, incubate the scaffold with secondary antibody prepared in 1% BSA in a closed, moist chamber for 1.5 h at RT.
12. Wash three times with PBS.
13. Counterstain scaffold with 300 nM DAPI for 30 min to 1 h at RT. Keep protected from light in a closed, moist chamber.
14. Remove DAPI solution, wash scaffolds three times with PBS, and store in PBS supplemented with 50% glycerol for up to 4 weeks at 4 °C.
15. Before imaging, cut each half lengthwise into two additional pieces, so the arch of the lumen is minimized for better imaging. Leave in PBS until ready to image. Place the scaffold strips with the lumen facing down on microscope slides during imaging.
16. Image using a confocal microscope at 20× magnification.

*3.1.12 Processing Infected 3D IEC-Based Scaffolds for Scanning Electron Microscopy*

1. Fix the infected and uninfected scaffolds in 2.5% glutaraldehyde as described [22].
2. Prepare the samples by progressive dehydration in a graded series of ethanol concentrations (30%, 50%, 75%, 95%, and twice in 100% for 30 min at each concentration).
3. Dry the samples by critical point drying with a liquid $CO_2$ dryer.
4. Coat with 5 nm of platinum/palladium using a sputter coater and image using a scanning electron microscope (e.g., Zeiss UltraPlus or Zeiss Ultra 55 (Carl Zeiss SBE, LLC, Thornwood, NY) at a voltage of 2–3 kV).

*3.1.13 Quantification of* C. parvum *Infection in 3D IEC-Based Scaffolds [7]*

Extraction of RNA from Scaffolds

1. At each time point, remove the luminal contents from each scaffold and reserve.
2. Add 40 μL of trypsin–EDTA solution into the lumen and incubate the scaffold for 10 min at 37 °C, 5% $CO_2$.
3. Collect the luminal contents following trypsinization, then flip the scaffold 180° and repeat trypsinization step with an additional 40 μL of trypsin–EDTA solution. Collect all the cells released and reserve.
4. Combine all reserved contents and centrifuge at 16,000 × *g* for 2 min at 4 °C.
5. At each time point, extract RNA from three infected scaffolds using the RNeasy Plus Mini Kit per manufacturer's protocol.
6. Remove contaminating genomic DNA using TURBO DNase Kit (Thermo Fisher Scientific, Inc.) per manufacturer's protocol.
7. Quantify RNA yield and purity using a NanoDrop spectrophotometer (Thermo Fisher Scientific, Inc.).

*3.1.14 RT-qPCR and Statistical Analysis*

Infection in IEC monolayers can be monitored by reverse transcription, quantitative PCR (RT-qPCR) using *C. parvum* 18S rRNA specific gene primers [19], and QuantiTect SYBR Green Master Mix.

1. For each sample, use 5 ng of RNA to synthesize cDNA using standard kits.
2. Prepare qPCR reactions in 96-well PCR plates and perform qPCR using a standard machine.
3. Reaction mixtures are heated to 95 °C for 15 min and then subjected to 40 thermal cycles (94 °C for 30 s, 52 °C for 30 s, and 72 °C for 30 s) of PCR amplification.

4. After amplification, perform melting curve analysis between 55 °C and 95 °C to assess the specificity of the reactions.
5. Perform three replicate reactions for each sample, and repeat the qPCR assay twice for a total of six technical replicates.
6. Determine the 18S cDNA copy number by comparison with a standard curve obtained by qPCR of DNA extracted from $10^2$ to $10^7$ oocysts using QIAamp DNA Mini Kit as described previously [7].
7. Since each *C. parvum* genome contains five copies of the 18S rRNA gene and each oocyst contains four sporozoites [36], the oocyst number is multiplied by 20 to obtain the approximate total number of 18S ribosomal cDNA copies.
8. Analyze the data statistically using the Kruskal–Wallis test followed by the Dunn's multiple comparisons test.

### 3.2 Human Intestinal Enteroid (hIE)-Based Models for *Cryptosporidium* Infection (Fig. 2)

#### 3.2.1 Growth and Maintenance of hIEs

1. Passage hIEs at a 1:2–1:3 ratio every 6–7 days.
2. Aspirate medium from wells without disturbing the Matrigel plug containing hIEs.
3. Add 500 μL of ice-cold complete medium to each well, and mechanically break up Matrigel plug by gently pipetting up and down with a micropipette (100–1000 μL) pipet (*see* **Note 2**).
4. Further disrupt the hIEs by passing two to three times through a 25G × 5/8″ needle attached to a 1 mL syringe. Transfer the contents of up to ten wells into a 15 mL conical tube. Add an additional two volumes of cold complete medium to dissolve the Matrigel.
5. Centrifuge tubes at 200 × *g* for 5 min at 4 °C and discard supernatant.
6. Resuspend the cell pellet in enough Matrigel to give 30 μL/well. Use cold pipet tips to prevent the Matrigel from congealing and avoid introducing bubbles into the suspension.
7. Using cold micropipette (20–200 μL) tips, take 30 μL of Matrigel suspension, and place in the center of each well in a 24-well plate.
8. Transfer plate to a 37 °C $CO_2$ incubator for 5–10 min to allow the Matrigel to set.
9. Add 500 μL of growth medium at RT to each well of hIEs, and culture in a 37 °C, 5% $CO_2$ incubator for 6–7 days, changing the growth medium every other day (Mondays, Wednesdays, and Fridays) (*see* **Note 4**).
10. If downstream analysis or infection is desired, culture hIEs in growth medium for 4 days, then switch to differentiation medium for 3–5 days.
11. Change the differentiation medium every other day until use.

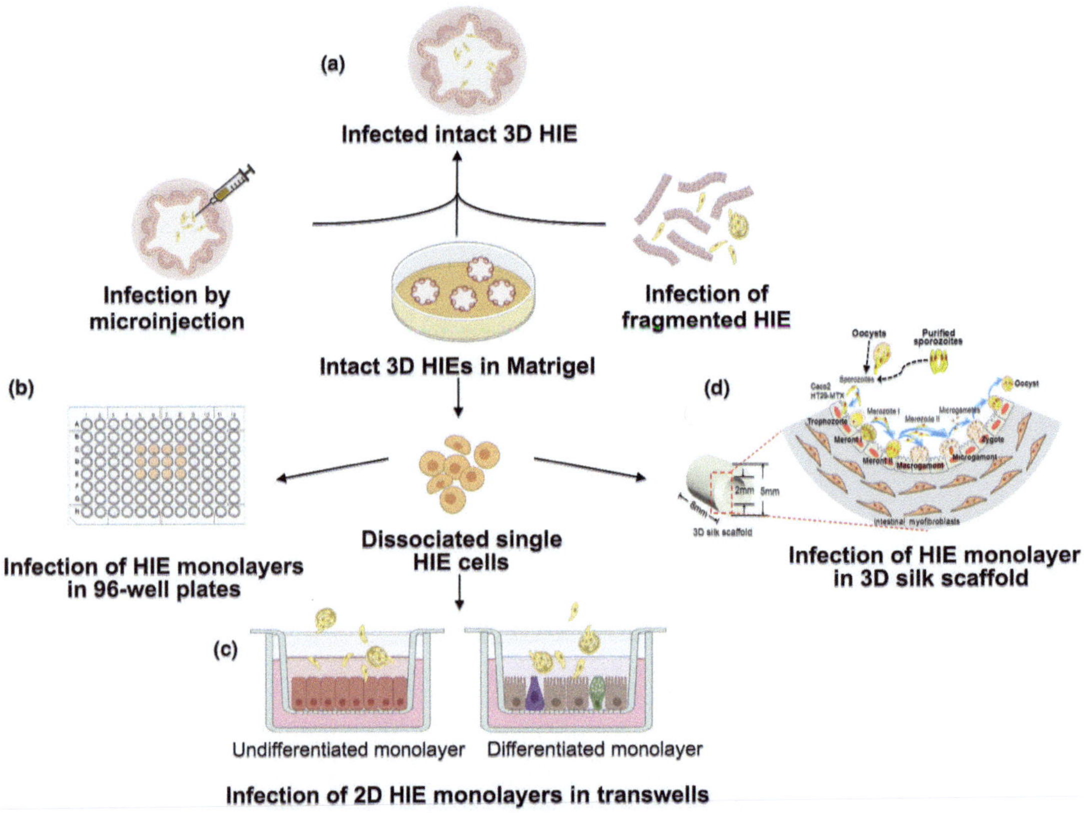


**Fig. 2** Human intestinal enteroid (hIE) models of *C. parvum* infection. (**a**) Intact 3D hIEs grown in Matrigel are fragmented, incubated with *C. parvum*, and re-plated in fresh Matrigel. Alternatively, hIEs are dissociated into single cells and grown as 2D monolayers in (**b**) microtiter plates or (**c**) transwell filters and infected apically with *C. parvum* oocysts or filtered sporozoites. (**d**) 3D silk scaffolds with luminal hIE monolayers are infected by introducing *C. parvum* oocysts or purified sporozoites into the lumen. (Figure reproduced with permission from reference [37]. Figure created in part with BioRender.com)

#### 3.2.2 Cryogenic Preservation and Revival of hIEs

Cryogenic Preservation of hIEs

1. After 6–7 days of culture, hIEs can be prepared for long-term storage in liquid nitrogen.
2. Aspirate the medium from wells without disturbing the Matrigel plug. Add 500 μL of cold complete medium to each well, and mechanically break up the Matrigel by passing through a micropipette (100–1000 μL) tip several times.
3. Transfer the contents of up to ten wells into an ice-cold 15 mL conical tube, and add an additional two volumes of cold complete medium to dissolve the Matrigel.
4. Centrifuge tube at 200 × *g* for 5 min at 4 °C and discard the supernatant.

5. Resuspend the pellet in enough freezing medium to give 500 μL for every 2 wells of hIEs that were processed. Place 500 μL into each cryovial. For easy identification, use color-coded vials for different lines of hIEs.
6. Transfer cryovials into a cell freezing container and place overnight at −80 °C. The next day transfer vials into liquid nitrogen for long-term storage.

Reviving Frozen Stocks of hIEs from Liquid Nitrogen

1. Thaw enough Matrigel overnight at 4 °C to give 120 μL per cryovial to be thawed.
2. Add 10 mL of complete medium into a 15 mL conical tube and keep on ice.
3. Take vial(s) from liquid nitrogen and immediately hold under tap water at RT until contents are thawed.
4. Transfer the contents in the vial(s) to the 15 mL tube containing 10 mL of ice-cold complete medium.
5. Centrifuge hIEs at 200 × *g* for 5 min at 4 °C, and discard the supernatant without disturbing the pellet.
6. Using cold micropipette tips, resuspend the pellet in 120 μL of Matrigel per cryovial, and place 30 μL in the center of as many wells as needed of a 24-well plate.
7. Carefully place plate in a 37 °C, 5% $CO_2$ incubator for 5–10 min to allow Matrigel to set.
8. Add 500 μL of growth medium at RT to each well of hIEs, and culture in a 37 °C, 5% $CO_2$ incubator for 6–7 days, changing the growth medium every other day (*see* **Note 4**).

### 3.3 Growth and Infection of 3D hIEs in Culture

#### 3.3.1 Growth of 3D hIEs in Culture

Grow and maintain hIE cultures as in Subheading 3.2.

#### 3.3.2 C. parvum Infection of 3D hIEs in Culture

Preparing oocysts for infection

1. Take sufficient oocysts to provide $5 \times 10^5$ oocysts for every well of hIEs to be infected. Include an extra 10% oocysts to account for losses.
2. Centrifuge oocysts at 16,000 × *g* for 3 min at 4 °C in a microcentrifuge and discard the supernatant.
3. Resuspend the pellet in 1 mL of 10% commercial bleach in PBS and incubate on ice for 10 min.

4. Centrifuge the oocysts at 16,000 × *g* for 3 min at 4 °C in a microcentrifuge.
5. Resuspend the pellet in 1 mL of PBS and centrifuge at 16,000 × *g* for 3 min at 4 °C. Repeat this wash step one more time.
6. Resuspend oocysts in 1 mL of 0.75% sodium taurocholate in PBS and incubate for 10 min at 15 °C.
7. Centrifuge at 16,000 × *g* for 3 min at 4 °C and discard the supernatant.
8. Resuspend in 1 mL of ice-cold complete medium and centrifuge at 16,000 × *g* for 3 min at 4 °C. Repeat this wash step one more time.
9. Resuspend the final oocyst pellet in 200 μL of complete medium.

Infection of 3D hIEs

1. Prepare a solution of 0.5 mg/mL pancreatin in complete medium solution.
2. For each well of hIEs to be infected, use two wells of 4-day-old differentiated or undifferentiated hIE cultures (*see* **Note 5**).
3. Aspirate medium from 4-day-old hIE cultures without disturbing the Matrigel plug, and add 500 μL of cold complete medium to each well.
4. Mechanically break up the Matrigel by pipetting up and down with a micropipette (100–1000 μL). Add an additional two volumes of ice-cold complete medium to dissolve the Matrigel. Transfer hIE suspension to a chilled 15 mL conical tube.
5. Centrifuge the tube at 200 × *g* for 5 min at 4 °C. Discard the supernatant and add more complete medium to wash the pellet. Repeat centrifugation step.
6. Carefully discard the supernatant without disturbing the hIE pellet.
7. Resuspend the pellet in the 200 μL of pre-treated oocyst suspension.
8. Add 200 μL of 0.5 mg/mL pancreatin in complete medium to the hIE/oocyst suspension, bringing the final concentration of pancreatin to 0.25 mg/mL.
9. Further break open the hIEs by pipetting up and down ~20 times with a micropipette (100–1000 μL).
10. Incubate the 15 mL tube containing the oocyst/hIE mix for 2 h at 37 °C to allow for excystation and attachment of the parasite to hIEs. Leave the cap loose to allow for gas exchange.

11. After 2 h of incubation, centrifuge the tube at 200 × *g* for 5 min at 4 °C and discard the supernatant.
12. Resuspend the pellet into enough Matrigel to give 30 μL per well of a 24-well plate. Add 30 μL droplets to each well and incubate plate for 5–10 min at 37 °C to allow Matrigel to set.
13. Add 500 μL of differentiation medium per well and incubate the plate in a 37 °C, 5% $CO_2$ incubator.
14. Harvest the hIEs at different time points post-infection to isolate RNA or make OCT blocks. Two wells of hIEs are required for preparing one OCT block.

#### *3.3.3 Processing of Infected 3D hIEs for Frozen Sections and Immunofluorescence Assays (IFA)*

Preparing OCT blocks for cryosectioning

1. Wash hIE cultures to be processed with 500 μL of ice-cold PBS.
2. Add 500 μL of ice-cold PBS per well and transfer to a 1.5 mL microcentrifuge tube.
3. Centrifuge tube(s) at 200 × *g* for 5 min at 4 °C, and carefully discard the supernatant without disturbing the pellet.
4. Add 500 μL of freshly made 4% PFA to the tube(s) and incubate for 20 min at RT.
5. After incubation, wash the pellet twice with 500 μL of cold PBS.
6. Resuspend the pellet in 50 μL of 0.2% methylene blue in PBS and incubate for 30 min at RT.
7. Add PBS to fill up the tube and centrifuge at 200 × *g* for 5 min at 4 °C.
8. Wash the pellet with PBS twice more or until the excess methylene blue is no longer present in the wash, and discard the supernatant.
9. Add a drop of OCT compound to the center of a cryomold.
10. Submerge a plastic disposable inoculation loop into the OCT, and then carefully scoop out the hIE pellet from the tube (*see* **Note 6**).
11. Transfer the whole clump of cells into the center droplet of the mold, place the mold onto dry ice immediately, and wait for the droplet to solidify. Then, add more OCT to fill up the mold.
12. Immediately store the molds overnight at −80 °C or until ready for sectioning.
13. Using a cryostat, prepare 5 μm sections from the OCT block for IFA.
14. Store blocks and prepared slides containing cut sections at − 80 °C until ready for IFA.

#### *3.3.4 Processing Sections of Infected 3D hIEs for Immunofluorescence Assays (IFA)*

Cryosections on slides can be processed by IFA to identify *C. parvum*-infected cells as described previously [34] and to examine other structural and nonstructural cellular components.

1. Prepare the blocking/permeabilization solution (0.5% Triton X-100, 5% normal goat serum in PBS) and the staining solution (0.05% Tween 20, 5% normal goat serum in PBS).
2. Remove slides from −80 °C and let them thaw for 30 min at RT before starting IFA.
3. Using a hydrophobic barrier (PAP) pen, draw a circle around the hIE sections and allow to dry. Solutions will be placed within these circles during staining.
4. Enough solution to completely cover the section without spilling over the PAP pen edge is needed (~75–100 μL for 1 section, ~150 μL for 2 sections next to each other).
5. Add blocking/permeabilization solution to each slide and incubate for 1 h at 37 °C. After blocking the sections, a smaller volume is necessary to cover them (~75 μL for a single section).
6. Place slides in a covered moist chamber.
7. Use staining solution to prepare the desired dilutions of primary antibodies. Cover the cryosections with primary antibody solution but avoid flooding the slide.
8. Place slides in a covered moist chamber and incubate overnight at 4 °C.
9. Wash the tissue sections three times with wash buffer by placing slides into a glass slide chamber, adding the wash buffer, and gently letting them rock on a rocking platform or rotator for ~5 min each wash.
10. Add the desired secondary antibodies diluted in staining solution.
11. Incubate the slides for 2 h at RT protected from light.
12. Wash tissue sections three times in wash buffer as in **Step 9**.
13. Place a drop of mounting medium over the section and seal with a coverslip. Secure the coverslip to the glass slide with clear nail polish.
14. Proceed for imaging by differential interference contrast (DIC) or fluorescence microscopy.

#### *3.3.5 Processing Infected 3D hIEs for Paraffin Embedding and Immunohistochemistry*

Where desired, immunohistochemistry can also be employed for staining sections of fixed hIEs as described previously described [38].

*3.3.6 Quantification of* C. parvum *Infection in 3D hIEs*

1. Extract RNA from 3D hIEs as described in Subheading 3.1.
2. Perform RT-qPCR and analyze the data as in Subheading 3.1.

### 3.4 3D Silk Scaffold-Based Model System Using hIEs

*3.4.1 Preparing 3D Silk Scaffolds for Seeding*

Prepare 3D silk scaffolds following the steps in Subheading 3.1 "Preparing Silk Scaffolds for Seeding."

*3.4.2 Seeding of hIEs onto 3D Silk Scaffolds*

Cell separation and dissociation of hIEs

1. For seeding one scaffold, collect three wells of hIEs (each containing at least 300–400 hIEs) cultured in growth medium for 6–7 days.
2. Carefully aspirate medium around the Matrigel plug containing hIEs from each well, leaving the plug intact.
3. Add 500 μL of cold 0.5 mM EDTA, and mechanically break up the Matrigel by gently pipetting up and down ten times with a micropipette (100–1000 μL) to help release hIEs from the Matrigel.
4. Transfer entire contents of up to ten wells into a chilled 15 mL tube kept on ice.
5. Centrifuge hIEs at 200 × *g* for 5 min at 4 °C and discard the supernatant.
6. Resuspend the pellet in 0.5 mL of trypsin–EDTA and incubate for 4 min at 37 °C.
7. Inactivate the trypsin by adding 1 mL of DMEM containing 10% FBS.
8. Dissociate the hIEs by vigorously pipetting up and down ~50 times using a micropipette (100–1000 μL). Pipet against the side of the tube to avoid making bubbles while dissociating the hIEs.
9. Place a 40 μm cell strainer on top of a 50 mL conical tube, and wet the membrane by passing 1 mL of DMEM containing 10% FBS through it.
10. Using a micropipette (100–1000 μL), pass the cells through the cell strainer and into the 50 mL conical tube.
11. Discard the cell strainer containing any cell clumps, and centrifuge the cells that passed through the cell strainer at 200 × *g* for 5 min at 4 °C.
12. Resuspend the cell pellet in an appropriate volume of growth media containing 10 μM Y-27632.

#### 3.4.3 Cell Seeding in 3D Silk Scaffolds

1. Remove petri dish containing collagen type I-coated scaffold from the $CO_2$ incubator.
2. Using forceps, slide out the Teflon-coated wires, and remove any excess solution or debris from the lumen using a micropipette (20–200 μL).
3. Add 35 μL of the homogenous hIE cell suspension into the lumen of each scaffold (*see* **Note 7**).
4. Incubate petri dish containing scaffold at 37 °C, 5% $CO_2$ for at least 45 min to allow cells to attach.
5. Repeat **Steps 1–12** of "Cell separation and dissociation of hIEs" above to be able to seed the other half of the scaffold.
6. Flip the scaffold over, and seed the second half by adding 35 μL of cell suspension into the lumen of each scaffold (ensure cell suspension is homogenous before adding).
7. Incubate the scaffold at 37 °C, 5% $CO_2$ for 1 h to allow cell attachment before transferring them into a 24-well plate using forceps.
8. Add 1.5 mL of growth medium with 10 μM Y-27632 to each well containing a seeded scaffold.
9. For downstream analysis or infection of scaffolds, switch to differentiation medium after 24 h. Incubate for 3–5 days, changing the differentiation media every other day until ready for infection.

#### 3.4.4 Infection of 3D hIE-Based Silk Scaffolds with Oocysts and Purified Sporozoites

See Subheading 3.1.

#### 3.4.5 Processing of Infected 3D hIE-Based Scaffolds for Confocal Microscopy

See Subheading 3.1.11.

#### 3.4.6 Processing of Infected 3D hIE-Based Scaffolds for Scanning Electron Microscopy

See Subheading 3.1.12.

#### 3.4.7 Quantification of C. parvum Infection in 3D hIE-Based Scaffolds

See Subheading 3.1.13.

### 3.5 Screening for *Cryptosporidium* Interventions in hIE-Derived Monolayers in Multi-Well Tissue Culture Plates

#### 3.5.1 Preparation of hIEs for Growth as Monolayers on Multi-Well Plates

1. Coat each well of a 96-well tissue culture plate to be seeded with 100 μL of ice-cold 33 μg/mL collagen type IV (*see* **Note 8**).
2. Incubate plate for 1.5–2 h at 37 °C, 5% $CO_2$.
3. Wash hIE cultures grown in 24-well plates with 500 μL of cold 0.5 mM EDTA in PBS, disrupt Matrigel plug with a micropipette (100–1000 μL) tip, and transfer to a 15 mL conical tube.
4. Centrifuge at 300 × *g* for 5 min at 4 °C and discard the supernatant.
5. Wash the pellet in 500 μL of trypsin–EDTA and incubate for 4 min at 37 °C. Use 1 mL of trypsin–EDTA when pooling six to ten wells of hIEs.
6. Add 1 mL of complete medium +10% FBS to inactivate trypsin.
7. Vigorously pipet the cell suspension ~50 times using a micropipette (100–1000 μL) to further dissociate the cells.
8. Place a 40 μm cell strainer over a 50 mL conical tube, and wash membrane with 1 mL of complete medium +10% FBS.
9. Pass cell suspension through strainer using a micropipette (100–1000 μL) and discard the strainer.
10. Dilute a ~10 μL aliquot of hIEs 1:10–1:20 in PBS and determine cell count using a hemocytometer.
11. Take the needed volume of cells to give $1 \times 10^5$ cells per well of a 96-well plate, centrifuge at 400 × *g* for 5 min at RT, and discard the supernatant (*see* **Note 9**).
12. Resuspend the pellet in enough growth medium supplemented with 10 μM Y-27632 to give a final volume of 100 μL per well.
13. Remove collagen type IV solution from wells.
14. Add 100 μL of cell suspension to each well.
15. 48 h later, remove growth medium from wells and replace with 100 μL of differentiation medium.
16. Change the differentiation medium every other day.

#### 3.5.2 Infection of hIE-Derived Monolayers with Transgenic *C. parvum* and Treatment with Interventions

1. After 4 days in differentiation medium, hIE-derived monolayers are ready to be infected. Place transgenic *C. parvum* oocysts (5000 per well of a 96-well plate) in a 1.5 mL microcentrifuge tube (*see* **Note 10**).
2. Centrifuge the tube at 16,000 × *g* for 3 min at 4 °C in a microcentrifuge and discard the supernatant.
3. Resuspend the pellet in 1 mL of 10% v/v commercial bleach in PBS and incubate on ice for 10 min.
4. Centrifuge the tube at 16,000 × *g* for 3 min at 4 °C in a microcentrifuge and discard the supernatant.

5. Resuspend the pellet in 1 mL of PBS.
6. Centrifuge the tube at 16,000 × *g* for 3 min at 4 °C in a microcentrifuge and discard the supernatant.
7. Resuspend the pellet in 1 mL of 0.75% w/v sodium taurocholate in PBS, and incubate at 15 °C for 10 min in a benchtop hot/cold dry block incubator.
8. Centrifuge the tube at 16,000 × *g* for 3 min at 4 °C in a microcentrifuge and discard the supernatant.
9. Resuspend the pellet in 1 mL of PBS.
10. Centrifuge the tube at 16,000 × *g* for 3 min at 4 °C in a microcentrifuge and discard the supernatant.
11. Resuspend the pellet in enough differentiation medium to give a final volume of 100 μL per well.
12. Add 100 μL to each well and allow the infection to proceed for 2 h at 37 °C, 5% $CO_2$.
13. Remove the medium from the wells and wash once with warm PBS to remove any sporozoites that did not invade.
14. Add 100 μL of differentiation medium containing the drugs, antibodies, or biologics to be tested at the desired concentrations to each well, and incubate for 72 h at 37 °C, 5% $CO_2$ (*see* **Note 11**).
15. Change the differentiation medium containing test compounds every 48 h.

#### 3.5.3 Luciferase Assay to Monitor Infection

1. After 72 h of incubation at 37 °C, 5% $CO_2$, remove the medium from all the wells of the 96-well plate using a multichannel pipettor (*see* **Note 12**).
2. Add 100 μL of Nano-Glo Luciferase Assay Buffer to each well.
3. Incubate the plate for 10 min at 37 °C, 5% $CO_2$ to lyse the cells.
4. Remove the plate from the incubator and add 100 μL of Nano-Glo Luciferase Assay Substrate mixed 1:50 with Assay Buffer to each well, bringing the final volume to 200 μL per well. Keep the reagents and plate protected from light.
5. Mix the samples thoroughly while avoiding bubble formation and transfer to a white 96-well plate.
6. Measure luminescence using a microplate reader (*see* **Note 13**).

#### 3.5.4 Toxicity Testing

1. After 4 days in differentiation medium, treat hIE monolayers in 96-well plate(s) with the compound(s) to be tested at the desired concentrations, and incubate at 37 °C, 5% $CO_2$.
2. 72 h posttreatment, add 15 μL of CellTiter 96 MTT Assay Dye Solution to each well of the plate using a multichannel pipettor.

3. Incubate the plate at 37 °C, 5% $CO_2$ for 1–4 h to allow MTT to be reduced to formazan (*see* **Note 14**).
4. Remove the plate from the incubator, and add 100 μL of CellTiter 96 MTT Assay Solubilization Solution to each well.
5. Incubate the plate at RT for 1 h (*see* **Note 15**).
6. Mix the samples thoroughly while avoiding bubble formation.
7. Measure absorbance at 570 nm using a microplate reader. A reference wavelength of 630–750 nm may be used. The absorbance may also be measured at a wavelength range of 550–600 nm.

### 3.6 2D Model for Growth of Polarized hIE Monolayers on Transwell Inserts (Fig. 2)

#### 3.6.1 Preparation of Transwells for Seeding

1. Place the transwell inserts in a 24-well plate (*see* **Note 16**).
2. Coat the upper compartment of each transwell insert with 100 μL of 33 μg/mL collagen type IV solution.
3. Add 300 μL of water to the 24-well plates containing the transwell inserts (lower compartment).
4. Incubate the 24-well plate with transwells for 1 h at 37 °C, 5% $CO_2$.
5. After 1 h, remove the liquid from the upper and lower compartments of the transwells.
6. Incubate the 24-well plate with transwells for 30 min at 37 °C, 5% $CO_2$ to allow them to dry.

#### 3.6.2 Growth of Differentiated hIE Monolayers on Transwells

1. Collect hIEs from a 24-well plate. Each well should have at least 50 hIEs cultured in growth medium for 6–7 days.
2. Remove the growth medium from all wells, and place 500 μL of cold 0.5 mM EDTA in every other well. Work on ice or at 4 °C.
3. Pipet up and down to break up the Matrigel.
4. Transfer the contents into the next well; break up the Matrigel in that well and so on.
5. Place the cells in a 50 mL conical tube on ice.
6. After hIEs from all the wells have been collected, centrifuge the tube at 300 × *g* for 5 min at 4 °C, and discard the supernatant.
7. Add 1 mL of warm trypsin–EDTA for every 4 wells of hIEs collected in **Step 1**, and incubate for 4 min at 37 °C, 5% $CO_2$.
8. At this point, remove the transwells from the incubator.
9. Inactivate the trypsin by adding an equal amount of complete medium containing 10% FBS to the cells.
10. Dissociate the cells by pipetting vigorously up and down 50 times using a micropipette (100–1000 μL).

11. Place a 40 μm cell strainer on top of a 50 mL conical tube. Wet the cell strainer membrane with complete medium containing 10% FBS.
12. Pipet the cells onto the strainer. The cells should pass through by gravity. Discard the strainer and any material that remained in it. Centrifuge the cells at 300 × *g* for 5 min at RT and discard the supernatant.
13. Resuspend the hIE cell pellet in 3 mL of growth medium.
14. Using a hemocytometer, count the total number of cells.
15. Add 600 μL of growth medium containing 10 μM Y27632 into the lower compartment of the transwell (i.e., the well of the 24-well plate containing the transwell insert).
16. Add 5–7.5 × $10^5$ total cells to the upper compartment of each transwell, and bring the volume of growth medium containing 10 μM Y27632 to 300 μL for each upper compartment of the transwell. Incubate the cells at 37 °C, 5% $CO_2$.
17. The next day, measure the transepithelial electrical resistance (TEER) of the monolayer.
18. Two days from the original seeding time, change the growth medium containing 10 μM Y27632 to an equal volume of differentiation medium. Change the differentiation medium every other day for 4 days.

#### 3.6.3 Growth of Undifferentiated hIE Monolayers on Transwells

1. Grow hIEs for 7 days as described above in "Growth of differentiated hIE monolayers on transwells" (**Steps 1–17**).
2. Continue to grow the hIEs on transwells in growth medium for 6 days.

#### 3.6.4 C. parvum Infection of Differentiated and Undifferentiated hIE Monolayers

1. Resuspend the desired number of oocysts in 1 mL of PBS.
2. Centrifuge the oocysts at 16,000 × *g* for 3 min at 4 °C.
3. Discard the supernatant and keep the oocysts on ice.
4. Add 1 mL of 10% bleach in PBS and incubate on ice for 10 min.
5. Centrifuge at 16,000 × *g* for 3 min at 4 °C and discard the supernatant.
6. Wash the pellet twice with 1 mL of PBS by centrifugation at 16,000 × *g* for 3 min at 4 °C.
7. Discard the supernatant, resuspend the pellet in 1 mL of 0.75% sodium taurocholate, and incubate for 10 min at 15 °C.
8. Centrifuge at 16,000 × *g* for 3 min at 4 °C and discard the supernatant.
9. Wash the pellet in 1 mL of PBS by centrifugation at 16,000 × *g* for 3 min at 4 °C two times.

10. Discard the supernatant, and resuspend the oocysts in growth medium (for undifferentiated hIE monolayers) or differentiation medium (for differentiated hIE monolayers). Count the number of oocysts using a hemocytometer.
11. Add oocysts to the upper compartment of the transwells at a multiplicity of infection (MOI) of 0.1–1 (*see* **Note 17**).

*3.6.5 Monitoring Monolayer Integrity Via Transepithelial Electrical Resistance (TEER) Measurements*

Perform all TEER measurements in a sterile environment using a Millicell-ERS probe (EMD Millipore, Burlington, MA) according to the manufacturer's instructions. Before measurements, make sure the voltohmmeter and electrodes are calibrated according to the manufacturer's instructions.

1. Place the electrodes in 70% ethanol for 10 min to sanitize. Allow the hIE-derived monolayers grown on the transwells to come to RT. Keep some growth medium at RT ready for rinsing the electrodes. Lastly, have a transwell (with medium only) ready.
2. Rinse the electrode in the growth medium at RT, and carefully place the long side of the electrode in the bottom compartment and the short side in the top compartment of the transwell. Measure the resistance (ohms) of the blank sample. Keep your hands steady to get a reproducible reading (*see* **Note 18**).
3. Measure the TEER of the first well; rinse the electrodes in the growth medium at RT and move on to the next well. Do this for every sample in the 24-well plate (*see* **Note 18**).
4. After all the TEER values have been recorded, place the 24-well plates containing transwells back in the incubator, or take desired time points for downstream assays (IFA, RT-qPCR, or Western blots).
5. TEER values are multiplied by 0.33 to account for the transwell growth area and expressed as ohms/cm$^2$.
6. After use, place the electrodes in 70% ethanol for 10 min, then dry and store in a clean area.

*3.6.6 Quantification of* C. parvum *Infection of hIE-Derived Monolayers by RT-qPCR*

1. At the desired time point post-infection, remove the medium from one transwell to a 1.5 mL microcentrifuge tube.
2. Add 100 μL of trypsin–EDTA to the upper compartment of the same transwell and incubate for 1–2 min at 37 °C.
3. Pipet up and down a few times to loosen the monolayer and transfer to the same microcentrifuge tube.
4. Place another 100 μL of trypsin–EDTA into the same transwell, scrape the membrane carefully using a micropipette (100–1000 μL) tip, and transfer to same microcentrifuge tube.
5. Add 1 mL of complete medium +10% FBS to inactivate the trypsin in the microcentrifuge tube.

6. Centrifuge at 5000 rpm for 5 min at 4 °C and discard the supernatant.
7. Resuspend pellet in 1 mL of PBS, centrifuge again, and discard the supernatant.
8. Store the pellet at −80 °C.
9. Extract RNA and quantify infection by RT-qPCR using a standard curve as described in Subheading 3.1 "Quantification of *C. parvum* infection in 3D hIEs."

#### 3.6.7 Quantitation of C. parvum Infection by Image Analysis

1. If obtaining sufficient RNA for RT-qPCR quantitation of *C. parvum* infection in the 2D model is a limitation, infection can be quantified by image analysis as described [31].
2. Obtain Z-stack images of the infected hIE monolayers following IFA and confocal microscopy, and process the images using Image J software [39]. Set a threshold to identify primary anti-*C. parvum* antibody-stained intracellular *C. parvum* stages by size (1–5 mM). The Image J software then automatically counts the total number of parasite stages in the image. Repeat this process for ten representative images at each time point and determine the mean.

#### 3.6.8 Processing hIE-derived Monolayers for IFA and Confocal Microscopy

1. For IFA and confocal microscopy, wash each transwell three times with 300 μL of PBS.
2. Add 300 μL of freshly made 4% PFA to the top and bottom chambers of the transwells.
3. Incubate for 15 min at RT.
4. After 15 min, remove the PFA from the transwells and wash three times with 300 μL of PBS.
5. Incubate with 300 μL of 0.25% Triton X-100 in PBS for 10 min at RT. Wash the transwells twice with 300 μL 0.1% Tween 20 in PBS at RT.
6. Wash the transwells twice with PBS.
7. Leave the transwells overnight at 4 °C in 300 μL of 5% NGS in both the top and lower chambers.
8. Wash the transwells twice with 300 μL of PBS.
9. For detecting markers of IEC differentiation, use mAbs against chromogranin A (enteroendocrine cells), lysozyme (Paneth cells), mucin 2 (MUC-2, goblet cells), and sucrose isomaltose (SI, enterocytes), as well as a mAb to Lgr5 as a marker of stem cells, as described [31].
10. For detecting markers of tight or adherens junction proteins, use mAbs to ZO-1 or E-cadherin respectively, as described [31].

11. For detecting and quantifying *C. parvum* infection and co-staining with markers of IEC differentiation or of tight or adherens junctions, use primary antibodies such as murine mAb 4E9 [40] or rabbit anti-gp40 [41, 31].
12. Dilute the primary antibody in 1% NGS and add 100 μL to the top chamber of the transwells. Add 300 μL of PBS to the bottom chamber of the transwells.
13. Incubate for 1 h at RT (or overnight at 4 °C, depending on antibody specifications).
14. Wash the transwells three times with 1 mL of PBS.
15. Incubate hIEs with the appropriate secondary antibodies: goat anti-rabbit IgG (Alexa Fluor 647) or goat anti-mouse IgG (Alexa Fluor 594), respectively. Dilute the secondary antibody in 1% NGS, and add 100 μL to the top chamber and 300 μL of PBS to the bottom chamber of the transwells.
16. Incubate for 1–2 h at RT depending on antibody specifications.
17. For double staining of infected hIEs with antibodies to SI and Lgr5, incubate PFA-fixed and permeabilized monolayers with a primary antibody to *C. parvum* and a mAb to Lgr5 overnight for 18 h at 4 °C as described [31].
18. Following three washes with PBS, stain nuclei with Hoechst 33342 and F-actin with Fluorescein Phalloidin as described [31].
19. Next, add a drop of Fluoro-Gel II (with DAPI) to a glass slide.
20. Use a clean blade and fine forceps to cut out the membranes of the transwells carefully. Place the transwell membranes facing up onto the slide, and add another drop of Fluoro-Gel II (with DAPI) onto the membrane.
21. Place a cover slip on the membrane, and use Kimwipes or paper towel to clean around the edges of the slide.
22. Seal the edges of the cover slip with clear nail polish.
23. Store slides in the dark at 4 °C until ready to be imaged via confocal microscopy.

#### 3.6.9 Image Analysis of Tight and Adherens Junctions Using Tight Junction Organizational Rate (TiJOR) and Intercellular Junction Organization Quantification (IJOQ)

1. Quantify the disruption of tight junctions, as previously described [42, 31].
2. Using Image J software [39], draw a polygon on the confocal image at the level of the ZO-1 network and extend to successive larger polygons. The software calculates the size of the polygon and the number of ZO-1 peaks it crosses using the TiJOR algorithm.
3. Quantify the disruption of adherens junctions using intercellular junction organization quantification (IJOQ), an automated algorithm that analyzes and quantifies the degree of

organizational disruption of intercellular junctions as previously described [43, 31]. Disruption is measured by tracing linear paths through the image and recording the frequency of intercellular junctions detected while tracing these paths using the algorithm.

*3.6.10 hIE Monolayer Height Measurements*

1. Measure hIE monolayer heights [44] as described previously [31].
2. Obtain Z stacks (0.5 mM per stack) of the hIE monolayer and create orthogonal views using Image J [39]. Draw a line from the basolateral to the apical surface of the monolayers and measure the length. Obtain three measurements for each image and record the mean as the monolayer height.

## 4 Notes

1. Conditioned media from L-WNT3a, R-spondin 1, Noggin, and L-WRN cells can be aliquoted and stored at −20 °C for 1 month. Avoid freeze and thaw cycles. Growth medium can be kept at −20 °C for up to 2 weeks. Complete medium and differentiation medium can be kept at 4 °C for up to 1 month.
2. Matrigel is stored long term at −20 °C. When working with Matrigel, thaw aliquots overnight at 4 °C and keep on ice to avoid a temperature increase. Matrigel will quickly solidify at warmer temperature. Therefore, when routinely passaging cells using Matrigel, all tubes and pipet tips must be pre-chilled at 4 °C before plating and kept on ice during plating.
3. Make one single cut down the center (avoid chopping motions). Make the second cut on the opposite side, so now you have 2 lengths with an open lumen that can be stained with different antibodies as required.
4. Newly thawed hIEs are initially slow to grow. Passage them one or two times after thawing before setting up an experiment with them.
5. Either undifferentiated or differentiated hIEs may be used for infections. Infection in differentiated hIEs lasts only for 5 days, whereas infection of undifferentiated hIEs lasts up to 3 weeks.
6. Submerging the inoculation loop into the OCT first helps to keep the hIEs together and prevents them from adhering to the loop.
7. Each scaffold requires 35 μL of cell suspension per side (top and bottom) of the lumen. For example, for seeding six scaffolds, resuspend the cell pellet in 420 μL of growth medium containing 10 μM Y-27632.

8. Stock collagen type IV is prepared as a 1 mg/mL solution in 0.6% acetic acid, which is then diluted 1:30 in water.
9. Use one well of a 7-day-old hIE culture for every three wells of 96-well plate to be seeded; final seeding density should be approximately $1 \times 10^5$ cells per well. Do not pool more than ten wells per 15 mL conical tube.
10. A transgenic *C. parvum* line that expresses a nanoluciferase reporter gene is used for infections. Purified transgenic oocysts are kept in PBS supplemented with antibiotic–antimycotic (10,000 U/mL penicillin, 10,000 μg/mL streptomycin, and 25 μg/mL amphotericin B) at 4 °C for up to 3 months.
11. 20 μM nitazoxanide may be used as a positive control (the only drug that is FDA-approved for treatment of *C. parvum* in immunocompetent hosts [45]). Make a nitazoxanide stock solution at a concentration of 65 mM in DMSO and dilute with differentiation medium as needed.
12. Use a multichannel pipettor and reagent reservoirs for all subsequent steps in this section.
13. The luminescence emitted is directly proportional to the number of parasites in the well.
14. The amount of formazan formation is proportional to the number of live cells in the well. Decreased formazan formation is indicative of cytotoxicity.
15. Incubation at 37 °C will accelerate solubilization and is recommended for assays with a large quantity of formazan formation. After solubilization, plates may be stored at 4 °C for several days.
16. Transwell inserts can be made of polycarbonate or polyethylene with pore sizes of 0.4 μm or 3 μm. Use 0.4 μm polycarbonate transwell inserts.
17. Be sure to not cross-contaminate uninfected wells with infected wells.
18. When samples are switched from growth medium to differentiation medium, be sure to use differentiation medium to rinse the electrodes.

## Acknowledgments

Work in the authors' laboratory was supported by NIH grants U19AI131126, R21AI120932, and R21AI128342 and by the Bill and Melinda Gates Foundation (OPP1164543) to HW.

**Competing Interest Statement** The authors declare no conflicts of interest relevant to the content of this chapter.

## References


1. Checkley W, White AC Jr, Jaganath D, Arrowood MJ, Chalmers RM, Chen X, Fayer R, Griffiths JK, Guerrant RL, Hedstrom L, Huston CD, Kotloff KL, Kang G, Mead JR, Miller M, Petri WA Jr, Priest JW, Roos DS, Striepen B, Thompson RC, Ward HD, Van Voorhis WA, Xiao L, Zhu G, Houpt ER (2015) A review of the global burden, novel diagnostics, therapeutics, and vaccine targets for cryptosporidium. Lancet Infect Dis 15(1): 85–94. https://doi.org/10.1016/S1473-3099(14)70772-8
2. Levine MM, Nasrin D, Acacio S, Bassat Q, Powell H, Tennant SM, Sow SO, Sur D, Zaidi AKM, Faruque ASG, Hossain MJ, Alonso PL, Breiman RF, O'Reilly CE, Mintz ED, Omore R, Ochieng JB, Oundo JO, Tamboura B, Sanogo D, Onwuchekwa U, Manna B, Ramamurthy T, Kanungo S, Ahmed S, Qureshi S, Quadri F, Hossain A, Das SK, Antonio M, Saha D, Mandomando I, Blackwelder WC, Farag T, Wu Y, Houpt ER, Verweiij JJ, Sommerfelt H, Nataro JP, Robins-Browne RM, Kotloff KL (2020) Diarrhoeal disease and subsequent risk of death in infants and children residing in low-income and middle-income countries: analysis of the GEMS case-control study and 12-month GEMS-1A follow-on study. Lancet Glob Health 8(2):e204–e214. https://doi.org/10.1016/S2214-109X(19)30541-8
3. Bourli P, Eslahi AV, Tzoraki O, Karanis P (2023) Waterborne transmission of protozoan parasites: a review of worldwide outbreaks - an update 2017-2022. J Water Health 21(10): 1421–1447. https://doi.org/10.2166/wh.2023.094
4. Castellanos-Gonzalez A, Cabada MM, Nichols J, Gomez G, White AC Jr (2013) Human primary intestinal epithelial cells as an improved in vitro model for Cryptosporidium parvum infection. Infect Immun 81(6): 1996–2001. https://doi.org/10.1128/IAI.01131-12
5. Varughese EA, Bennett-Stamper CL, Wymer LJ, Yadav JS (2014) A new in vitro model using small intestinal epithelial cells to enhance infection of Cryptosporidium parvum. J Microbiol Methods 106:47–54. https://doi.org/10.1016/j.mimet.2014.07.017
6. Miller CN, Josse L, Brown I, Blakeman B, Povey J, Yiangou L, Price M, Cinatl J Jr, Xue WF, Michaelis M, Tsaousis AD (2018) A cell culture platform for cryptosporidium that enables long-term cultivation and new tools for the systematic investigation of its biology. Int J Parasitol 48(3–4):197–201. https://doi.org/10.1016/j.ijpara.2017.10.001
7. DeCicco RePass MA, Chen Y, Lin Y, Zhou W, Kaplan DL, Ward HD (2017) Novel bioengineered three-dimensional human intestinal model for long-term infection of Cryptosporidium parvum. Infect Immun 85:3. https://doi.org/10.1128/IAI.00731-16
8. Morada M, Lee S, Gunther-Cummins L, Weiss LM, Widmer G, Tzipori S, Yarlett N (2016) Continuous culture of Cryptosporidium parvum using hollow fiber technology. Int J Parasitol 46(1):21–29. https://doi.org/10.1016/j.ijpara.2015.07.006
9. Bhalchandra S, Cardenas D, Ward HD (2018) Recent breakthroughs and ongoing limitations in *cryptosporidium* research. F1000Res 7. https://doi.org/10.12688/f1000research.15333.1
10. Larregieu CA, Benet LZ (2013) Drug discovery and regulatory considerations for improving in silico and in vitro predictions that use Caco-2 as a surrogate for human intestinal permeability measurements. AAPS J 15(2): 483–497. https://doi.org/10.1208/s12248-013-9456-8
11. Ben-David U, Siranosian B, Ha G, Tang H, Oren Y, Hinohara K, Strathdee CA, Dempster J, Lyons NJ, Burns R, Nag A, Kugener G, Cimini B, Tsvetkov P, Maruvka YE, O'Rourke R, Garrity A, Tubelli AA, Bandopadhayay P, Tsherniak A, Vazquez F, Wong B, Birger C, Ghandi M, Thorner AR, Bittker JA, Meyerson M, Getz G, Beroukhim R, Golub TR (2018) Genetic and transcriptional evolution alters cancer cell line drug response. Nature 560(7718):325–330. https://doi.org/10.1038/s41586-018-0409-3
12. Sato T, Vries RG, Snippert HJ, van de Wetering M, Barker N, Stange DE, van Es JH, Abo A, Kujala P, Peters PJ, Clevers H (2009) Single Lgr5 stem cells build crypt-villus structures in vitro without a mesenchymal niche. Nature 459(7244):262–265. https://doi.org/10.1038/nature07935
13. Kovbasnjuk O, Zachos NC, In J, Foulke-Abel J, Ettayebi K, Hyser JM, Broughman JR, Zeng XL, Middendorp S, de Jonge HR, Estes MK, Donowitz M (2013) Human enteroids: preclinical models of non-inflammatory diarrhea. Stem Cell Res Ther 4(Suppl 1):S3. https://doi.org/10.1186/scrt364
14. Zachos NC, Kovbasnjuk O, Foulke-Abel J, In J, Blutt SE, de Jonge HR, Estes MK, Donowitz M (2016) Human enteroids/colonoids

and intestinal organoids functionally recapitulate normal intestinal physiology and pathophysiology. J Biol Chem 291(8):3759–3766. https://doi.org/10.1074/jbc.R114.635995

15. Almeqdadi M, Mana MD, Roper J, Yilmaz OH (2019) Gut organoids: mini-tissues in culture to study intestinal physiology and disease. Am J Physiol Cell Physiol 317(3):C405–C419. https://doi.org/10.1152/ajpcell.00300.2017
16. Rouch JD, Scott A, Lei NY, Solorzano-Vargas RS, Wang J, Hanson EM, Kobayashi M, Lewis M, Stelzner MG, Dunn JC, Eckmann L, Martin MG (2016) Development of functional microfold (M) cells from intestinal stem cells in primary human Enteroids. PLoS One 11(1):e0148216. https://doi.org/10.1371/journal.pone.0148216
17. Foulke-Abel J, In J, Kovbasnjuk O, Zachos NC, Ettayebi K, Blutt SE, Hyser JM, Zeng XL, Crawford SE, Broughman JR, Estes MK, Donowitz M (2014) Human enteroids as an ex-vivo model of host-pathogen interactions in the gastrointestinal tract. Exp Biol Med (Maywood) 239(9):1124–1134. https://doi.org/10.1177/1535370214529398
18. Zachos NC, Kovbasnjuk O, Foulke-Abel J, In J, Blutt SE, De Jonge HR, Estes MK, Donowitz M (2015) Human enteroids/colonoids and intestinal organoids functionally recapitulate normal intestinal physiology and pathophysiology. J Biol Chem. https://doi.org/10.1074/jbc.R114.635995
19. Yu H, Hasan NM, In JG, Estes MK, Kovbasnjuk O, Zachos NC, Donowitz M (2017) The contributions of human mini-intestines to the study of intestinal physiology and pathophysiology. Annu Rev Physiol 79:291–312. https://doi.org/10.1146/annurev-physiol-021115-105211
20. Ramani S, Crawford SE, Blutt SE, Estes MK (2018) Human organoid cultures: transformative new tools for human virus studies. Curr Opin Virol 29:79–86. https://doi.org/10.1016/j.coviro.2018.04.001
21. Roh TT, Chen Y, Rudolph S, Gee M, Kaplan DL (2021) InVitro models of intestine innate immunity. Trends Biotechnol 39(3):274–285. https://doi.org/10.1016/j.tibtech.2020.07.009
22. Chen Y, Zhou W, Roh T, Estes MK, Kaplan DL (2017) In vitro enteroid-derived three-dimensional tissue model of human small intestinal epithelium with innate immune responses. PLoS One 12(11):e0187880. https://doi.org/10.1371/journal.pone.0187880
23. Chen Y, Lin Y, Davis KM, Wang Q, Rnjak-Kovacina J, Li C, Isberg RR, Kumamoto CA, Mecsas J, Kaplan DL (2015) Robust bioengineered 3D functional human intestinal epithelium. Sci Rep 5:13708. https://doi.org/10.1038/srep13708
24. Heo I, Dutta D, Schaefer DA, Iakobachvili N, Artegiani B, Sachs N, Boonekamp KE, Bowden G, Hendrickx APA, Willems RJL, Peters PJ, Riggs MW, O'Connor R, Clevers H (2018) Modelling cryptosporidium infection in human small intestinal and lung organoids. Nat Microbiol 3(7):814–823. https://doi.org/10.1038/s41564-018-0177-8
25. Lahree A, Gilbert L (2024) Development of organoids to study infectious host interactions. Methods Mol Biol 2742:151–164. https://doi.org/10.1007/978-1-0716-3561-2_12
26. Nikolaev M, Mitrofanova O, Broguiere N, Geraldo S, Dutta D, Tabata Y, Elci B, Brandenberg N, Kolotuev I, Gjorevski N, Clevers H, Lutolf MP (2020) Homeostatic mini-intestines through scaffold-guided organoid morphogenesis. Nature 585(7826):574–578. https://doi.org/10.1038/s41586-020-2724-8
27. Bhalchandra S, Lamisere H, Honorine W (2020) Intestinal organoid/enteroid-based models for cryptosporidium. Curr Opin Microbiol 58:124–129
28. Dutta D, Heo I, O'Connor R (2019) Studying cryptosporidium infection in 3D tissue-derived human organoid culture systems by microinjection. J Vis Exp 151. https://doi.org/10.3791/59610
29. Korwin-Mihavics BR, Dews EA, di Genova BM, Huston CD (2023) Organoid-based in vitro systems to model Cryptosporidium parvum infection in 2D and 3D. bioRxiv. https://doi.org/10.1101/2023.09.29.560165
30. Hariss F, Delbeke M, Guyot K, Zarnitzky P, Ezzedine M, Certad G, Meresse B (2023) Cytotoxic innate intraepithelial lymphocytes control early stages of cryptosporidium infection. Front Immunol 14:1229406. https://doi.org/10.3389/fimmu.2023.1229406
31. Lamisere H, Bhalchandra S, Kane AV, Zeng XL, Mo D, Adams W, Estes MK, Ward HD (2022) Differential response to the course of Cryptosporidium parvum infection and its impact on epithelial integrity in differentiated versus undifferentiated human intestinal Enteroids. Infect Immun 90(11):e0039722. https://doi.org/10.1128/iai.00397-22
32. Zou WY, Blutt SE, Crawford SE, Ettayebi K, Zeng XL, Saxena K, Ramani S, Karandikar UC, Zachos NC, Estes MK (2019) Human intestinal Enteroids: new models to study gastrointestinal virus infections. Methods Mol Biol

1576:229–247. https://doi.org/10.1007/7651_2017_1

33. Heijmans J, van Lidth de Jeude JF, Koo BK, Rosekrans SL, Wielenga MC, van de Wetering M, Ferrante M, Lee AS, Onderwater JJ, Paton JC, Paton AW, Mommaas AM, Kodach LL, Hardwick JC, Hommes DW, Clevers H, Muncan V, van den Brink GR (2013) ER stress causes rapid loss of intestinal epithelial stemness through activation of the unfolded protein response. Cell Rep 3(4): 1128–1139. https://doi.org/10.1016/j.celrep.2013.02.031
34. Vinayak S, Pawlowic MC, Sateriale A, Brooks CF, Studstill CJ, Bar-Peled Y, Cipriano MJ, Striepen B (2015) Genetic modification of the diarrhoeal pathogen *Cryptosporidium parvum*. Nature 523(7561):477–480. https://doi.org/10.1038/nature14651
35. Lendner M, Daugschies A (2014) Cryptosporidium infections: molecular advances. Parasitology 141(11):1511–1532. https://doi.org/10.1017/S0031182014000237
36. Le Blancq SM, Khramtsov NV, Zamani F, Upton SJ, Wu TW (1997) Ribosomal RNA gene organization in Cryptosporidium parvum. Mol Biochem Parasitol 90(2):463–478
37. Bhalchandra S, Lamisere H, Ward H (2020) Intestinal organoid/enteroid-based models for cryptosporidium. Curr Opin Microbiol 58:124–129. https://doi.org/10.1016/j.mib.2020.10.002
38. Saxena K, Blutt SE, Ettayebi K, Zeng XL, Broughman JR, Crawford SE, Karandikar UC, Sastri NP, Conner ME, Opekun AR, Graham DY, Qureshi W, Sherman V, Foulke-Abel J, In J, Kovbasnjuk O, Zachos NC, Donowitz M, Estes MK (2015) Human intestinal enteroids: a new model to study human rotavirus infection, host restriction, and pathophysiology. J Virol 90(1):43–56. https://doi.org/10.1128/JVI.01930-15
39. Collins TJ (2007) ImageJ for microscopy. BioTechniques 43(1 Suppl):25–30. https://doi.org/10.2144/000112517
40. Cevallos AM, Bhat N, Verdon R, Hamer DH, Stein B, Tzipori S, Pereira ME, Keusch GT, Ward HD (2000) Mediation of Cryptosporidium parvum infection in vitro by mucin-like glycoproteins defined by a neutralizing monoclonal antibody. Infect Immun 68(9): 5167–5175
41. O'Connor RM, Wanyiri JW, Cevallos AM, Priest JW, Ward HD (2007) Cryptosporidium parvum glycoprotein gp40 localizes to the sporozoite surface by association with gp15. Mol Biochem Parasitol 156(1):80–83
42. Terryn C, Sellami M, Fichel C, Diebold MD, Gangloff S, Le Naour R, Polette M, Zahm JM (2013) Rapid method of quantification of tight-junction organization using image analysis. Cytometry A 83(2):235–241. https://doi.org/10.1002/cyto.a.22239
43. Mo D, Xu S, Rosa JP, Hasan S, Adams W (2022) Dynamic python-based method provides quantitative analysis of intercellular junction organization during S. pneumoniae infection of the respiratory epithelium. Front Cell Infect Microbiol 12:865528. https://doi.org/10.3389/fcimb.2022.865528
44. Noel G, Baetz NW, Staab JF, Donowitz M, Kovbasnjuk O, Pasetti MF, Zachos NC (2017) A primary human macrophage-enteroid co-culture model to investigate mucosal gut physiology and host-pathogen interactions. Sci Rep 7:45270. https://doi.org/10.1038/srep45270
45. Cabada MM, White AC Jr (2010) Treatment of cryptosporidiosis: do we know what we think we know? Curr Opin Infect Dis 23(5): 494–499. https://doi.org/10.1097/QCO.0b013e32833de052

# Chapter 22

# *Cryptosporidium* Oocyst Purification Using Discontinuous Gradient Centrifugation

**Michael J. Arrowood**


## Abstract

Many laboratory studies in cryptosporidial research require a source of purified oocysts. Sources can include experimentally infected laboratory animals or from samples collected from naturally infected animals and from clinical cases of human cryptosporidiosis. Purification of oocysts can be accomplished with readily available laboratory equipment including table-top centrifuges and microcentrifuges. Following purification, oocysts can be stored in antibiotic-supplemented buffers or in 2.5% aqueous potassium dichromate for over 6 months. Ultimately, oocyst viability and infectivity declines to less than 10% after 1 year, so if isolates are expected to be maintained, serial passage in a suitable host at ≤6-month intervals is recommended. Oocysts purified as described in this chapter are suitable for animal infection studies, cell culture studies, and a wide range of molecular biological studies, environmental studies, drug testing, and disinfection studies.


**Key words** *Cryptosporidium*, Oocyst, Purification, Isolation, Sucrose, Sheather's solution, Cesium chloride, Centrifugation, Gradient

## 1 Introduction

This method describes a bulk stool processing method for cryptosporidial oocyst isolation and purification using discontinuous sucrose and cesium chloride gradients (*see* Fig. 1). The method is scalable, having been adapted for small-scale purification of individual mouse stool samples [1], as well as bulk stool processing from experimentally infected bovine calves [2, 3]. Using essentially the same conditions and reagents at the proportions described below, most stool samples can be processed with success. A known caveat is that clinical human stools can vary significantly in composition, affecting the efficiency of the purification process. Oocysts purified using these gradient techniques are suitable for immediate use or further purification and/or sorting using flow cytometry-based methods as well as immunomagnetic bead separation. This method does not include a "defatting" step using an organic solvent (e.g.,

Jan R. Mead and Michael J. Arrowood (eds.), *Cryptosporidium: Methods and Protocols*, Methods in Molecular Biology, vol. 2978, https://doi.org/10.1007/978-1-0716-4824-7_22,

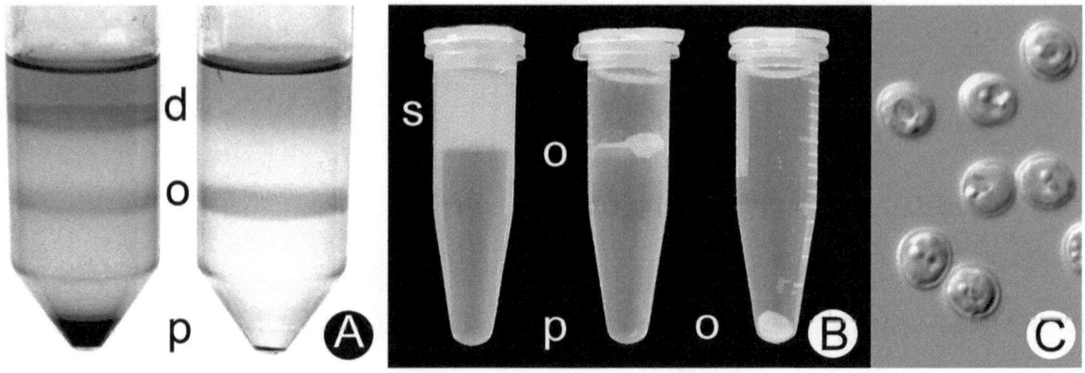


**Fig. 1** Panels A and B illustrate the progressive gradient techniques used to purify *C. parvum* oocysts. Panel A shows primary (left) and secondary (right) discontinuous sucrose gradients after centrifugation with light debris (d) at the interface between the sample and the PEST 1:4 layer, the oocysts (o) at the interface between the PEST 1:4 and PEST 1:2 layers, and the pellet (p) of dense debris. Panel B shows examples of the cesium chloride step gradients: the first tube (left) shows the sucrose gradient-purified oocyst sample (s) layered over the cesium chloride column before centrifugation; the second tube (middle) shows the same tube after centrifugation with the oocysts (o) at the interface between the sample and the cesium chloride and a small pellet (p) of dense debris at the bottom; the third tube (right) shows the purified pellet of oocysts (o) after the saline wash step. Panel C shows a differential interference contrast (DIC) view of representative *C. parvum* oocysts purified using the sucrose and cesium chloride gradient techniques

ethyl acetate, similar to methods for stool sedimentation before O&P evaluation), although some laboratories include it. If organic solvents are used, the useful shelf life of the oocysts may be shortened. If oocysts are to be sanitized before experimental use (e.g., "bleach" treated), we recommend delaying this step until just before use, as the treatment is likely to shorten the shelf life of the oocysts. Also, consider "sanitized" oocysts to be partially compromised and may not be suitable for disinfection studies, as the sanitizing step is likely to alter the oocyst wall and make oocysts more susceptible to the experimental disinfectant treatment.

This updated chapter includes the use of the classic Sheather's solution-based sucrose gradients, as described in the previous edition, but also now includes recommended alternative sucrose gradient solutions (incorporating phenoxyethanol instead of phenol) for reasons of improved safety in preparation, handling, and disposal (*see* **Note 1**).

## 2 Materials

1. Large glass or plastic funnels.
2. Stainless-steel sieves: size 60 and 20 mesh.
3. Stainless-steel or dishwasher safe buckets (sized to match sieve diameter).

4. Stainless-steel forceps.
5. Small plastic pan.
6. Plastic spatulas.
7. Ring-stand and holder (if using funnel).
8. Small wash bottles: saline, deionized water.
9. Paper towels.
10. Polypropylene bottles (1 L and 4 L).
11. Repeater pipet.
12. 250 mL polycarbonate centrifuge bottles (transparent).
13. 200 mL polypropylene conical centrifuge bottles (translucent).
14. 50 mL polypropylene conical centrifuge tubes (transparent).
15. 5, 10, 50, and 100 mL disposable pipets.
16. 3 mL disposable transfer pipet.
17. Micropipettes and tips.
18. Deionized water ($dH_2O$).
19. Phosphate-buffered saline 1X (PBS): 0.025 M, pH 7.2.
20. PBS 10X concentration (0.025 M, pH 7.2 or 0.01 M, pH 7.4).
21. Saline: 0.85% NaCl w/v aqueous.
22. Saline supplemented with 0.1% Tween 80 (Saline/Tween).
23. Antibiotic mix: e.g., Penicillin, Streptomycin, Fungizone, prepared at 1X working concentration in PBS.
24. Potassium dichromate ($K_2Cr_2O_7$) solutions: 2.5% and 5.0% w/v in deionized water.
25. Tween 80 or Tween 20.
26. Sheather's sucrose solution: 500 g sucrose, 320 mL $H_2O$, 9 mL aqueous phenol (85%).
27. PEST (Phenoxyethanol | Sucrose | Tween with PBS) gradient solutions (*see* Table 1):
28. STP (Sheather's | Tween | PBS) gradient solutions (*see* Table 2):
29. Antifoam Y-30 emulsion.
30. Bobbitt valves.
31. Aspirator bottles (1 L) with tubulation (bottom hose outlet), fitted with Bobbitt valve and shortened pipet (use silicone tubing) (*see* Fig. 2).
32. Vacuum aspirator flask (4 L) with vacuum flask trap (typically 1 L or 500 mL) outfitted with disposable Pasteur pipet inserted in silicone tubing connected to glass tubing in rubber stopper of aspirator flask.

**Table 1**
**Preparation of PEST Solutions (Phenoxyethanol I Sucrose I Tween with PBS)**

| Component | PEST 1:2 (1.103 g/l density) | PEST 1:4 (1.064 g/l density) |
|---|---|---|
| $H_2O$ | 750 mL | 810 mL |
| PBS (10X stock) | 84 mL | 90 mL |
| Sucrose | 262 g | 157 g |
| 2-Phenoxyethanol | 3 mL | 3 mL |
| Tween 80 (*see* **Note 2**) | 10 mL | 10 mL |

Thoroughly mix the solutions, starting in the order listed and store at RT. For larger volumes of these working solutions, simply multiply each component by the appropriate factor. Make sure solutions are thoroughly mixed and, as necessary, verify density (using a hydrometer) before use

**Table 2**
**Preparation of STP Solutions (Sheather's Solution I Tween I PBS)**

| Component | STP 1:2 (1.103 g/l density) | STP 1:4 (1.064 g/l density) |
|---|---|---|
| Sheather's Solution | 300 mL | 200 mL |
| PBS (1X stock) | 600 mL | 800 mL |
| Tween 80 (*see* **Note 2**) | 9 mL | 9 mL |

Thoroughly mix the solutions, starting in the order listed and store at RT. For larger volumes of these working solutions, simply multiply each component by the appropriate factor. Make sure solutions are thoroughly mixed and, as necessary, verify density (using a hydrometer) before use

33. Hydrophobic in-line filter for vacuum aspirator to protect vacuum system from contamination.
34. Carboy (~20 L) with spigot and Bobbitt valve, filled with deionized water ($dH_2O$) (*see* Fig. 2).
35. Hot water bath set at ~55 °C (if decontaminating dishwasher is not available).
36. 5 gallon or larger tank for heat disinfection of waste (or equivalent).
37. Stir plate and stir bar (large, suitable for the 5 gallon tank used to disinfect waste).
38. Immersion heater or heat source suitable for use with disinfection tank.
39. Heat controller for immersion heater (typically includes a temperature sensor immersed in the disinfection tank with fail-safe circuitry if overheating is detected).
40. Timer for heating circuit (each system will need to be calibrated to achieve the target temperature and to hold it for the necessary time to achieve inactivation (i.e., ~55 °C for $\geq$1 h).

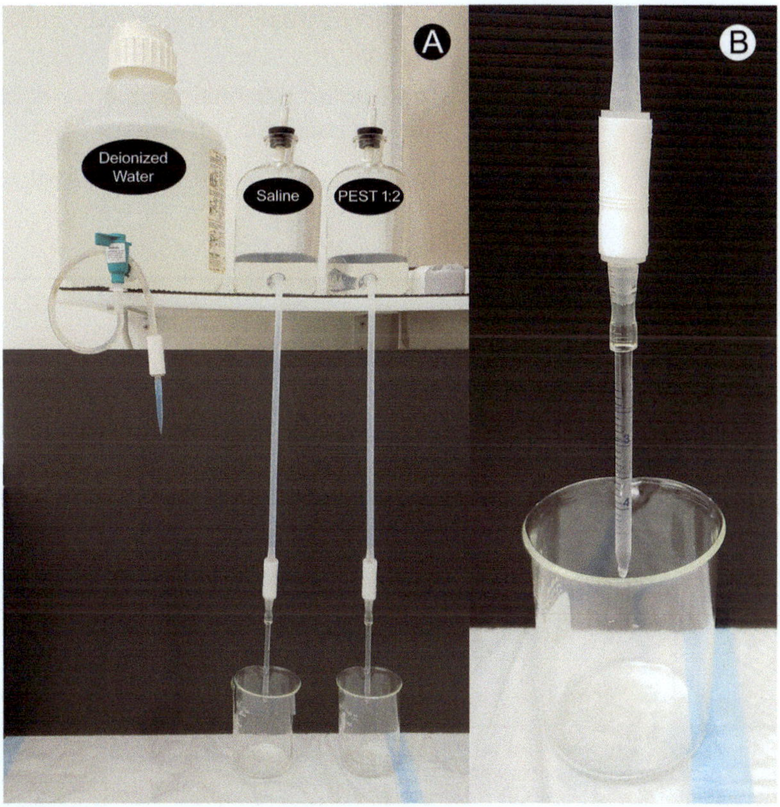


**Fig. 2** Assembled reagents used in the purification of cryptosporidial oocysts. Reservoirs are located on shelf above the workbench, so reagents can be delivered to the bottles and tubes at bench height. A solid black background is included to provide contrast behind tubing and Bobbitt valves; however, during use, the workbench should be ventilated (ideally a laminar flow workbench or equivalent). The carboy on the left is filled with deionized water and is used for sample dilution and washing of gradient components from samples. Note that the Bobbitt valve can be fitted with a disposable 1 mL micropipette tip for small-volume use (cesium chloride gradients). Two aspirator bottles with tubulations (bottom hose outlets) are fitted with Bobbitt valves and shortened pipets (using silicone tubing). Saline is used for washing gradient components from samples and for sample suspension and storage. The PEST 1:2 reservoir is used to underlay the PEST 1:4 volume in gradient bottles or tubes

41. 5–15 gallon polypropylene barrels (or otherwise suitable) for disposal of heat-treated chemical waste.
42. Disposable counter pads (absorbent with plastic-backed, waterproof shell).
43. Cesium chloride solution (sp. gr. 1.15): 21.8 g CsCl, 103.3 mL $dH_2O$.
44. Table-top centrifuge (swing bucket rotor, refrigerated, must reach ~3200× *g*).

45. Microcentrifuge (refrigerated, must reach ~13,200 rpm or 16,000× *g*).
46. Appropriate personal protective equipment (gloves, lab coat, safety glasses, etc.).
47. Biosafety Level 2 workbench (laminar flow system recommended).

## 3 Methods

*See* Fig. 3 for a schematic flowchart of the purification strategy.

### 3.1 Safety Recommendations When Working with Fecal Samples Containing Oocysts

1. Gloves, safety glasses, and a lab coat are to be worn at all times. Consider wearing a full-face splash shield if working on an open counter.
2. When working with fluids and tissues containing oocysts, dispensing bottles with undiluted commercial hydrogen peroxide

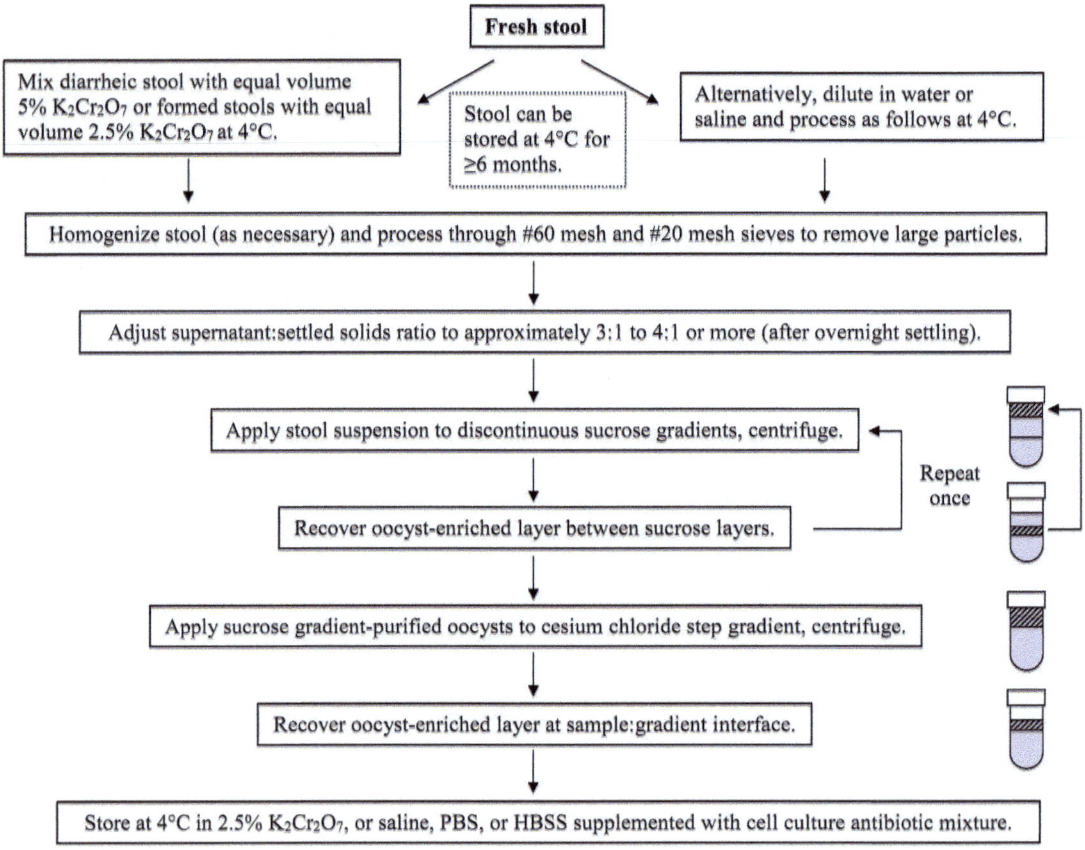


**Fig. 3** Schematic flowchart summarizing the purification strategy for *Cryptosporidium* oocysts for use in laboratory studies

(identified on bottle as 3% hydrogen peroxide or "10 vol" hydrogen peroxide) should be available in case of spills.

3. All instruments that come in contact with oocyst-contaminated materials should be placed in a stainless-steel pan to be autoclaved or sanitized in the appropriate cycle in the laboratory dishwasher. Instruments that cannot be autoclaved may be immersed in a heated water bath (~55 °C) for 60 min or placed in 3% hydrogen peroxide for a period of 30 min (minimum to 2 h maximum) before washing. Wash with a commercial disinfecting detergent before rinsing with tap and finally deionized water.
4. All disposable labware that comes into contact with oocyst-contaminated tissues or fluids should be placed in autoclave bags and autoclaved as per local guidelines prior to disposal.
5. In case of spills, use absorbent disposable paper towels to remove contaminating material and transfer to autoclave bag. Flood contaminated areas with 3% hydrogen peroxide for a minimum of 30 min (add additional hydrogen peroxide as necessary to prevent drying). Excess hydrogen peroxide and contaminating fluid can be absorbed with disposable paper towels and transferred to an autoclave bag. Allow surfaces to thoroughly dry before use.

### 3.2 Sieving Stool Samples (See Note 3)

1. Cover workspace with plastic disposable counter pad (plastic-backed pad) to protect counter. Assemble the sieving apparatus by inserting the #20 sieve (850 μm mesh) into a small (~1 gallon) stainless-steel bucket ("primary sieve") and the #60 (250 μm mesh) sieve on a second stainless-steel bucket ("secondary sieve"). Place the bucket in the large plastic tray (if used) to catch spills (*see* Fig. 4).
2. Pour the "raw" stool suspension/mixture through the screen into the bucket. Stop pouring when the liquid stops flowing freely through the screen or when the bucket is approximately 3/4 "full."
3. Use a plastic scraper and deionized water sprayed from a wash bottle to gently push small particulate matter through the sieve. Large matter is removed with the spatula and discarded into the hazardous waste container.
4. If the sieve becomes clogged with solid material, replace with a clean sieve. Continue sieving until finished, or pause while the sieves are cleaned in the dishwasher.
5. Continue sieving the feces until the day's collection has been processed through the #20 sieve. Transfer the sieve to the sanitizing dishwasher for processing/cleaning.

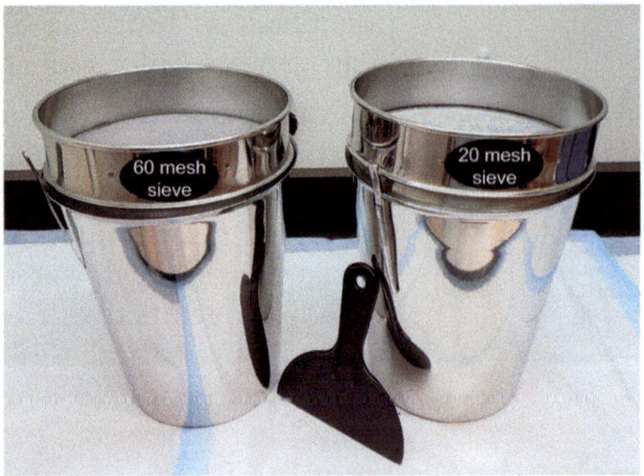


**Fig. 4** Example of stainless-steel buckets paired with stainless-steel sieves for sieving stool samples. Note the use of forceps to create a break at the interface between the buckets and the sieves (prevents vapor seal/lock during sieving; otherwise, sample spattering can occur when lifting the sieves). If buckets have pouring spouts, the forceps are generally not needed. A disposable plastic spatula is used to assist sieving of stool samples (along with water or saline spray bottles)

6. Rinse out the original feces bottle with a small amount of saline and set aside (do NOT pour bleach inside; these bottles will be cleaned and decontaminated in the dishwasher).
7. Pour the #20 sieved feces through the #60 sieve into the bucket. Stop pouring when the liquid stops flowing freely through the screen or when the bucket is approximately 3/4 "full."
8. When the feces stops flowing freely through the #60 sieve, use the plastic spatula and saline sprayed from the wash bottle to gently push small particulate material through. Large matter is removed with the spatula and discarded into the biohazard waste container.
9. If sieve becomes clogged with solid material, repeat cleaning procedure as in **step 3**.
10. Continue sieving the feces until the day's collection has been processed through the #60 sieve. Transfer the sieve to the dishwasher (*see* **step 4**).
11. If the sieved fecal sample is to be condensed immediately, go to the next section (*see* Subheading 3.3). If the sample is to be stored temporarily, transfer the sieved feces to a large bottle (typically a 1 gallon wide-mouth polypropylene bottle), label it as "sieved," and identify the sample appropriately (e.g., *C. parvum* isolate, calf ID, and date post-infection). Store bottle(s) at 4 °C in the laboratory refrigerator.

12. Stainless-steel buckets are cleaned and sanitized in the laboratory dishwasher.
13. Insert spatulas, scrapers, and other tools in the laboratory dishwasher for cleaning and disinfection.
14. If contaminated, discard counter pad(s) into the biohazard waste container. Replace with new disposable pad(s).
15. If the countertop is soiled, wipe it with paper towels to remove bulk waste. Spray/flood the cleaned surface with 3% hydrogen peroxide, and allow peroxide to stand for ≥30 min (do not allow it to dry; reapply peroxide as necessary). Dry, rinse with tap water, and dry again.

### 3.3 Condensing Sieved Stool Samples

1. If previously stored, mix bottle(s) of raw, sieved feces to suspend fecal sediment.
2. Pour approximately 700–750 mL into each 750 mL centrifuge bottle, and centrifuge at a ~4000 rpm (3200× *g*) in refrigerated (if available) table-top centrifuge (with swing buckets) for 10 min at 4 °C.
3. Aspirate supernatant into vacuum flask or pour/decant supernatant into waste tank. If a large volume of sieved stool remains to be condensed, add sieved stool suspension to the bottles (on top of the pellet(s)) and centrifuge as before. Aspirate/pour/decant supernatant as before. Depending on pellet volumes, pool pellets into one or more centrifuge bottles and dilute to 750 mL with deionized water.
4. Centrifuge as before for 10 min. Aspirate or decant supernatant. Pool and disperse all pellets from one day's collection (or individual sample) into a single 1 L bottle, and dilute to approximately 800 mL with 2.5% potassium dichromate or 0.85% saline. Label bottle with appropriate identifying details as before, including sieved status. Store sieved and condensed feces at 4 °C (*see* **Note 4**).
5. Disinfect pooled liquid waste using heat (*see* Subheading 3.7).

### 3.4 Primary Oocyst Purification Using Discontinuous Sucrose Gradients (See Notes 5 and 6)

1. Four 250 mL bottles equal one primary run (compared with the original tube method employing 50 mL tubes [2]). Use four 250 mL transparent polycarbonate bottles per gradient centrifugation run.
2. Dispense 80 mL of PEST 1:4 into each 250 mL bottle using a graduated cylinder. If not already marked, mark the bottle with a permanent marker to indicate the top (meniscus) of the 1:4 layer.
3. Carefully dispense 80 mL of PEST 1:2 under the 1:4 layer using the Bobbitt valve/pipet connected to the PEST 1:2 reservoir (Fig. 2). The Bobbitt valve allows precise flow control

when underlaying the gradient solutions, allowing a sharp interface with minimal mixing of the PEST solutions. If not already marked, mark the bottle to indicate the "new" top (meniscus) of the 1:4 layer overlaying the 1:2 layer.

4. Mix (or vortex) fecal sample(s) well. Using a 50 mL disposable pipet and an automatic pipet aid, carefully transfer/overlay 40 mL of the fecal sample(s) onto prepared gradients in each of the four bottles. This step will require care to avoid mixing the sample into the PEST 1:4 layer.
5. Centrifuge the bottles at 1000× *g* (~2050 RPM) in a refrigerated, table-top centrifuge for 25 min at 4 °C.
6. Using a vacuum flask assembly (with antifoam), carefully aspirate the upper sample layer (orange if dichromate, clear if saline) down to approximately halfway into the 1:4 layer.
7. Collect the oocyst-enriched layer (1:4/1:2 interface) by pipetting (100 mL disposable pipet) the supernatant of each bottle down to about 1/2 cm from the pellet (~95–100 mL). Transfer/dispense the liquid into one 200 mL conical polypropylene bottle (1 conical bottle for each 250 mL gradient bottle). Repeat for each bottle. Set gradient bottles aside: at the end of purification run(s), residual supernatant and pellets will be aspirated from the original bottles into the waste flask. Place empty original gradient bottles into the dishwasher (or hot water bath) for disinfection/cleaning.
8. Fill conical bottles containing oocyst-enriched fractions to the ~190–200 mL mark (shoulder of the bottle) with $dH_2O$. Balance the bottles in the centrifuge carrier adapters (add $dH_2O$ or saline as necessary), cap tightly, and mix.
9. Centrifuge bottles at 1500× *g* (minimum) up to 3000× *g* (preferred) for 10 min at 4 °C.
10. Aspirate supernatants into a waste flask, and transfer pellets from two conical bottles into one 50 mL conical centrifuge tube. Repeat for the second pair of conical bottles.
11. Fill each tube to the 45–50 mL mark with $dH_2O$, mix thoroughly, and centrifuge for 10 min at 1500× *g* (minimum) up to 3000× *g* (latter preferred) in a refrigerated table-top centrifuge at 4 °C.
12. Aspirate supernatants, add a small volume (~5 mL) of saline or 2.5% $K_2Cr_2O_7$, thoroughly disperse/resuspend pellets using a disposable transfer pipet, and dilute to 40 mL with 0.85% saline or 2.5 % potassium dichromate. If processing calf stool, consider labeling each tube with these data: source ID/day post-infection/primary run ID/current date/initials and the primary symbol. For example:

**6IA7 C 1° 8/15/18 XXX**

6 = passage number for isolate
IA = Iowa isolate designation
7 = day post-infection that identifies this subsample
C = tube contents represent primary run C
1° = tube contents a product of a primary gradient purification
8/15/18 = current date (date of gradient run)
XXX = initials of technician who performed gradient run

13. Store labeled oocyst preparations at 4 °C.
14. Enumerate oocyst concentration using hemocytometer (*see* Subheading 3.6). Record tube ID and oocyst concentration in the logbook. (Oocyst numbers are recorded as **NUMBER** × $10^6$/mL).
15. Disinfect pooled liquid waste using heat (*see* Subheading 3.7).

### 3.5 Secondary Oocyst Purification Using Discontinuous Sucrose Gradients (See Notes 5 and 6)

1. Select four tubes of primary sucrose gradient-purified oocysts for secondary purification. Be careful to note the sample ID and primary gradient run date on the tube. If stored in 2.5% $K_2Cr_2O_7$, mix the tubes thoroughly and centrifuge for 10 min at 3000× *g* in the refrigerated table-top centrifuge. Aspirate supernatants and thoroughly disperse/resuspend pellets into a small volume (~ 5 mL) of saline using a disposable transfer pipet (3 mL). Dilute oocyst suspensions to 40 mL with Saline/Tween.
2. Dispense 80 mL of PEST 1:4 into each 250 mL bottle using a graduated cylinder. If not already marked, mark the bottle with a permanent marker to indicate the top (meniscus) of the 1:4 layer.
3. Carefully dispense 80 mL of PEST 1:2 under the 1:4 layer using the Bobbitt valve/pipet connected to the PEST 1:2 reservoir (Fig. 2). If not already marked, mark the bottle to indicate the "new" top (meniscus) of the 1:4 layer overlaying the 1:2 layer.
4. Overlay the sucrose gradients each with 40 mL oocyst suspension (one tube/bottle, and note each bottle accordingly). Tracking bottles can be simplified if using bottle adapters with labels.
5. Centrifuge bottles at 1000× *g* (2050 RPM) in the refrigerated, table-top centrifuge for 25 min at 4 °C.
6. Using a vacuum flask assembly, aspirate the upper sample layer (typically clear if suspended in saline) down to approximately halfway into the 1:4 layer.

7. Collect the oocyst-enriched layer (1:4/1:2 interface) by pipetting (100 mL pipet) the supernatant of each bottle down to about 1/2 cm from the pellet (~95–100 mL). Transfer/dispense the liquid into one 175 mL conical polypropylene bottle (1 conical bottle for each 250 mL gradient bottle). Repeat for each bottle. Aspirate pellets from the original bottles into the waste flask. Place original gradient bottles into the dishwasher for disinfection/cleaning.
8. Fill bottles containing oocyst-enriched fractions to the ~190–200 mL mark (shoulder of the bottle) with $dH_2O$. Balance the bottles in the centrifuge carrier adapters (add $dH_2O$ or saline as necessary), cap tightly, and mix.
9. Centrifuge bottles at 1500× *g* up to 3000× *g* (preferred) for 10 min at 4 °C.
10. Aspirate supernatants and transfer pellets from two conical bottles into one 50 mL polypropylene tube. Repeat for the second pair of conical bottles.
11. Fill each tube to the 45 mL mark with $dH_2O$ or saline, mix thoroughly, and centrifuge at 1500× *g* up to 3000× *g* (preferred) for 10 min at 4 °C.
12. Aspirate supernatants, add a small volume (~5 mL) of saline or 2.5% $K_2Cr_2O_7$, thoroughly disperse/resuspend pellets using a disposable transfer pipet, and dilute to 40 mL with 0.85% saline or 2.5 % $K_2Cr_2O_7$. Each tube will now be labeled with these data: calf ID/day post-infection/primary run source ID/current date/initials and the secondary symbol. For example:

    **6IA7 C/D 2° 8/15/18 XXX**

    IA = Iowa isolate designation
    7 = day post-infection that identifies this subsample
    C/D = tube contents represent pooled material from primary runs C & D
    2° = tube contents a product of a secondary gradient purification
    8/15/18 = current date (date of gradient run)
    XXX = initials of technician who performed gradient run
13. Store labeled oocyst preparations at 4 °C.
14. Enumerate oocyst concentration using hemocytometer (*see* Subheading 3.6). Record tube ID and oocyst concentration in the logbook. (Oocyst numbers are recorded as **NUMBER** × $10^6$/mL).
15. Disinfect pooled liquid waste using heat (*see* Subheading 3.7).

### 3.6 Tertiary Oocyst Purification Using Discontinuous Cesium Chloride Gradients (See Notes 5 and 6)

1. Select four tubes of secondary sucrose gradient-purified oocysts for cesium chloride purification. Be careful to note the sample ID and secondary Sheather's run date on the tube. Mix tubes thoroughly, and centrifuge for 10 min at 1500× *g* (minimum) up to 3000× *g* (preferred) in the table-top centrifuge. If oocysts were stored in 2.5% potassium dichromate, wash out dichromate with saline or PBS using centrifugation (do not apply dichromate to the top of the CsCl gradient; otherwise, dichromate crystals may form). If oocysts were stored in saline or similar buffer, aspirate supernatant and resuspend each pellet into 3 mL of saline.
2. Prepare and label six 1.7 mL clear siliconized (low binding) graduated microcentrifuge tubes for each tube of secondary oocyst preparations.
3. Dispense 1 mL of CsCl gradient solution into each 1.7 mL siliconized microcentrifuge tube. (This is simplified by using a "repeater" pipet).
4. Carefully overlay each CsCl gradient with 0.5 mL of the saline/oocyst suspension, and close the cap tightly.
5. Centrifuge the tubes at ~13,200 rpm (16,000× *g*) for 3 min at 4 °C using the high-speed microcentrifuge. If available, use minimal-to-mild braking during deceleration.
6. Carefully remove the microcentrifuge tubes and collect the oocyst-containing layer at the interface between the sample and the CsCl layer: e.g., using a micropipettor, draw 1.0 mL from the top of the tube (collecting the original 0.5 mL saline and 0.5 mL of the CsCl layer).
7. Transfer each 1.0 mL oocyst suspension into a clean 1.7 microcentrifuge tube.
8. Dilute each tube to ~1.5 mL with saline or deionized $H_2O$. Cap the tubes and mix by inverting several times before centrifugation. Centrifuge the tubes at ~13,200 rpm (16,000× *g*) for 3 min at 4 °C using the microcentrifuge.
9. Aspirate and discard supernatants using the vacuum aspirator assembly. Dispense 0.5 mL of saline into the first tube of each set of six tubes. Carefully resuspend pellets with only 0.5 mL of saline or deionized $H_2O$, and pool each set of six microcentrifuge tube suspensions into one microcentrifuge tube. Rinse each set of tubes with another 0.5 mL of saline or deionized $H_2O$, and pool with the first 0.5 mL suspension. There should be slightly more than 1 mL of suspension in each of the four tubes. Dilute suspensions to the 1.5 mL mark with saline or deionized $H_2O$. Cap tubes and mix by inverting several times before centrifugation.

10. Centrifuge the tubes at ~13,200 rpm (16,000× *g*) for 3 min at 4 °C using the microcentrifuge.
11. Aspirate and discard supernatant. If there are any orange dichromate crystals, resuspend the pellets in 1.0 mL of saline and repeat centrifugation. Wash the pellets carefully until crystals are gone.
12. Resuspend each pellet to 1.0 mL with saline or PBS supplemented with antibiotics mix, or use 2.5% potassium dichromate. Each tube will now be labeled with these data: sample ID/day post-infection/secondary run source ID/current date/initials, and the "tertiary" symbol. For example:

**6IA7 C/D 3° 8/15/18 XXX**

6 = passage number for isolate
IA = Iowa isolate designation
7 = day post-infection that identifies this subsample
C/D = tube contents represent pooled material from primary runs C & D
3° = tube contents a product of a secondary gradient purification
8/15/18 = current date (date of gradient run)
XXX = initials of technician who performed gradient run

13. Store labeled oocyst preparations at 4 °C.
14. Enumerate oocyst concentration using hemocytometer (*see* Subheading 3.6). Record tube ID and oocyst concentration in the logbook. (Oocyst numbers are recorded as **NUMBER** × $10^8$/mL).
15. Disinfect pooled liquid waste using heat (*see* Subheading 3.7).

### 3.7 Heat Disinfection of Liquid Waste Contaminated with Oocysts (See Note 7)

1. Pour supernatant/pellet waste from the centrifuge bottles or vacuum aspirator waste flask into the bulk waste container.
2. Pool waste until the volume reaches the minimum for heat inactivation or has been supplemented with tap water to reach the minimum volume (e.g., 5 gallons—the minimum volume that covers the working surface of the immersion heating element).
3. Assemble the Teflon-coated heating rod (or equivalent heat source), temperature controller set to ~55 °C, magnetic stir plate (or equivalent), appropriate container top, and power timer for the waste tank. *Ensure the waste sample is mixed during heat inactivation to maintain uniform heat throughout the waste volume.*
4. Disinfect the supernatant/pellet suspension with appropriate heat treatment. For example, a 5 gallon tank outfitted with a

500 watt immersion heater takes approximately 3 h to inactivate 5 gallons of waste (this includes the time for the liquid to reach the target temperature (~55 °C) and remain there for at least 1 h). Inactivated waste can be transferred after the heating element has been off for at least 30 min (to allow the heating element to cool). After cooling, transfer the disinfected waste into an appropriate disposal container (e.g., 5–15 gallon polypropylene drum).

5. Dispose of waste appropriately. Waste with potassium dichromate or cesium chloride will require hazardous chemical waste disposal.

### *3.8 Enumerating Oocysts Using the Hemocytometer (See Fig. 5)*

1. Make appropriate dilution of oocyst sample in saline or deionized water (e.g. undilute to 1/5 dilution of primary gradient purified oocysts, 1/10 to 1/25 dilution of secondary gradient purified oocysts, or 1/50 to 1/100 dilution of tertiary gradient purified oocysts). Dilution may be adjusted up or down as necessary.
2. Place 30 mm × 40 mm coverslip over the center position of the hemocytometer. Carefully pipet diluted oocyst suspension into the counting chamber. Do not overfill. Allow cell suspension to settle (~30 s to 1 min).
3. Use phase contrast or bright field settings. Locate the center portion of the ruled area of the counting chamber with the 10X objective and adjust focus.

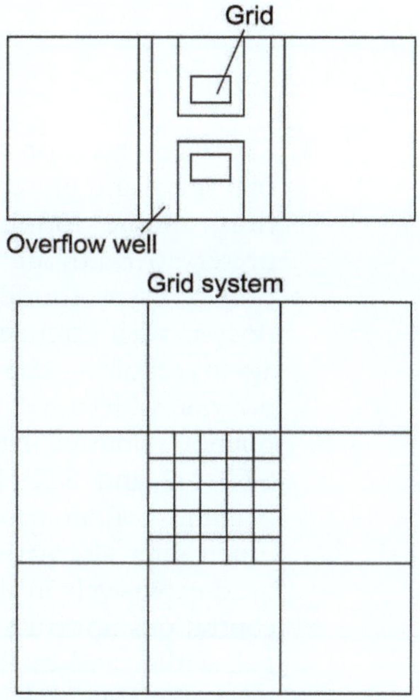


**Fig. 5** Schematic illustration of hemocytometer and hemocytometer grid system.

4. Count the number of oocysts using the 40X objective within the center square of the grid area's 9 large squares (Fig. 5, *grid system center*). Count the oocysts in or touching the top and left borders of each square. Do not count oocysts touching the right or bottom borders of each square. Typically, count the number of oocysts in 5 small squares within the center large square (*see* Fig. 5, *grid square 5 × 5 in center*). A minimum of 77 oocysts should be counted within the 5 squares to give an acceptable approximation of the number of cells/mL. For cell counts <77, use a lower dilution of oocyst suspension or count all 25 squares (adjust formula below as necessary).
5. Calculate the number of oocysts/mL using the following formula:

where:

dilution factor of oocyst suspension = $r$ (inverse of dilution, e.g. inverse of $1/5 = 5$).

dilution factor of hemocytometer = $10^4$.

$$\textbf{(\#of oocysts counted in 5 small squares}^{*}\ \textbf{5)}^{*}\ r^{*}\ \textbf{10}^{4}\ \textbf{= \#oocysts/mL}$$

6. Discard disposable cover glass into "Sharps" bottle, rinse hemocytometer with 70% alcohol and tap water, wipe dry with disposable "wipe" or air-dry, or immerse in hot water bath (55 °C for 1 h). Alternatively, immerse hemocytometer in 3% hydrogen peroxide for ≥30 min, rinse with deionized water and dry.

## 4 Notes

1. Sheather's solution typically contains phenol as an antimicrobial agent and preservative (and to prevent sucrose crystallization). Some formulations incorporate formaldehyde as a preservative although these are not recommended for purification of cryptosporidial oocysts (formaldehyde can inactivate oocysts with extended contact). Phenol is a known hazardous agent, requiring safe handling and disposal (https://www.epa.gov/sites/default/files/2016-09/documents/phenol.pdf). If phenol is omitted in the Sheather's solution used to prepare the STP 1:2 and STP 1:4 gradient solutions, they will quickly become contaminated with fungal and bacterial agents. A much safer alternative to phenol is phenoxyethanol which is used extensively in the cosmetics industry in products at concentrations up to 1%. Phenoxyethanol is antibacterial, antifungal, with a well-established safety profile and, furthermore, is biodegradable [4–6]. Phenoxyethanol is used and preferred in

anatomical embalming (cadavers) and tissue preservation applications [7]. Sucrose gradient solutions prepared using phenoxyethanol improve on phenol-based gradient solutions as being safer to prepare, handle, and for disposal after use. Follow local, national, and institutional guidelines on safe handling and disposal of mixed wastes.

2. Using a positive displacement pipet or syringe, add Tween 80 gradually with constant stirring until dissolved. Store at room temperature in sealed bottles.

3. The typical pattern is to sieve a portion of the raw stool through successive sieves (~3+ liters per centrifuge run) and fill centrifuge bottles for condensing. Repeat until all stool material has been sieved and condensed. If sieves fit tightly on buckets (i.e. there are no air space breaks, e.g. pouring spout), add forceps/tweezers or other spacer to achieve an air gap (*see* Fig. 4), otherwise sieves may form a liquid/vapor seal during sieving and cause spattering/contamination of the workspace when lifted to break the seal.

4. Rule of thumb regarding suspension of fecal pellet: wait several hours or overnight and check the ratio of settled solids to total volume. The settled solids should represent approximately 1/4 to 1/5 the total volume of 2.5% potassium dichromate or 0.85% saline. Adjust volume as necessary to achieve target ratio. Recovery efficiencies improve with added dilution but require more gradient runs to complete purification.

5. Add a few drops of Antifoam Y-30 to the 4 L vacuum aspirator flask before starting a gradient purification run to prevent excessive foaming during the vacuum aspiration steps. For bulk purification runs, a typical sequence of three consecutive four-bottle runs will generate just less than 4 L of waste which will fit in the aspirator flask. If antifoam is not used, foaming may cause a portion of the waste to be drawn into the small, vacuum trap flask between the aspirator flask and the vacuum supply. Be sure to have a hydrophobic in-line filter installed after the trap and before the vacuum valve to prevent contamination of the vacuum supply. Consider using a disposable 50 mL tube as a temporary "docking station" for disposable Pasteur pipets attached to the 4 L vacuum aspirator flask via silicone tubing. The 50 mL tube and pipet can be changed with each sample or succession of gradient runs. Carefully balance the bottles in the centrifuge bucket adapters to assure the centrifuge runs will not be interrupted by an "out-of-balance" error or imbalance that causes mixing of bottle contents (prepared gradients) during acceleration/deceleration.

6. While we track each gradient run to monitor purification performance, the method described can be simplified by pooling

oocyst pellets together at the end of each gradient process (i.e., all of the primary sucrose gradient pellets can be pooled together after all of the stool material is processed). This way, only one count will need to be made before proceeding to the next gradient technique. Some laboratories may wish to keep oocysts separate based on the day shed post-infection. Alternatively, oocysts from an entire experimental shedding period may be pooled together into a single "lot" for subsequent studies. We have not found significant differences among oocyst generated between days 5 and 12 post-infection from experimentally infected bovine calves.

7. For safety reasons, heating systems should have overheating circuitry that shuts off power in the event of overheating. The use of a timer with a manual "ON" feature and an automatic "OFF" feature reduces the risk of heating cycles running too long if left unattended. Heating cycles can be timed routinely to accomplish the inactivation within a window of time for a given container/volume/heat source. A temperature of 55 °C and a duration of 1 h are used to minimize the risk of handling large volumes of hot liquid waste (e.g., autoclaving) while also ensuring cryptosporidial oocysts are inactivated. Although pasteurization temperatures are recognized to inactivate oocysts, published data and our own quality control data support the use of the target temperature for disinfection [8–10].

## Acknowledgments

The author is retired from service at the Centers for Disease Control and Prevention. The use of trade names is for identification only and does not imply endorsement by the Public Health Service or the US Department of Health and Human Services. The findings and conclusions in this report are those of the author and do not necessarily represent the official position of the Centers for Disease Control and Prevention.

**Competing Interest Statement** The authors declare no conflicts of interest relevant to the content of this chapter.

### References

1. Arrowood MJ, Hurd MR, Mead JR (1995) A new method for evaluating experimental cryptosporidial parasite loads using immunofluorescent flow cytometry. J Parasitol 81(3): 404–409
2. Arrowood MJ, Sterling CR (1987) Isolation of *Cryptosporidium* oocysts and sporozoites using discontinuous sucrose and isopycnic Percoll gradients. J Parasitol 73:314–319
3. Arrowood MJ, Donaldson K (1996) Improved purification methods for calf-derived

*Cryptosporidium parvum* oocysts using discontinuous sucrose and cesium chloride gradients. J Eukaryot Microbiol 43(5):S 89

4. Dreno B, Zuberbier T, Gelmetti C, Gontijo G, Marinovich M (2019) Safety review of phenoxyethanol when used as a preservative in cosmetics. J Eur Acad Dermatol Venereol 33 (Suppl 7):15–24. https://doi.org/10.1111/jdv.15944
5. Expert Panel (1990) 9 Final Report on the Safety Assessment of Phenoxyethanol. J Am Coll Toxicol 9(2):259–277. https://doi.org/10.3109/10915819009078737
6. Krowka J, Loretz L, Brzuska K, Almeida JF, Diehl M, Gonsior SJ, Johnson A, Sellam S (2014) Phenoxyethanol as a safe and important preservative in personal care. Cosmetics & Toiletries 129:24–27
7. Crosado B, Loffler S, Ondruschka B, Zhang M, Zwirner J, Hammer N (2020) Phenoxyethanol-based embalming for anatomy teaching: an 18 years' experience with Crosado embalming at the University of Otago in New Zealand. Anat Sci Educ 13(6): 778–793. https://doi.org/10.1002/ase.1933
8. Fujino T, Matsui T, Kobayashi F, Haruki K, Yoshino Y, Kajima J, Tsuji M (2002) The effect of heating against *Cryptosporidium* oocysts. J Vet Med Sci 64(3):199–200
9. Harp JA, Fayer R, Pesch BA, Jackson GJ (1996) Effect of pasteurization on infectivity of *Cryptosporidium parvum* oocysts in water and milk. Appl Environ Microbiol 62(8): 2866–2868
10. Anderson BC (1985) Moist heat inactivation of *Cryptosporidium* sp. Am J Public Health 75: 1433–1434

# INDEX



Jan R. Mead and Michael J. Arrowood (eds.), *Cryptosporidium: Methods and Protocols*, Methods in Molecular Biology, vol. 2978, https://doi.org/10.1007/978-1-0716-4824-7,

## D



## E



## F



## G



## H



## I



## J



## K



## L



## M

## T



## W



## Z